BUSINESS/SCIENCE/TECHNOLOGY DIVISION
CHICAGO PUBLIC LIBRARY
400 SOUTH STATE STREET
CHICAGO, IL 60605

"Four legs good, two legs better."
>A modified version of the
>Animal Farm's Constitution.

"Two logs good, p logs better."
>The original Constitution
>of mathematicians.

RANDOM WALK IN RANDOM AND NON-RANDOM ENVIRONMENTS

Second Edition

RANDOM WALK IN RANDOM AND NON-RANDOM ENVIRONMENTS

Second Edition

Pál Révész

Technische Universität Wien, Austria
Technical University of Budapest, Hungary

World Scientific

NEW JERSEY • LONDON • SINGAPORE • BEIJING • SHANGHAI • HONG KONG • TAIPEI • CHENNAI

Published by

World Scientific Publishing Co. Pte. Ltd.
5 Toh Tuck Link, Singapore 596224
USA office: 27 Warren Street, Suite 401-402, Hackensack, NJ 07601
UK office: 57 Shelton Street, Covent Garden, London WC2H 9HE

Library of Congress Cataloging-in-Publication Data
Random walk in random and non-random environments / Pál Révész.--2nd ed.
   p. cm.
  Includes bibliographical references and indexes.
  ISBN 981-256-361-X (alk. paper)
   1. Random walks (Mathematics). I. Title.

QA274.73 .R48 2005
519.2'82--dc22

                                     2005045536

British Library Cataloguing-in-Publication Data
A catalogue record for this book is available from the British Library.

Copyright © 2005 by World Scientific Publishing Co. Pte. Ltd.

All rights reserved. This book, or parts thereof, may not be reproduced in any form or by any means, electronic or mechanical, including photocopying, recording or any information storage and retrieval system now known or to be invented, without written permission from the Publisher.

For photocopying of material in this volume, please pay a copying fee through the Copyright Clearance Center, Inc., 222 Rosewood Drive, Danvers, MA 01923, USA. In this case permission to photocopy is not required from the publisher.

Printed in Singapore by Mainland Press

Preface to the First Edition

"I did not know that it was so dangerous to drink a beer with you. You write a book with those you drink a beer with," said Professor Willem Van Zwet, referring to the preface of the book Csörgő and I wrote (1981) where it was told that the idea of that book was born in an inn in London over a beer. In spite of this danger Willem was brave enough to invite me to Leiden in 1984 for a semester and to drink quite a few beers with me there. In fact I gave a seminar in Leiden, and the handout of that seminar can be considered as the very first version of this book. I am indebted to Willem and to the Department of Leiden for a very pleasant time and a number of useful discussions.

I wrote this book in 1987-89 in Vienna (Technical University) partly supported by Fonds zur Förderung der Wissenschaftlichen Forschung, Project Nr. P6076. During these years I had very strong contact with the Mathematical Institute of Budapest. I am especially indebted to Professors E. Csáki and A. Földes for long conversations which have a great influence on the subject of this book. The reader will meet quite often with the name of P. Erdős, but his role in this book is even greater. Especially most results of Part II are fully or partly due to him, but he had a significant influence even on those results that appeared under my name only.

Last but not least, I have to mention the name of M. Csörgő, with whom I wrote about 30 joint papers in the last 15 years, some of them strongly connected with the subject of this book.

Vienna, 1989.
P. Révész
Technical University of Vienna
Wiedner Hauptstrasse 8-10/107
A–1040 Vienna
Austria

Preface to the Second Edition

If you write a monograph on a new, just developing subject, then in the next few years quite a number of brand-new papers are going to appear in your subject and your book is going to be outdated. If you write a monograph on a very well-developed subject in which nothing new happens, then it is going to be outdated already when it is going to appear. In 1989 when I prepared the First Edition of this book it was not clear for me that its subject was already overdeveloped or it was a still developing area. A year later Erdős told me that he had been surprised to see how many interesting, unsolved problems had appeared in the last few years about the very classical problem of coin-tossing (random walk on the line). In fact Erdős himself proposed and solved a number of such problems.

I was happy to see the huge number of new papers (even books) that have appeared in the last 16 years in this subject. I tried to collect the most interesting ones and to fit them in this Second Edition. Many of my friends helped me to find the most important new results and to discover some of the mistakes in the First Edition.

My special thanks to E. Csáki, M. Csörgő, A. Földes, D. Khoshnevisan, Y. Peres, Q. M. Shao, B. Tóth, Z. Shi.

Vienna, 2005.

Contents

Preface to the First Edition ... v

Preface to the Second Edition ... vii

Introduction ... xv

I. SIMPLE SYMMETRIC RANDOM WALK IN $\mathbb{Z}^1$

Notations and abbreviations ... 3

1 Introduction of Part I ... 9
  1.1 Random walk ... 9
  1.2 Dyadic expansion ... 10
  1.3 Rademacher functions ... 10
  1.4 Coin tossing ... 11
  1.5 The language of the probabilist ... 11

2 Distributions ... 13
  2.1 Exact distributions ... 13
  2.2 Limit distributions ... 19

3 Recurrence and the Zero-One Law ... 23
  3.1 Recurrence ... 23
  3.2 The zero-one law ... 25

4 From the Strong Law of Large Numbers to the Law of Iterated Logarithm ... 27
  4.1 Borel–Cantelli lemma and Markov inequality ... 27
  4.2 The strong law of large numbers ... 28
  4.3 Between the strong law of large numbers and the law of iterated logarithm ... 29
  4.4 The LIL of Khinchine ... 31

5 Lévy Classes ... 33
  5.1 Definitions ... 33
  5.2 EFKP LIL ... 34
  5.3 The laws of Chung and Hirsch ... 39
  5.4 When will S_n be very large? ... 39

	5.5 A theorem of Csáki	41

6 Wiener Process and Invariance Principle 47
6.1 Four lemmas 47
6.2 Joining of independent random walks 49
6.3 Definition of the Wiener process 51
6.4 Invariance Principle 52

7 Increments 57
7.1 Long head-runs 57
7.2 The increments of a Wiener process 66
7.3 The increments of S_N 77

8 Strassen Type Theorems 83
8.1 The theorem of Strassen 83
8.2 Strassen theorems for increments 90
8.3 The rate of convergence in Strassen's theorems 92
8.4 A theorem of Wichura 95

9 Distribution of the Local Time 97
9.1 Exact distributions 97
9.2 Limit distributions 103
9.3 Definition and distribution of the local time
of a Wiener process 104

10 Local Time and Invariance Principle 109
10.1 An invariance principle 109
10.2 A theorem of Lévy 111

11 Strong Theorems of the Local Time 117
11.1 Strong theorems for $\xi(x,n)$ and $\xi(n)$ 117
11.2 Increments of $\eta(x,t)$ 119
11.3 Increments of $\xi(x,n)$ 123
11.4 Strassen type theorems 124
11.5 Stability 126

12 Excursions 135
12.1 On the distribution of the zeros of a random walk 135
12.2 Local time and the number of long excursions
(Mesure du voisinage) 141
12.3 Local time and the number of high excursions 146
12.4 The local time of high excursions 147
12.5 How many times can a random walk reach its maximum? . 152

13 Frequently and Rarely Visited Sites — 157
13.1 Favourite sites 157
13.2 Rarely visited sites 161

14 An Embedding Theorem — 163
14.1 On the Wiener sheet 163
14.2 The theorem 164
14.3 Applications 168

15 A Few Further Results — 171
15.1 On the location of the maximum of a random walk 171
15.2 On the location of the last zero 175
15.3 The Ornstein–Uhlenbeck process and
a theorem of Darling and Erdős 179
15.4 A discrete version of the Itô formula 183

16 Summary of Part I — 187

II. SIMPLE SYMMETRIC RANDOM WALK IN $\mathbb{Z}^d$

Notations — 191

17 The Recurrence Theorem — 193

18 Wiener Process and Invariance Principle — 203

19 The Law of Iterated Logarithm — 207

20 Local Time — 211
20.1 $\xi(0,n)$ in $\mathbb{Z}^2$ 211
20.2 $\xi(n)$ in $\mathbb{Z}^d$ 218
20.3 A few further results 220

21 The Range — 221
21.1 The strong law of large numbers 221
21.2 CLT, LIL and Invariance Principle 225
21.3 Wiener sausage 226

22 Heavy Points and Heavy Balls — 227
22.1 The number of heavy points 227
22.2 Heavy balls 236
22.3 Heavy balls around heavy points 239
22.4 Wiener process 240

23 Crossing and Self-crossing   241

24 Large Covered Balls   245
24.1 Completely covered discs centered in the origin of $\mathbb{Z}^2$ 245
24.2 Completely covered disc in $\mathbb{Z}^2$ with arbitrary centre 263
24.3 Almost covered disc centred in the origin of $\mathbb{Z}^2$ 264
24.4 Discs covered with positive density in $\mathbb{Z}^2$ 265
24.5 Completely covered balls in $\mathbb{Z}^d$ 272
24.6 Large empty balls....................... 277
24.7 Summary of Chapter 24 280

25 Long Excursions   281
25.1 Long excursions in $\mathbb{Z}^2$ 281
25.2 Long excursions in high dimension 284

26 Speed of Escape   287

27 A Few Further Problems   293
27.1 On the Dirichlet problem 293
27.2 DLA model 296
27.3 Percolation 297

III. RANDOM WALK IN RANDOM ENVIRONMENT

Notations   301

28 Introduction   303

29 In the First Six Days   307

30 After the Sixth Day   311
30.1 The recurrence theorem of Solomon 311
30.2 Guess how far the particle is going away in an RE 313
30.3 A prediction of the Lord 314
30.4 A prediction of the physicist................. 326

31 What Can a Physicist Say About the Local Time $\xi(0,n)$?   329
31.1 Two further lemmas on the environment 329
31.2 On the local time $\xi(0,n)$ 330

32 On the Favourite Value of the RWIRE   337

33 A Few Further Problems   345
    33.1 Two theorems of Golosov 345
    33.2 Non-nearest-neighbour random walk 347
    33.3 RWIRE in $\mathbb{Z}^d$ 348
    33.4 Non-independent environments 350
    33.5 Random walk in random scenery 350
    33.6 Random environment and random scenery 353
    33.7 Reinforced random walk 353

References   357

Author Index   375

Subject Index   379

Introduction

The first examinee is saying: Sir, I did not have time enough to study everything but I learned very carefully the first chapter of your handout.

Very good – says the professor – you will be a great specialist. You know what a specialist is. A specialist knows more and more about less and less. Finally he knows everything about nothing.

The second examinee is saying: Sir, I did not have enough time but I read your handout without taking care of the details. Very good – answers the professor – you will be a great polymath. You know what a polymath is. A polymath knows less and less about more and more. Finally he knows nothing about everything.

Recalling this old joke and realizing that the biggest part of this book is devoted to the study of the properties of the simple symmetric random walk (or equivalently, coin tossing) the reader might say that this is a book for specialists written by a specialist. The most trivial plea of the author is to say that this book does not tell *everything* about coin tossing and even the author does not know *everything* about it. Seriously speaking I wish to explain my reasons for writing such a book.

You know that the first probabilists (Bernoulli, Pascal, etc.) investigated the properties of coin tossing sequences and other simple games only. Later on the progress of the probability theory went into two different directions:

(i) to find newer and deeper properties of the coin tossing sequence,

(ii) to generalize the results known for a coin tossing sequence to more complicated sequences or processes.

Nowadays the second direction is much more popular than the first one. In spite of this fact this book mostly follows direction (i).

I hope that:

(a) using the advantage of the simple situation coming from concentrating on coin tossing sequences, the reader becomes familiar with the problems, results and partly the methods of proof of probability theory, especially those of the limit theorems, without suffering too much from technical tools and difficulties,

(b) since the random walk (especially in $\mathbb{Z}^d$) is the simplest mathematical model of the Brownian motion, the reader can find a simple way to the problems (at least to the classical problems) of statistical physics,

(c) since it is nearly impossible to give a more or less complete picture of the properties of the random walk without studying the analogous properties of the Wiener process, the reader can find a simple way to the study of the stochastic processes and should learn that it is impossible to go deeply in direction (i) without going a bit in direction (ii),

(d) any reader having any degree in math can understand the book, and reading the book can get an overall picture about random phenomena, and the readers having some knowledge in probability can get a better overview of the recent problems and results of this part of the probability theory,

(e) some parts of this book can be used in any introductory or advanced probability course.

The main aim of this book is to collect and compare the results – mostly strong theorems – which describe the properties of a simple symmetric random walk. The proofs are not always presented. In some cases more proofs are given, in some cases none. The proofs are omitted when they can be obtained by routine methods and when they are too long and too technical. In both cases the reader can find the exact reference to the location of the (or of a) proof.

"The earth was without form and void, and darkness was upon the face of the deep."

The First Book of Moses

I. SIMPLE SYMMETRIC RANDOM WALK IN $\mathbb{Z}^1$

Notations and abbreviations

Notations

General notations

1. $X_1, X_2, \ldots$ is a sequence of independent, identically distributed random variables with
$$\mathbf{P}\{X_1 = 1\} = \mathbf{P}\{X_1 = -1\} = 1/2.$$

2. $S_0 = 0, S_n = S(n) = X_1 + X_2 + \cdots + X_n$ $(n = 1, 2, \ldots)$. $\{S_n\}$ is the (simple symmetric) random walk.

3. $M_n^+ = M^+(n) = \max_{0 \leq k \leq n} S_k,$

   $M_n^- = M^-(n) = -\min_{0 \leq k \leq n} S_k,$

   $M_n = M(n) = \max_{0 \leq k \leq n} |S_k| = \max(M_n^+, M_n^-),$

   $M_n^* = M_n^+ + M_n^-,\ Y_n = M_n^+ - S_n.$

4. $\{W(t); t \geq 0\}$ is a Wiener process (cf. Section 6.3).

5. $m^+(t) = \sup_{0 \leq s \leq t} W(s),$

   $m^-(t) = -\inf_{0 \leq s \leq t} W(s),$

   $m(t) = \sup_{0 \leq s \leq t} |W(s)| = \max(m^+(t), m^-(t))\ (t \geq 0),$

   $m^*(t) = m^+(t) + m^-(t),$
   $y(t) = m^+(t) - W(t).$

6. $b_n = b(n) = (2n \log \log n)^{-1/2},$

   $\gamma_n = \gamma(n, a) = \left(2a \left(\log \dfrac{n}{a} + \log \log n\right)\right)^{-1/2}.$

7. $[x]$ is the largest integer less than or equal to x.

8. $f(n) \gg g(n) \leftrightarrow g(n) = o(f(n)) \leftrightarrow \lim_{n \to \infty} \dfrac{f(n)}{g(n)} = \infty.$

9. $g(n) = O(f(n)) \leftrightarrow 0 < \liminf_{n\to\infty} \dfrac{f(n)}{g(n)} \leq \limsup_{n\to\infty} \dfrac{f(n)}{g(n)} < \infty.$

10. $f(n) \approx g(n) \leftrightarrow \lim_{n\to\infty} \dfrac{f(n)}{g(n)} = 1.$

11. Sometimes we use the notation $f(n) \sim g(n)$ without any exact mathematical meaning, just saying that $f(n)$ and $g(n)$ are close to each other in some sense.

12. $\Phi(x) = \dfrac{1}{\sqrt{2\pi}} \int_{-\infty}^{x} e^{-u^2/2} du$ is the standard normal distribution function.

13. $N \in N(m, \sigma) \leftrightarrow \mathbf{P}\{\sigma^{-1}(N-m) < x\} = \Phi(x).$

14. $\#\{\ldots\} = |\{\ldots\}|$ is the cardinality of the set in the bracket.

15. $\mathbb{R}^d$ resp. $\mathbb{Z}^d$ is the d-dimensional Euclidean space resp. its integer grid.

16. $B = B^d$ is the set of Borel-measurable sets of $\mathbb{R}^d$.

17. $\lambda(\cdot)$ is the Lebesgue measure on $\mathbb{R}^d$.

18. $\log_p$ ($p = 1, 2, \ldots$) is p-th iterated of log and lg resp. $\lg_p$ is the logarithm resp. p-th iterated of the logarithm of base 2.

19. Let $\{U_n\}$ and $\{V_n\}$ be two sequences of random variables. $\{U_n,\ n = 1, 2, \ldots\} \stackrel{\mathcal{D}}{=} \{V_n\ n = 1, 2, \ldots\}$ if the finite dimensional distributions of $\{U_n\}$ are equal to the corresponding finite dimensional distributions of $\{V_n\}$.

Notations to the increments

1. $I_1(n, a) = \max\limits_{0 \leq k \leq n-a} (S_{k+a} - S_k),$

2. $I_2(n, a) = \max\limits_{0 \leq k \leq n-a} |S_{k+a} - S_k|,$

3. $I_3(n, a) = \max\limits_{0 \leq k \leq n-a} \max\limits_{0 \leq j \leq a} (S_{k+j} - S_k),$

4. $I_4(n, a) = \max\limits_{0 \leq k \leq n-a} \max\limits_{0 \leq j \leq a} |S_{k+j} - S_k|,$

5. $I_5(n, a) = \min\limits_{0 \leq k \leq n-a} \max\limits_{0 \leq j \leq a} |S_{k+j} - S_k|,$

6. $J_1(t,a) = \sup_{0 \le s \le t-a} (W(s+a) - W(s))$,

7. $J_2(t,a) = \sup_{0 \le s \le t-a} |W(s+a) - W(s)|$,

8. $J_3(t,a) = \sup_{0 \le s \le t-a} \sup_{0 \le u \le a} (W(s+u) - W(s))$,

9. $J_4(t,a) = \sup_{0 \le s \le t-a} \sup_{0 \le u \le a} |W(s+u) - W(s)|$,

10. $J_5(t,a) = \inf_{0 \le s \le t-a} \sup_{0 \le u \le a} |W(s+u) - W(s)|$,

11. Z_n is the largest integer for which $I_1(n, Z_n) = Z_n$, i.e. Z_n is the length of the longest run of pure heads in n Bernoulli trials.

Notations to the Strassen-type theorems

1. $s_n(x) = b_n \left(S_{[nx]} + \left(x - \dfrac{[nx]}{n} \right) X_{[nx]+1} \right)$ $(0 \le x \le 1)$,

2. $w_t(x) = b_t W(tx)$ $(0 \le x \le 1,\ t > 0)$,

3. $C(0,1)$ is the set of continuous functions defined on the interval $[0,1]$,

4. $\mathcal{S}(0,1)$ is the Strassen's class, containing those functions $f(\cdot) \in C(0,1)$ for which $f(0) = 0$ and $\int_0^1 (f'(x))^2 dx \le 1$.

Notations to the local time

1. $\xi(x, n) = \#\{k : 0 < k \le n,\ S_k = x\}$ $(x = 0, \pm 1, \pm 2, \ldots,\ n = 1, 2, \ldots)$ is the local time of the random walk $\{S_k\}$. For any $A \subset \mathbb{Z}^1$ we define the occupation time $\Xi(A, n) = \sum_{x \in A} \xi(x, n)$.

2. $\eta(x, t)$ $(-\infty < x < +\infty,\ t \ge 0)$ is the local time of $W(\cdot)$ (cf. Section 9.3).

3. $H(A, t) = \lambda\{s : 0 \le s \le t,\ W(s) \in A\}$ $(A \subset \mathbb{R}^1$ is a Borel set, $t \ge 0)$ is the occupation time of $W(\cdot)$ (cf. Section 9.3).

4. Consider those values of k for which $S_k = 0$. Let these values in increasing order be $0 = \rho_0 < \rho_1 < \rho_2 < \ldots$, i.e. $\rho_1 = \min\{k : k > 0,\ S_k = 0\}$, $\rho_2 = \min\{k : k > \rho_1,\ S_k = 0\}, \ldots, \rho_n = \min\{k : k > \rho_{n-1},\ S_k = 0\}, \ldots$

5. For any $x = 0, \pm 1, \pm 2, \ldots$ consider those values of k for which $S_k = x$. Let these values in increasing order be $0 < \rho_1(x) < \rho_2(x) < \ldots$ i.e. $\rho_1(x) = \min\{k : k > 0, S_k = x\}, \rho_2(x) = \min\{k : k > \rho_1(x), S_k = x\}, \ldots, \rho_n(x) = \min\{k : k > \rho_{n-1}(x), S_k = x\} \ldots$ Clearly $\rho_i(0) = \rho_i$. In case of a Wiener process define $\rho_u^* = \inf\{t : t \geq 0, \eta(0,t) \geq u\}$.

6. $\xi(n) = \max_x \xi(x, n)$.

7. $\eta(t) = \sup_x \eta(x,t)$.

8. The random sequences

$$E_1 = \{S_0, S_1, \ldots, S_{\rho_1}\}, \; E_2 = \{S_{\rho_1}, S_{\rho_1+1}, \ldots, S_{\rho_2}\}, \ldots$$

are called the first, second, ... excursions (away from 0) of the random walk $\{S_k\}$.

9. The random sequences

$$E_1(x) = \{S_{\rho_1(x)}, S_{\rho_1(x)+1}, \ldots, S_{\rho_2(x)}\},$$
$$E_2(x) = \{S_{\rho_2(x)}, S_{\rho_2(x)+1}, \ldots, S_{\rho_3(x)}\}, \ldots$$

are called the first, second, ... excursions away from x of the random walk $\{S_k\}$.

10. For any $t > 0$ let $\alpha(t) = \sup\{\tau : \tau < t, W(\tau) = 0\}$ and $\beta(t) = \inf\{\tau : \tau > t, W(\tau) = 0\}$. Then the path $\{W_t(s); \alpha(t) \leq s \leq \beta(t)\}$ is called an excursion of $W(\cdot)$.

11. ζ_n is the number of those terms of $S_1, S_2, \ldots, S_n$ which are positive or which are equal to 0 but the preceding term of which is positive.

12. $\Theta(n) = \#\{k : 1 \leq k \leq n, \; S_{k-1}S_{k+1} < 0\}$ is the number of crossings.

13. $R(n) = \max\{k : k > 1$ for which there exists a $0 < j < n - k$ such that $\xi(0, j+k) = \xi(0,j)\}$ is the length of the longest zero-free interval.

14. $r(t) = \sup\{s : s > 0$ for which there exists a $0 < u < t - s$ such that $\eta(0, u + s) = \eta(0, u)\}$.

15. $\Psi(n) = \max\{k : 0 \leq k \leq n, \; S_k = 0\}$ is the location of the last zero up to n.

16. $\psi(t) = \sup\{s : 0 < s \leq t, \; W(s) = 0\}$.

17. $\hat{R}(n) = \max\{k : k > 1$ for which there exists a $0 < j < n - k$ such that $M^+(j+k) = M^+(j)\}$ is the length of the longest flat interval of M_k^+ up to n.

18. $\hat{r}(t) = \sup\{s : s > 0$ for which there exists a $0 < u < t - s$ such that $m^+(u+s) = m^+(u)\}$.

19. $R^*(n) = \max\{k : k > 1$ for which there exists a $0 < j < n - k$ such that $M(j+k) = M(j)\}$.

20. $r^*(t) = \sup\{s : s > 0$ for which there exists a $0 < u < t - s$ such that $m(u+s) = m(u)\}$.

21. $\mu(n)$ is the location of the maximum of the absolute value of a random walk $\{S_k\}$ up to n, i.e. $\mu(n)$ is defined by $S(\mu(n)) = M(n)$ and $\mu(n) \leq n$. If there are more integers satisfying the above conditions then the smallest one will be considered as $\mu(n)$.

22. $\mathcal{M}(t) = \inf\{s : 0 < s \leq t$ for which $W(s) = m(t)\}$.

23. $\mu^+(n) = \inf\{k : 0 \leq k \leq n$ for which $S(k) = M^+(n)\}$.

24. $\mathcal{M}^+(t) = \inf\{s : 0 < s \leq t$ for which $W(s) = m^+(t)\}$.

25. $\chi(n)$ is the number of those places where the maximum of the random walk $S_0, S_1, \ldots, S_n$ is reached, i.e. $\chi(n)$ is the largest positive integer for which there exists a sequence of integers $0 \leq k_1 < k_2 < \ldots < k_{\chi(n)} \leq n$ such that

$$S(k_1) = S(k_2) = \cdots = S(k_{\chi(n)}) = M^+(n).$$

Abbreviations

1. r.v. = random variable,

2. i.i.d.r.v.'s = independent, identically distributed r.v.'s,

3. LIL = law of iterated logarithm,

4. UUC, ULC, LUC, LLC, AD, QAD (cf. Section 5.1),

5. i.o. = infinitely often,

6. a.s. = almost surely.

Chapter 1

Introduction of Part I

The problems and results of the theory of simple symmetric random walk in $\mathbb{Z}^1$ can be presented using different languages. The physicist will talk about random walk or Brownian motion on the line. (We use the expression "Brownian motion" in this book only in a non-well-defined physical sense and we will say that the simple symmetric random walk or the Wiener process are mathematical models of the Brownian motion.) The number theorist will talk about dyadic expansions of the elements of $[0,1]$. The people interested in orthogonal series like to formulate the results in the language of Rademacher functions. The gambler will talk about coin tossing and his gain. And a probabilist will consider independent, identically distributed random variables and the partial sums of those.

Mathematically speaking all of these formulations are equivalent. In order to explain the grammar of these languages in this Introduction we present a few of our notations and problems using the different languages. However, later on mostly the "language of the physicist and that of the probabilist" will be used.

1.1 Random walk

Consider a particle making a random walk (Brownian motion) on the real line. Suppose that the particle starts from $x = 0$ and moves one unit to the left with probability $1/2$ and one unit to the right with probability $1/2$ during one time unit. In the next step it moves one step to the left or to the right with equal probabilities independently from its location after the first step. Continuing this procedure we obtain a random walk that is the simplest mathematical model of the linear Brownian motion.

Let S_n be the location of the particle after n steps or in time n. This model clearly implies that

$$\mathbf{P}\{S_{n+1} = i_{n+1} \mid S_n = i_n, S_{n-1} = i_{n-1}, \ldots, S_1 = i_1, S_0 = i_0 = 0\}$$
$$= \mathbf{P}\{S_{n+1} = i_{n+1} \mid S_n = i_n\} = 1/2 \tag{1.1}$$

where $i_0 = 0, i_1, i_2, \ldots, i_n, i_{n+1}$ is a sequence of integers with $|i_1 - i_0| = |i_2 - i_1| = \ldots = |i_{n+1} - i_n| = 1$. It is also natural to ask: how far does the particle go away (resp. going away to the right or to the left) during the first n steps. It means that we consider

$$M_n = \max_{0 \leq k \leq n} |S_k| \quad \text{resp.} \quad M_n^+ = \max_{0 \leq k \leq n} S_k \quad \text{or} \quad M_n^- = -\min_{0 \leq k \leq n} S_k.$$

1.2 Dyadic expansion

Let x be any real number in the interval $[0,1]$ and consider its dyadic expansion

$$x = 0, \varepsilon_1 \varepsilon_2 \ldots = \sum_{k=1}^{\infty} \varepsilon_k 2^{-k}$$

where $\varepsilon_i = \varepsilon_i(x)$ $(i = 1, 2, \ldots)$ is equal to 0 or 1. In fact

$$\varepsilon_i = [2^i x] \pmod 2.$$

Observe that

$$\lambda\{x: \ \varepsilon_{j_1}(x) = \delta_1, \ \varepsilon_{j_2}(x) = \delta_2, \ldots, \varepsilon_{j_n}(x) = \delta_n\} = 2^{-n} \quad (1.2)$$

where $1 \leq j_1 < j_2 < \ldots < j_n; n = 1, 2, \ldots; \delta_1, \delta_2, \ldots, \delta_n$ is an arbitrary sequence of 0's and +1's and λ is the Lebesgue measure. Let $S_0 = S_0(x) = 0$ and $S_n = S_n(x) = n - 2\sum_{i=1}^{n} \varepsilon_i(x)$ $(n = 1, 2, \ldots)$. Then (1.2) implies

$$\lambda\{x: \ S_{n+1} = i_{n+1}, \ S_n = i_n, \ldots, S_1 = i_1, \ S_0 = i_0\} = 2^{-(n+1)} \quad (1.3)$$

where $i_0 = 0$, $i_1, i_2, \ldots, i_{n+1}$ is a sequence of integers with $|i_1 - i_0| = |i_2 - i_1| = \ldots = |i_{n+1} - i_n| = 1$. Clearly (1.3) is equivalent to (1.1). Hence any theorem proved for a random walk can be translated to a theorem on dyadic expansion.

A number theorist is interested in the frequency $N_n(x) = \sum_{i=1}^{n} \varepsilon_i(x)$ of the ones among the first n digits of $x \in [0, 1]$. Since $N_n(x) = (n - S_n(x))/2$ any theorem formulated for S_n implies a corresponding theorem for $N_n(x)$.

1.3 Rademacher functions

In the theory of orthogonal series the following sequence of functions is well-known. Let

$$r_1(x) = \begin{cases} 1 & \text{if } x \in [0, 1/2), \\ -1 & \text{if } x \in [1/2, 1], \end{cases}$$

$$r_2(x) = \begin{cases} 1 & \text{if } x \in [0, 1/4) \cup [1/2, 3/4), \\ -1 & \text{if } x \in [1/4, 1/2) \cup [3/4, 1], \end{cases}$$

$$r_3(x) = \begin{cases} 1 & \text{if } x \in [0, 1/8) \cup [1/4, 3/8) \cup [1/2, 5/8) \cup [3/4, 7/8), \\ -1 & \text{if } x \in [1/8, 1/4) \cup [3/8, 1/2) \cup [5/8, 3/4) \cup [7/8, 1], \ldots \end{cases}$$

An equivalent definition, by dyadic expansion, is
$$r_n(x) = 1 - 2\varepsilon_n(x).$$

The functions $r_1(x), r_2(x), \ldots$ are called Rademacher functions. It is a sequence of orthonormed functions, i.e.
$$\int_0^1 r_i(x)r_j(x)dx = \begin{cases} 1 & \text{if } i = j, \\ 0 & \text{if } i \neq j. \end{cases}$$

Observe that
$$\lambda\{x : r_{j_1}(x) = \delta_1,\ r_{j_2}(x) = \delta_2, \ldots, r_{j_n}(x) = \delta_n\} = 2^{-n} \quad (1.4)$$
where $1 \leq j_1 < j_2 < \ldots < j_n$ $(n = 1, 2, \ldots)$; $\delta_1, \delta_2, \ldots, \delta_n$ is an arbitrary sequence of $+1$'s and -1's and λ is the Lebesgue measure. Putting $S_0 = S_0(x) = 0$ and $S_n = S_n(x) = \sum_{i=1}^n r_i(x)$ $(n = 1, 2, \ldots)$ we obtain (1.3).

1.4 Coin tossing

Two gamblers (A and B) are tossing a coin. A wins one dollar if the tossing results in a head and B wins one dollar if the result is tail. Let S_n be the amount gained by A (in dollars) after n tossings. (Clearly S_n can be negative and $S_0 = 0$ by definition.) Then S_N satisfies (1.1) if the game is fair, i.e. the coin is regular.

1.5 The language of the probabilist

Let $X_1, X_2, \ldots$ be a sequence of i.i.d.r.v.'s with
$$\mathbf{P}\{X_i = 1\} = \mathbf{P}\{X_i = -1\} = 1/2 \quad (i = 1, 2, \ldots),$$
i.e.
$$\mathbf{P}\{X_{j_1} = \delta_1, X_{j_2} = \delta_2, \ldots, X_{j_n} = \delta_n\} = 2^{-n} \quad (1.5)$$
where $1 \leq j_1 < j_2 < \ldots < j_n$, $(n = 1, 2, \ldots)$ and $\delta_1, \delta_2, \ldots, \delta_n$ is an arbitrary sequence of $+1$'s and -1's. Let
$$S_0 = 0 \quad \text{and} \quad S_n = \sum_{k=1}^n X_k \quad (n = 1, 2, \ldots).$$
Then (1.5) implies that $\{S_n\}$ is a Markov chain, i.e.
$$\mathbf{P}\{S_{n+1} = i_{n+1} \mid S_n = i_n, S_{n-1} = i_{n-1}, \ldots, S_1 = i_1, S_0 = i_0 = 0\}$$
$$= \mathbf{P}\{S_{n+1} = i_{n+1} \mid S_n = i_n\} = 1/2 \quad (1.6)$$
where $i_0 = 0, i_1, i_2, \ldots, i_n, i_{n+1}$ is a sequence of integers with $|i_1 - i_0| = |i_2 - i_1| = \ldots = |i_{n+1} - i_n| = 1$.

Chapter 2

Distributions

2.1 Exact distributions

A trivial combinatorial argument gives

THEOREM 2.1

$$\mathbf{P}\{S_{2n} = 2k\} = \binom{2n}{n-k} 2^{-2n} \tag{2.1}$$

where $k = -n, -n+1, \ldots, n; \ n = 1, 2, \ldots,$

$$\mathbf{P}\{S_{2n+1} = 2k+1\} = \binom{2n+1}{n-k} 2^{-2n-1} \tag{2.2}$$

where $k = -n-1, -n, \ldots, n; \ n = 1, 2, \ldots$. (2.1) and (2.2) together give

$$\mathbf{P}\{S_n = k\} = \begin{cases} \binom{n}{(n-k)/2} 2^{-n} & \text{if } k \equiv n \pmod{2}, \\ 0 & \text{otherwise} \end{cases} \tag{2.3}$$

where $k = -n, -n+1, \ldots, n; \ n = 1, 2, \ldots$. Further, for any $n = 1, 2, \ldots$, $t \in \mathbb{R}^1$ we have

$$\mathbf{E}S_n = 0, \quad \mathbf{E}S_n^2 = n, \quad \mathbf{E}\exp(tS_n) = \left(\frac{e^t + e^{-t}}{2}\right)^n. \tag{2.4}$$

The following inequality (Bernstein inequality) can also be obtained by elementary methods:

THEOREM 2.2 (cf. e.g. Rényi, 1970/B, p. 387).

$$\mathbf{P}\left\{\left|\frac{S_n}{n}\right| \geq \varepsilon\right\} \leq 2\exp\left(-\frac{2n\varepsilon^2}{(1+2\varepsilon)^2}\right)$$

for any $n = 1, 2, \ldots$ and $0 < \varepsilon \leq 1/4$.

For later reference we present also a slightly more general form of the Bernstein inequality.

THEOREM 2.3 (cf. Rényi, 1970/B, p. 387). *Let $X_1^*, X_2^*, \ldots$ be a sequence of i.i.d.r.v.'s with*

$$\mathbf{P}\{X_1^* = 1\} = 1 - \mathbf{P}\{X_1^* = 0\} = p.$$

Then for any $0 < \varepsilon \le pq$ we have

$$\mathbf{P}\left\{\left|\frac{S_n^*}{n} - p\right| \ge \varepsilon\right\} \le 2\exp\left(-\frac{n\varepsilon^2}{2pq\left(1 + \frac{\varepsilon}{2pq}\right)^2}\right)$$

where $S_n^* = X_1^* + X_2^* + \cdots + X_n^*$ and $q = 1 - p$.

A slightly more precise form of the above Theorem is the so-called

LARGE DEVIATION THEOREM (cf. Durrett, 1991, p. 61). *Let $0 < a < 1$. Then*

$$\lim_{n\to\infty} n^{-1}\log \mathbf{P}\{S_n^* \ge na\} = -\frac{a}{2}\log\frac{q(1+a)}{p(1-a)} + \frac{1}{2}\log\frac{4pq}{1-a^2}.$$

THEOREM 2.4 (cf. e.g. Rényi, 1970/A, p. 233).

$$p_{n,k} = \mathbf{P}\{M_n^+ = k\}$$
$$= \binom{n}{[(n-k)/2]} 2^{-n} \quad (k = 0, 1, 2, \ldots, n;\ n = 1, 2, \ldots), \tag{2.5}$$

and for any $t \in \mathbb{R}^1$

$$\psi_{2n}(t) = \mathbf{E}\exp(tM_{2n}^+)$$
$$= 2^{-2n}\sum_{k=0}^{2n}\binom{2n}{[(2n-k)/2]}e^{kt} \le (1+e^t)\left(\frac{e^t + e^{-t}}{2}\right)^{2n}. \tag{2.6}$$

Proof 1 of (2.5). (Rényi, 1970/A, p. 233). Let

$$\bar{M}_n^+ = \max_{1 \le k \le n+1} \sum_{j=2}^k X_j.$$

Then

$$p_{n+1,k} = \mathbf{P}\{X_1 = 1, \bar{M}_n^+ = k-1\} + \mathbf{P}\{X_1 = -1, \bar{M}_n^+ = k+1\}$$
$$= \frac{1}{2}(p_{n,k-1} + p_{n,k+1}) \quad (k \ge 1). \tag{2.7}$$

Similarly for $k = 0$

$$p_{n+1,0} = \mathbf{P}\{X_1 = -1, \bar{M}_n^+ \leq 1\} = \frac{1}{2}(p_{n,1} + p_{n,0}). \tag{2.8}$$

Since $p_{1,0} = p_{1,1} = 1/2$ we get (2.5) from (2.7) and (2.8) by induction.

Proof 2 of (2.5). Clearly

$$\mathbf{P}\{M_n^+ \geq k\} = \mathbf{P}\{S_n \geq k\} + \mathbf{P}\{S_n < k, M_n^+ \geq k\}$$

$$= 2^{-n} \sum_{\substack{j=k \\ j \equiv n \pmod{2}}}^{n} \binom{n}{\frac{n-j}{2}} + \mathbf{P}\{S_n < k, M_n^+ \geq k\}.$$

Let

$$\rho_1(k) = \min\{l : S_l = k\},$$

$$S_l^{(k)} = \begin{cases} S_l & \text{if } l \leq \rho_1(k), \\ k - \sum_{j=\rho_1(k)+1}^{l} X_j & \text{if } l > \rho_1(k), \end{cases}$$

i.e. $S_l^{(k)}$ for $l > \rho_1(k)$ is the reflection of S_l in the mirror $y = k$. (Hence the method of this proof is called *reflection principle*.) Then

$$\mathbf{P}\{M_n^+ \geq k\} = 2^{-n} \sum_{\substack{j=k \\ j \equiv n \pmod{2}}}^{n} \binom{n}{(n-j)/2} + \mathbf{P}\{S_n < k, \rho_1^{(k)} < n\}$$

$$= 2^{-n} \sum_{\substack{j=k \\ j \equiv n \pmod{2}}}^{n} \binom{n}{(n-j)/2} + \mathbf{P}\{S_n^{(k)} > k\}$$

$$= 2^{-n} \sum_{\substack{j=k \\ j \equiv n \pmod{2}}}^{n} \binom{n}{(n-j)/2} + 2^{-n} \sum_{\substack{j=k+1 \\ j \equiv n \pmod{2}}}^{n} \binom{n}{(n-j)/2}$$

$$= 2\mathbf{P}\{S_n > k\} + \mathbf{P}\{S_n = k\} = 2^{-n} \sum_{j=k}^{n} \binom{n}{[(n-j)/2]} \tag{2.9}$$

which proves (2.5).

(2.6) can be obtained by a direct calculation.

THEOREM 2.5 *For any integers $a \leq 0 \leq b$, $a < b$, $a \leq \nu \leq b$ we have*

$$p_n(a,b,\nu) = \mathbf{P}\{a < -M_n^- \leq M_n^+ < b, S_n = \nu\}$$
$$= \sum_{k=-\infty}^{\infty} q_n(\nu + 2k(b-a)) - \sum_{k=-\infty}^{\infty} q_n(2b - \nu + 2k(b-a)) \quad (2.10)$$

where

$$q_n(j) = \mathbf{P}\{S_n = j\} = \begin{cases} \binom{n}{(n-j)/2} 2^{-n} & \text{if } j \equiv n \pmod{2}, \\ 0 & \text{otherwise} \end{cases}$$

$(j = -n, -n+1, \ldots, n; \; n = 0, 1, \ldots).$

Proof. (Billingsley, 1968, p. 78). In case $n = 0$

$$p_0(a,b,\nu) = \begin{cases} 1 & \text{if } \nu = 0 \text{ and } a^2 + b^2 > 0, \\ 0 & \text{otherwise,} \end{cases}$$

and we obtain (2.10) easily. Assume that (2.10) holds for $n-1$ and for any a, b, ν satisfying the conditions of the Theorem. Now we prove (2.10) by induction. Note that $p_n(0, b, \nu) = p_n(a, 0, \nu) = 0$ and the same is true for the righthand side of (2.10) (since the terms cancel because $q_n(j) = q_n(-j)$). Hence we may assume that $a < 0 < b$. But in this case $a + 1 \leq 0$ and $b - 1 \geq 0$. Hence by induction (2.10) holds with parameters $n-1, a+1, b+1, \nu$ and $n-1, a-1, b-1, \nu$. We obtain (2.10) observing that

$$q_n(j) = \frac{1}{2} q_{n-1}(j-1) + \frac{1}{2} q_{n-1}(j+1),$$

and

$$p_n(a,b,\nu) = \frac{1}{2} p_{n-1}(a-1, b-1, \nu-1) + \frac{1}{2} p_{n-1}(a+1, b+1, \nu+1).$$

THEOREM 2.6 *For any integers $a \leq 0 \leq b$ and $a \leq u \leq \nu \leq b$ we have*

$$\mathbf{P}\{a < -M_n^- \leq M_n^+ < b, u < S_n < \nu\}$$
$$= \sum_{k=-\infty}^{\infty} \mathbf{P}\{u + 2k(b-a) < S_n < \nu + 2k(b-a)\}$$
$$- \sum_{k=-\infty}^{\infty} \mathbf{P}\{2b - \nu + 2k(b-a) < S_n < 2b - u + 2k(b-a)\}, \quad (2.11)$$

$$\mathbf{P}\{a < -M_n^- \le M_n^+ < b\}$$
$$= \sum_{k=-\infty}^{\infty} \mathbf{P}\{a + 2k(b-a) < S_n < b + 2k(b-a)\}$$
$$- \sum_{k=-\infty}^{\infty} \mathbf{P}\{b + 2k(b-a) < S_n < 2b - a + 2k(b-a)\} \qquad (2.12)$$

and

$$\mathbf{P}\{M_n < b\} = \sum_{k=-\infty}^{\infty} \mathbf{P}\{(4k-1)b < S_n < (4k+1)b\}$$
$$- \sum_{k=-\infty}^{\infty} \mathbf{P}\{(4k+1)b < S_n < (4k+3)b\}. \qquad (2.13)$$

(2.11) is a simple consequence of (2.10), (2.12) follows from (2.11) taking $u = a$, $\nu = b$ and (2.13) follows from (2.12) taking $a = -b$.

To evaluate the distribution of $I_i(n,a)$ ($i = 1,2,3,4,5$) seems to be very hard (cf. Notations to the Increments). However, we can get some information about the distribution of $I_1(n,a)$.

LEMMA 2.1 (Erdős – Révész, 1976).

$$p(n+j, n) := \mathbf{P}\{I_1(n+j, n) = n\} = \frac{j+2}{2^{n+1}} \quad (j = 0, 1, 2, \ldots, n).$$

Clearly $p(n+j, n)$ is the probability that a coin tossing sequence of length $n + j$ contains a pure-head-run of length n.

Proof. Let

$$A = \{I_1(n+j, n) = n\} \quad \text{and} \quad A_k = \{S_{k+n} - S_k = n\}.$$

Then

$$A = A_0 + \bar{A}_0 A_1 + \bar{A}_0 \bar{A}_1 A_2 + \cdots + \bar{A}_0 \bar{A}_1 \cdots \bar{A}_{j-1} A_j$$
$$= A_0 + \bar{A}_0 A_1 + \bar{A}_1 A_2 + \cdots + \bar{A}_{j-1} A_j.$$

Since $\mathbf{P}\{A_0\} = 2^{-n}$ and $\mathbf{P}\{\bar{A}_0 \bar{A}_1 \ldots \bar{A}_j A_{j+1}\} = 2^{-n-1}$ for any $j = 1, 2, \ldots$, we have the Lemma.

The next recursion can be obtained in a similar way.

LEMMA 2.2 *For any $j = 1, 2, \ldots$ we have*

$$\mathbf{P}\{I_1(2n+j,n) = n\}$$
$$= p(2n+j,n) = \frac{n+2}{2^{n+1}} + \left(1 - \frac{1}{2^n}\right)\frac{1}{2^{n+1}} + (1-p(n+1,n))\frac{1}{2^{n+1}}$$
$$+ (1-p(n+2,n))\frac{1}{2^{n+1}} + \ldots + (1-p(n+j-1,n))\frac{1}{2^{n+1}}.$$

In case $j \leq n$ we obtain

$$p(2n+j,n) = \frac{n+2+j}{2^{n+1}} - \frac{1}{2^{2n+3}}j(j+3).$$

In some cases it is worthwhile to have a less exact but simpler formula. For example, we have

LEMMA 2.3 (Deheuvels – Erdős – Grill – Révész, 1987).

$$(j+2)2^{-n-1} - (j+2)^2 2^{-2n-2} \leq \mathbf{P}\{I_1(n+j,n) = n\} \leq (j+2)2^{-n-1} \quad (2.14)$$

for any $n = 1, 2, \ldots; \ j = 1, 2, \ldots$.

The idea of the proof is the same as those of the above two lemmas. The details are omitted.

The exact distribution of Z_n (cf. Notations to the Increments) is also known, namely:

THEOREM 2.7 (Székely – Tusnády, 1979).

$$\mathbf{P}\{Z_n < s\} = 2^{-n} F_n^{(s)}$$

where

$$F_n^{(s)} = \sum_{k=1}^{n+1}\sum_{j=0}^{k}(-1)^j \binom{k}{j}\binom{n-sj}{k-1}$$
$$= \sum_{j=0}^{n}(-1)^j \left\{\binom{n-sj}{j-1}2^{n-j(s+1)+1} + \binom{n-sj}{j}2^{n-j(s+1)}\right\}.$$

Remark 1. Csáki, Földes and Komlós (1987) worked out a very general method to obtain inequalities like (2.14). Their method gives a somewhat weaker result than (2.14). However, their result is also strong enough to produce most of the strong theorems given later (cf. Section 7.3).

2.2 Limit distributions

Utilizing the Stirling formula

$$n! = \left(\frac{n}{e}\right)^n (2\pi n)^{1/2} \exp\left(\frac{\sigma_n}{12n}\right)$$

(where $0 < \sigma_n < 1$) and the results of Section 2.1, the following limit theorems can be obtained.

THEOREM 2.8 (e.g. Rényi, 1970/A, p. 208). *Assume that for some $0 < \varepsilon < 1/2$ the inequality $\varepsilon n < k < (1-\varepsilon)n$ is satisfied. Then*

$$\binom{n}{k} 2^{-n} = \frac{2^{nd(K)}}{(2\pi n K(1-K))^{1/2}} \left(1 + O\left(\frac{1}{n}\right)\right)$$

where $K = k/n$ and $d(K) = K \log 2K + (1-K) \log 2(1-K)$. If we also assume that $|k - n/2| = o(n^{2/3})$ then

$$\binom{n}{k} 2^{-n} = (\pi n/2)^{-1/2} \exp\left(-\frac{(k-n/2)^2}{n/2}(1+o(1))\right).$$

Especially

$$\binom{2n}{n} 2^{-2n} \approx (\pi n)^{-1/2}.$$

The next theorem is the so-called *Central Limit Theorem*.

THEOREM 2.9 (Gnedenko – Kolmogorov, 1954, §40).

$$\sup_x |\mathbf{P}\{n^{-1/2} S_n < x\} - \Phi(x)| \leq 2n^{-1/2}.$$

A stronger version of Theorem 2.9 is another form of the Large Deviation Theorem:

THEOREM 2.10 (e.g. Feller, 1966, p. 517).

$$\lim_{n\to\infty} \frac{\mathbf{P}\{n^{-1/2} S_n < -x_n\}}{\Phi(-x_n)} = \lim_{n\to\infty} \frac{\mathbf{P}\{n^{-1/2} S_n > x_n\}}{1 - \Phi(x_n)} = 1$$

provided that $0 < x_n = o(n^{1/6})$.

Theorem 2.10 can be generalized as follows:

THEOREM 2.11 (e.g. Feller, 1966, p. 517). *Let $X_1^*, X_2^*, \ldots$ be a sequence of i.i.d.r.v.'s with*

$$\mathbf{E}X_i^* = 0, \quad \mathbf{E}(X_i^*)^2 = 1, \quad \mathbf{E}(\exp(tX_i^*)) < \infty$$

for all t in some interval $|t| < t_0$. Then

$$\lim_{n \to \infty} \frac{\mathbf{P}\{n^{-1/2}S_n^* < -x_n\}}{\Phi(-x_n)} = \lim_{n \to \infty} \frac{\mathbf{P}\{n^{-1/2}S_n^* > x_n\}}{1 - \Phi(x_n)} = 1$$

provided that $0 < x_n = o(n^{1/6})$ where $S_n^* = X_1^* + X_2^* + \cdots + X_n^*$.

THEOREM 2.12 (e.g. Rényi, 1970/A, p. 234).

$$\lim_{n \to \infty} \mathbf{P}\{n^{-1/2}M_n^+ < x\} = \mathbf{P}\{|N| < x\} = 2\Phi(x) - 1$$

uniformly in $x \in \mathbb{R}^1$ where $N \in N(0,1)$. Further,

$$\lim_{n \to \infty} \mathbf{E}(n^{-1/2}M_n^+) = (2/\pi)^{1/2}.$$

THEOREM 2.13

$$\lim_{n \to \infty} \mathbf{P}\{n^{-1/2}M_n < x\} = G(x) = H(x)$$

uniformly in $x \in \mathbb{R}^1$. Further,

$$\lim_{n \to \infty} \frac{\mathbf{P}\{n^{-1/2}M_n > x_n\}}{1 - G(x_n)} = \lim_{n \to \infty} \frac{\mathbf{P}\{n^{-1/2}M_n < x_n^{-1}\}}{H(x_n^{-1})} = 1$$

where

$$G(x) = \frac{1}{\sqrt{2\pi}} \int_{-x}^{x} \sum_{k=-\infty}^{+\infty} (-1)^k \exp\left(-\frac{(u - 2kx)^2}{2}\right) du,$$

$$H(x) = \frac{4}{\pi} \sum_{k=0}^{\infty} \frac{(-1)^k}{2k+1} \exp\left(-\frac{\pi^2(2k+1)^2}{8x^2}\right)$$

provided that $0 < x_n = o(n^{1/6})$. Consequently for any $\varepsilon > 0$

$$\mathbf{P}\{n^{-1/2}M_n > x_n\} \geq (1-\varepsilon)(1 - G(x_n))$$
$$\geq 2(1-\varepsilon)(1 - \Phi(x_n)) \geq (1-\varepsilon)\sqrt{\frac{2}{\pi}} \left(\frac{1}{x_n} - \frac{1}{x_n^3}\right) e^{-x_n^2/2}, \quad (2.15)$$

DISTRIBUTIONS

$$\mathbf{P}\{n^{-1/2}M_n > x_n\} \le (1+\varepsilon)(1-G(x_n))$$
$$\le (1+\varepsilon)\left[1 - \frac{1}{\sqrt{2\pi}}\int_{-x_n}^{x_n} e^{-u^2/2}du + \frac{2}{\sqrt{2\pi}}\int_{x_n}^{3x_n} e^{-u^2/2}du\right]$$
$$\le (1+\varepsilon)\frac{4}{\sqrt{2\pi}}\frac{1}{x_n}e^{-x_n^2/2}, \tag{2.16}$$

$$\mathbf{P}\{n^{-1/2}M_n < x_n^{-1}\} \le (1+\varepsilon)H(x_n^{-1}) \le \frac{4(1+\varepsilon)}{\pi}\exp\left(-\frac{\pi^2}{8}x_n^2\right) \tag{2.17}$$

and

$$\mathbf{P}\{n^{-1/2}M_n < x_n^{-1}\} \ge (1-\varepsilon)H(x_n^{-1})$$
$$\ge \frac{4(1-\varepsilon)}{\pi}\left[\exp\left(-\frac{\pi^2}{8}x_n^2\right) - \frac{1}{3}\exp\left(-\frac{9\pi^2}{8}x_n^2\right)\right] \tag{2.18}$$

if $0 < x_n = o(n^{1/6})$ and n is large enough.

Remark 1. As we claimed $G(x) = H(x)$ however in Theorem 2.13 the asymptotic distribution in the form of $G(\cdot)$ is proposed to be used when x is large. When x is small, $H(\cdot)$ is more adequate.

Finally we present the limit distribution of Z_n.

THEOREM 2.14 (Földes, 1975, Goncharov, 1944). *For any positive integer k we have*

$$\mathbf{P}\{Z_n - [\lg N] < k\} = \exp(-2^{-(k+1)-\{\lg N\}}) + o(1)$$

where $\{\lg N\} = \lg N - [\lg N]$.

Remark 2. As we have mentioned earlier the above Theorems can be proved using the analogous exact theorems and the Stirling formula. Indeed this method (at least theoretically) is always applicable, but it often requires very hard work. Hence sometimes it is more convenient to use characteristic functions or other analytic methods.

Chapter 3

Recurrence and the Zero-One Law

3.1 Recurrence

One of the most classical strong theorems on random walk claims that the particle returns to the origin infinitely often with probability 1. That is

RECURRENCE THEOREM (Pólya, 1921).

$$\mathbf{P}\{S_n = 0 \quad i.o.\} = 1.$$

We present three proofs of this theorem. The first one is based on the following lemma:

LEMMA 3.1 *Let $0 \leq i \leq k$. Then for any $m \geq i$ we have*

$$\begin{aligned} &p(0,i,k) \\ &= \mathbf{P}\{\min\{j : j \geq m, S_j = 0\} < \min\{j : j \geq m, S_j = k\} \mid S_m = i\} \\ &= k^{-1}(k-i), \end{aligned} \quad (3.1)$$

i.e. the probability that a particle starting from i hits 0 before k is $k^{-1}(k-i)$.

Proof. Clearly we have

$$p(0,0,k) = 1, \quad p(0,k,k) = 0.$$

When the particle is located in i then it hits 0 before k if

(i) either it goes to $i-1$ (with probability $1/2$) and from $i-1$ goes to 0 before k (with probability $p(0, i-1, k)$),

(ii) or it goes to $i+1$ (with probability $1/2$) and from $i+1$ goes to 0 before k (with probability $p(0, i+1, k)$).

That is

$$p(0,i,k) = \frac{1}{2}p(0,i-1,k) + \frac{1}{2}p(0,i+1,k)$$

($i = 1, 2, \ldots, k-1$). Hence $p(0, i, k)$ is a linear function of i, being 1 in 0 and 0 in k, which implies (3.1).

Proof 1 of the Recurrence Theorem. Assume that $S_1 = 1$, say. By (3.1) for any $\varepsilon > 0$ there exists a positive integer $n_0 = n_0(\varepsilon)$ such that $p(0, 1, n) = 1 - 1/n \geq 1 - \varepsilon$ if $n \geq n_0$. Consequently the probability that the particle returns to 0 is larger than $1 - \varepsilon$ for any $\varepsilon > 0$. Hence the particle returns to 0 with probability 1 at least once. Having one return, the probability of a second return is again 1. In turn it implies that the particle returns to 0 infinitely often with probability 1.

Proof 2 of the Recurrence Theorem. Introduce the following notations

$$p_0 = 1,$$
$$p_{2k} = \mathbf{P}\{S_{2k} = 0\} = 2^{-2k}\binom{2k}{k} \quad (k = 1, 2, \ldots),$$
$$A_{2k} = \{S_{2k} = 0, \ S_{2k-2} \neq 0, \ S_{2k-4} \neq 0, \ldots, S_2 \neq 0\},$$
$$q_{2k} = \mathbf{P}\{A_{2k}\},$$
$$P(z) = \sum_{k=0}^{\infty} p_{2k} z^{2k} \quad (|z| < 1),$$
$$Q(z) = \sum_{k=1}^{\infty} q_{2k} z^{2k} \quad (|z| < 1).$$

(Note that q_{2k} is the probability of the event that the first return of the particle to the origin occurs in the $(2k)$-th step but not before.)

Since $p_{2k} \approx (\pi k)^{-1/2}$ (cf. Theorems 2.1 and 2.8) we have

$$\lim_{z \nearrow 1} P(z) = \infty.$$

Observe that

$$\{S_{2k} = 0\} = A_{2k} + \sum_{j=1}^{k-1} A_{2k-2j}\{S_{2k} = 0\}$$

and

$$\mathbf{P}\{A_{2k-2j}\{S_{2k} = 0\}\} = q_{2k-2j} p_{2j}.$$

Hence

(0) $p_0 = 1,$
(i) $p_2 = q_2,$
(ii) $p_4 = q_4 + q_2 p_2,$
(iii) $p_6 = q_6 + q_4 p_2 + q_2 p_4, \cdots$
(k) $p_{2k} = q_{2k} + q_{2k-2} p_2 + \cdots + q_2 p_{2k-2}, \cdots$

Multiplying the k-th equation by z^{2k} ($|z| < 1$) and summing up to infinity we obtain
$$P(z) = P(z)Q(z) + 1,$$
i.e.
$$Q(z) = 1 - \frac{1}{P(z)} \quad \text{and} \quad \lim_{z \nearrow 1} Q(z) = 1.$$
Since $Q(1) = \sum_{k=1}^{\infty} q_{2k} = 1$ is the probability that the particle returns to the origin at least once, we obtain the theorem.

3.2 The zero-one law

The above two proofs of the Recurrence Theorem are based on the fact that if
$$\mathbf{P}\{S_n = 0 \text{ at least for one } n\} = 1$$
then
$$\mathbf{P}\{S_n = 0 \text{ i.o.}\} = 1.$$
Similarly one can see that if $\mathbf{P}\{S_n = 0 \text{ at least for one } n\}$ were less than 1 then $\mathbf{P}\{S_n = 0 \text{ i.o.}\}$ would be equal to 0. Hence without any calculation one can see that $\mathbf{P}\{S_n = 0 \text{ i.o.}\}$ is equal to 0 or 1. Consequently in order to prove the Recurrence Theorem it is enough to prove that $\mathbf{P}\{S_n = 0 \text{ i.o.}\} > 0$.

In the study of the behaviour of the infinite sequences of independent r.v.'s we frequently realize that the probabilities of certain events can be only 0 or 1. Roughly speaking we have: let $Y_1, Y_2, \ldots$ be a sequence of independent r.v.'s. Then, if A is an event depending on $Y_n, Y_{n+1}, \ldots$ (but it is independent from $Y_1, Y_2, \ldots, Y_{n-1}$) for every n, it follows that the probability of A equals either 0 or 1. More formally speaking we have

ZERO-ONE LAW (Kolmogorov, 1933). *Let $Y_1, Y_2, \ldots$ be independent r.v.'s. Then if $A \subset \Omega$ is a set, measurable on the sample space of $Y_n, Y_{n+1}, \ldots$ for every n, it follows that*
$$\mathbf{P}\{A\} = 0 \quad \text{or} \quad \mathbf{P}\{A\} = 1.$$

Example 1. Let $Y_1, Y_2, \ldots$ be independent r.v.'s. Then $\sum_{i=1}^{\infty} Y_i$ converges a.s. or diverges a.s.

Having the zero-one law we present a third proof of the Recurrence Theorem. It is based on the following:

LEMMA 3.2 *For any $-\infty < a \leq b < +\infty$ we have*
$$\mathbf{P}\{\liminf_{n \to \infty} S_n = a\} = \mathbf{P}\{\limsup_{n \to \infty} S_n = b\} = 0.$$

Proof is trivial.

Proof 3 of the Recurrence Theorem. Lemma 3.2, the zero-one law and the fact that S_n is symmetrically distributed clearly imply that

$$\mathbf{P}\{\liminf_{n\to\infty} S_n = -\infty\} = \mathbf{P}\{\limsup_{n\to\infty} S_n = \infty\} = 1 \qquad (3.2)$$

which in turn implies the Recurrence Theorem.

Note that (3.2) is equivalent to the Recurrence Theorem. In fact the Recurrence Theorem implies (3.2) without having used the zero-one law.

Chapter 4

From the Strong Law of Large Numbers to the Law of Iterated Logarithm

4.1 Borel–Cantelli lemma and Markov inequality

The proofs of almost all strong theorems are based on different forms of the Borel–Cantelli lemma and those of the Markov inequality. Here we present the most important versions.

BOREL–CANTELLI LEMMA 1 *Let $A_1, A_2, \ldots$ be a sequence of events for which $\sum_{n=1}^{\infty} \mathbf{P}\{A_n\} < \infty$. Then*

$$\mathbf{P}\{\limsup_{n \to \infty} A_n\} = \mathbf{P}\left\{\prod_{n=1}^{\infty} \sum_{i=n}^{\infty} A_i\right\} = \mathbf{P}\{A_n \text{ i.o.}\} = 0,$$

i.e. with probability 1 only a finite number of the events A_n occur simultaneously.

BOREL–CANTELLI LEMMA 2 *Let $A_1, A_2, \ldots$ be a sequence of pairwise independent events for which $\sum_{n=1}^{\infty} \mathbf{P}\{A_n\} = \infty$. Then*

$$\mathbf{P}\{\limsup_{n \to \infty} A_n\} = 1,$$

i.e. with probability 1 an infinite number of the events A_n occur simultaneously.

BOREL–CANTELLI LEMMA 2*(Spitzer, 1964). *Let $A_1, A_2, \ldots$ be a sequence of events for which*

$$\sum_{n=1}^{\infty} \mathbf{P}\{A_n\} = \infty \quad \text{and} \quad \liminf_{n \to \infty} \frac{\sum_{k=1}^{n} \sum_{i=1}^{n} \mathbf{P}\{A_k A_i\}}{\left(\sum_{k=1}^{n} \mathbf{P}\{A_k\}\right)^2} \leq C \quad (C \geq 1).$$

Then
$$\mathbf{P}\{\limsup_{n\to\infty} A_n\} \geq C^{-1}.$$

MARKOV INEQUALITY *Let X be a non-negative r.v. with $\mathbf{E}X < \infty$. Then for any $\lambda > 0$*
$$\mathbf{P}\{X \geq \lambda \mathbf{E}X\} \leq \frac{1}{\lambda}.$$

As a simple consequence of the Markov inequality we obtain

CHEBYSHEV INEQUALITY *Let X be an r.v. with $\mathbf{E}X^2 < \infty$. Then for any $\lambda > 0$*

$$\mathbf{P}\{|X-\mathbf{E}X| \geq \lambda(\mathbf{E}(X-\mathbf{E}X)^2)^{1/2}\} = \mathbf{P}\{(X-\mathbf{E}X)^2 \geq \lambda^2 \mathbf{E}(X-\mathbf{E}X)^2\} \leq \frac{1}{\lambda^2}.$$

Similarly we get

THEOREM 4.1 *Let X be an r.v. with $\mathbf{E}(\exp(tX)) < \infty$ for some $t > 0$. Then for any $\lambda > 0$ we have*

$$\mathbf{P}\{X \geq \lambda\} = \mathbf{P}\{\exp(tX) \geq e^{\lambda t}\} \leq \frac{\mathbf{E}e^{tX}}{e^{\lambda t}}.$$

Borel–Cantelli lemmas 1 and 2 and Markov inequality can be found practically in any probability book (see e.g. Rényi, 1970/B).

4.2 The strong law of large numbers

THEOREM OF BOREL (1909).

$$\lim_{n\to\infty} n^{-1} S_n = 0 \quad a.s. \tag{4.1}$$

Remark 1. Applying this theorem for dyadic expansion, for almost all $x \in [0, 1]$ we obtain $\lim_{n\to\infty} n^{-1} N_n(x) = 1/2$. In fact the original theorem of Borel was formulated in this form. Borel also observed that if instead of the dyadic expansion we consider t-adic expansion ($t = 2, 3, \ldots$) of $x \in [0, 1]$ and $N_n(x, s, t)$ ($s = 0, 1, 2, \ldots, t-1$, $t = 2, 3, \ldots$) is the number of s's among the first n digits of the t-adic expansion of x, then

$$\lim_{n\to\infty} \left| \frac{N_n(x, s, t)}{n} - \frac{1}{t} \right| = 0 \quad (s = 0, 1, 2, \ldots, t-1;\ t = 2, 3, \ldots) \tag{4.2}$$

for almost all x. Hence Borel introduced the following:

Definition. A number $x \in [0,1]$ is *normal* if for any $s = 0, 1, 2, \ldots, t-1$; $t = 2, 3, \ldots$ (4.2) holds.

The above result easily implies

THEOREM 4.2 (Borel, 1909).

Almost all $x \in [0,1]$ are normal.

It is interesting to note that in spite of the fact that almost all $x \in [0,1]$ are normal it is hard to find any concrete normal number.

Proof 1 of (4.1). (Gap method). Clearly (cf. (2.4))

$$\mathbf{E} n^{-1} S_n = 0, \quad \mathbf{E} n^{-2} S_n^2 = n^{-1}.$$

Hence by Chebyshev inequality for any $\varepsilon > 0$

$$\mathbf{P}\{|n^{-1} S_n| \geq \varepsilon\} \leq n^{-1} \varepsilon^{-2}$$

and by Borel–Cantelli lemma 1

$$n^{-2} S_{n^2} \to 0 \quad \text{a.s.} \quad (n \to \infty).$$

Now we have to estimate the value of S_k for the k's lying in the gap, i.e. between n^2 and $(n+1)^2$. If $n^2 \leq k < (n+1)^2$ then

$$|k^{-1} S_k| = |n^{-2} S_{n^2} n^2 k^{-1} + k^{-1}(S_k - S_{n^2})| \leq |n^{-2} S_{n^2}| + k^{-1}((n+1)^2 - n^2).$$

Since both members of the right-hand side tend to 0, the proof is complete.

Proof 2 of (4.1). (Method of high moments). A simple calculation gives

$$\mathbf{E}(n^{-4} S_n^4) = n^{-3} + 6 \binom{n}{2} n^{-4} = O(n^{-2}),$$

and again the Markov inequality and the Borel–Cantelli lemma imply the theorem.

As we will see later on most of the proofs of the strong theorems are based on a joint application of the above two methods.

4.3 Between the strong law of large numbers and the law of iterated logarithm

The Theorem of Borel claims that the distance of the particle from the origin after n steps is $|S_n| = o(n)$ a.s. It is natural to ask whether a better rate can be obtained. In fact we have

THEOREM OF HAUSDORFF (1913). *For any $\varepsilon > 0$*

$$\lim_{n\to\infty} n^{-1/2-\varepsilon} S_n = 0 \quad a.s.$$

Proof. Let K be a positive integer. Then a simple calculation gives

$$\mathbf{E} S_n^{2K} = O(n^K).$$

Hence the Markov inequality implies

$$\mathbf{P}\{|S_n| \geq n^{1/2+\varepsilon}\} = \mathbf{P}\{S_n^{2K} \geq n^{K+\varepsilon K}\} \leq O(n^{-\varepsilon K}).$$

If K is so big that $\varepsilon K > 1$ then by the Borel–Cantelli lemma we obtain the theorem. (The method of high moments was applied.)

Similarly one can prove

THEOREM 4.3

$$\limsup_{n\to\infty} \frac{|S_n|}{n^{1/2}\log n} \leq 1 \quad a.s.$$

Proof. By (2.4) we have

$$\mathbf{E}\exp(n^{-1/2} S_n) \to e^{1/2} \quad (n \to \infty).$$

Hence

$$\mathbf{P}\{S_n \geq (1+\varepsilon) n^{1/2} \log n\}$$
$$= \mathbf{P}\{\exp(n^{-1/2} S_n) \geq \exp((1+\varepsilon)\log n) = n^{1+\varepsilon}\} \leq n^{-1-\varepsilon/2}$$

if n is large enough. Consequently

$$\limsup_{n\to\infty} \frac{S_n}{n^{1/2}\log n} \leq 1 \quad a.s.$$

and the statement of the theorem follows from the symmetry of S_n.

The best possible rate was obtained by Khinchine. His result is the so-called *Law of Iterated Logarithm* (LIL).

4.4 The LIL of Khinchine

THE LIL OF KHINCHINE (1923).

$$\limsup_{n\to\infty} b_n S_n = \limsup_{n\to\infty} b_n |S_n| = \limsup_{n\to\infty} b_n M_n$$
$$= \limsup_{n\to\infty} b_n M_n^+ = \limsup_{n\to\infty} b_n M_n^- = 1 \quad a.s.$$

where $b_n = (2n \log \log n)^{-1/2}$.

Proof. The proof will be presented in two steps. The first one gives an upper bound of $\limsup_{n\to\infty} b_n M_n$, the second one gives a lower bound of $\limsup_{n\to\infty} b_n S_n$. These two results combined imply the Theorem.

Step 1. We prove that for any $\varepsilon > 0$

$$\limsup_{n\to\infty} b_n M_n \leq 1 + \varepsilon \quad a.s.$$

By (2.16) we obtain

$$\mathbf{P}\{M_n \geq (1+\varepsilon)b_n^{-1}\} \leq \exp(-(1+\varepsilon)\log\log n) = (\log n)^{-1-\varepsilon} \quad (4.3)$$

if n is large enough. Let $n_k = [\Theta^k]$ ($\Theta > 1$). Then by the Borel–Cantelli lemma we get

$$M_{n_k} \leq (1+\varepsilon)b_{n_k}^{-1} \quad a.s.$$

for all but finitely many k. Let $n_k \leq n < n_{k+1}$. Then

$$M_n \leq M_{n_{k+1}} \leq (1+\varepsilon)b_{n_{k+1}}^{-1} \leq (1+2\varepsilon)b_{n_k}^{-1} \leq (1+2\varepsilon)b_n^{-1} \quad a.s.$$

provided that Θ is close enough to 1.

We obtain

$$\limsup_{n\to\infty} b_n T_n \leq 1$$

where T_n is any of S_n, $|S_n|$, M_n, M_n^+, M_n^-.

Observe that in this proof the gap method was used. However, to obtain inequality (4.3) it is not enough to evaluate the moments or the moment generating function of M_n (or that of S_n) but we have to use the stronger result of Theorem 2.13.

Step 2. Let $n_k = [\Theta^k]$ ($\Theta > 1$). Then for any $\varepsilon > 0$

$$\mathbf{P}\{b_{n_{k+1}}(S_{n_{k+1}} - S_{n_k}) \geq 1 - \varepsilon\} \geq O\left((\log n_{k+1})^{-1+\varepsilon/2}\right)$$

if k is large enough. Since the events $\{b_{n_{k+1}}(S_{n_{k+1}} - S_{n_k}) > (1-\varepsilon)\}$ are independent, we have by Borel–Cantelli lemma 2

$$b_{n_{k+1}}(S_{n_{k+1}} - S_{n_k}) > 1 - \varepsilon \qquad \text{i.o. a.s.}$$

Consequently
$$b_{n_{k+1}} S_{n_{k+1}} > 1 - \varepsilon - b_{n_{k+1}}|S_{n_k}|.$$

Applying the result proved in Step 1 we obtain

$$b_{n_{k+1}}|S_{n_k}| = b_{n_k}^{-1}|S_{n_k}|b_{n_k}b_{n_{k+1}} \leq (1+\varepsilon)b_{n_k}^{-1}b_{n_{k+1}} \leq \varepsilon \quad \text{a.s.}$$

if k and Θ are large enough. Hence $\limsup_{n\to\infty} b_n S_n \geq 1 - \varepsilon$ a.s. for any $\varepsilon > 0$, which implies the Theorem.

Note that the above Theorem clearly implies

$$\limsup_{n\to\infty} S_n = \infty, \quad \liminf_{n\to\infty} S_n = -\infty \qquad \text{a.s.},$$

which in turn implies the Recurrence Theorem of Section 3.1.

For later references we mention the following strong generalization of the LIL of Khinchine.

LIL OF HARTMAN–WINTNER (1941). *Let $Y_1, Y_2, \ldots$ be a sequence of i.i.d.r.v.'s with*
$$\mathbf{E}Y_1 = 0, \quad \mathbf{E}Y_1^2 = 1.$$

Then
$$\limsup_{n\to\infty} b_n(Y_1 + Y_2 + \cdots + Y_n) = 1 \quad a.s.$$

Remark 1. Strassen (1966) also investigated the case $\mathbf{E}Y_1^2 = \infty$. In fact he proved that if $Y_1, Y_2, \ldots$ is a sequence of i.i.d.r.v.'s with $\mathbf{E}Y_1 = 0$ and $\mathbf{E}Y_1^2 = \infty$ then

$$\limsup_{n\to\infty} b_n|Y_1 + Y_2 + \cdots + Y_n| = \infty \quad \text{a.s.}$$

Later Berkes (1972) has shown that this result of Strassen is the strongest possible one in the following sense: for any function $f(n)$ with $\lim_{n\to\infty} f(n) = 0$ there exists a sequence $Y_1, Y_2, \ldots$ of i.i.d.r.v.'s for which $\mathbf{E}Y_1 = 0$, $\mathbf{E}Y_1^2 = \infty$ and
$$\lim_{n\to\infty} b_n f(n)|Y_1 + Y_2 + \cdots + Y_n| = 0 \quad \text{a.s.}$$

Chapter 5

Lévy Classes

5.1 Definitions

The LIL of Khinchine tells us exactly (in certain sense) how far the particle can be from the origin after n steps. A trivial reformulation of the LIL is the following:
(i) for any $\varepsilon > 0$

$$S_n \leq (1+\varepsilon)b_n^{-1} \quad \text{a.s. for all but finitely many } n$$

and
(ii)
$$S_n \geq (1-\varepsilon)b_n^{-1} \quad \text{i.o. a.s.}$$

Having in mind this form of the LIL, Lévy asked how the class of those functions (or monotone increasing functions) $f(n)$ can be characterized for which

$$S_n < f(n) \quad \text{a.s.}$$

for all but finitely many n. (i) tells us that $(1+\varepsilon)b_n^{-1}$ is such a function for any $\varepsilon > 0$ and (ii) claims that $(1-\varepsilon)b_n^{-1}$ is not such a function. The LIL does not answer the question whether b_n^{-1} is such a function or not. However, one can prove that b_n^{-1} is not such a function but $(2n(\log \log n + 3/2 \log \log \log n))^{1/2}$ belongs to the mentioned class. In order to formulate the answer of Lévy's question introduce the following definitions.

Let $\{Y(t), t \geq 0\}$ be a stochastic process then

Definition 1. The function $a_1(t)$ belongs to the upper-upper class of $\{Y(t)\}$ ($a_1 \in \mathrm{UUC}(Y(t))$) if for almost all $\omega \in \Omega$ there exists a $t_0(\omega) > 0$ such that $Y(t) < a_1(t)$ if $t > t_0 = t_0(\omega)$.

Definition 2. The function $a_2(t)$ belongs to the upper-lower class of $\{Y(t)\}$ ($a_2 \in \mathrm{ULC}(Y(t))$) if for almost all $\omega \in \Omega$ there exists a sequence of positive numbers $0 < t_1 = t_1(\omega) < t_2 = t_2(\omega) < \ldots$ with $t_n \to \infty$ such that $Y(t_i) \geq a_2(t_i)$ ($i = 1, 2, \ldots$).

Definition 3. The function $a_3(t)$ belongs to the lower-upper class of $\{Y(t)\}$ ($a_3 \in \text{LUC}(Y(t))$) if for almost all $\omega \in \Omega$ there exists a sequence of positive numbers $0 < t_1 = t_1(\omega) < t_2 = t_2(\omega) < \ldots$ with $t_n \to \infty$ such that $Y(t_i) \leq a_3(t_i)$ ($i = 1, 2, \ldots$).

Definition 4. The function $a_4(t)$ belongs to the lower-lower class of $\{Y(t)\}$ ($a_4 \in \text{LLC}(Y(t))$) if for almost all $\omega \in \Omega$ there exists a $t_0 = t_0(\omega) > 0$ such that $Y(t) > a_4(t)$ if $t > t_0$.

Let $Y_1, Y_2, \ldots$ be a sequence of random variables then the four Lévy classes $\text{UUC}(Y_n)$, $\text{ULC}(Y_n)$, $\text{LUC}(Y_n)$, $\text{LLC}(Y_n)$ of $\{Y_n\}$ can be defined in the same way as it was done above for $Y(t)$.

We introduce two further definitions strongly connected with the above four definitions of the Lévy classes.

Definition 5. The process $Y(t)$ is asymptotically deterministic (AD) if there exist a function $a_1(t) \in \text{UUC}(Y(t))$ and a function $a_4(t) \in \text{LLC}(Y(t))$ such that $\lim_{t \to \infty} |a_4(t) - a_1(t)| = 0$.
Consequently
$$\lim_{t \to \infty} |a_4(t) - Y(t)| = \lim_{t \to \infty} |a_1(t) - Y(t)| = 0 \quad \text{a.s.}$$

Definition 6. The process $Y(t)$ is quasi AD (QAD) if there exist a function $a_1(t) \in \text{UUC}(Y(t))$ and a function $a_4(t) \in \text{LLC}(Y(t))$ such that $\limsup_{t \to \infty} |a_4(t) - a_1(t)| < \infty$.

The definition of AD (resp. QAD) sequences of r.v.'s can be obtained by a trivial reformulation of Definitions 5 (6).

Remark 1. Clearly $\text{UUC}(Y_n)$ (resp. $\text{UUC}(Y(t))$) is the complementer of $\text{ULC}(Y_n)$ (resp. $\text{ULC}(Y(t))$) and similarly $\text{LUC}(Y_n)$ (resp. $\text{LUC}(Y(t))$) is the complementer of $\text{LLC}(Y_n)$ (resp. $\text{LLC}(Y(t))$).

5.2 EFKP LIL

Now we formulate the celebrated Erdős (1942), Feller (1943, 1946), Kolmogorov–Petrowsky (1930–35) theorem.

EFKP LIL *The nondecreasing function $a(n) \in \text{UUC}(Y_n)$ if and only if*
$$I_1(a) := \sum_{n=1}^{\infty} \frac{a(n)}{n} \exp\left(-\frac{a^2(n)}{2}\right) < \infty$$

where Y_n is any of $n^{-1/2}S_n$, $n^{-1/2}|S_n|$, $n^{-1/2}M_n$, $n^{-1/2}M_n^+$, $n^{-1/2}M_n^-$.

This theorem completely characterises the $\text{UUC}(Y_n)$ if we take into consideration only nondecreasing functions and it implies

Consequence 1. For any $\varepsilon > 0$

$$S_n \leq (n(2\log\log n + (3+\varepsilon)\log\log\log n))^{1/2} \quad \text{a.s.} \tag{5.1}$$

for all but finitely many n. Further

$$S_n \geq (n(2\log\log n + 3\log\log\log n))^{1/2} \quad \text{i.o. a.s.} \tag{5.2}$$

Here we present the proof of Consequence 1 only instead of the proof of EFKP LIL (cf. Remark 1 at the end of Section 5.3).

Proof of Consequence 1. The proof will be presented in two steps.

Step 1. We prove that for any $\varepsilon > 0$ and for all but finitely many n

$$M_n \leq (n(2\log\log n + (3+\varepsilon)\log\log\log n))^{1/2} \quad \text{a.s.,} \tag{5.3}$$

which clearly implies (5.1). By (2.16) we obtain

$$\begin{aligned}
\mathbf{P}&\{M_n \geq (n(2\log_2 n + (3+\varepsilon)\log_3 n))^{1/2}\} \\
&\leq \frac{4(1+\varepsilon)}{\sqrt{2\pi}} \frac{1}{\sqrt{2\log_2 n}} \frac{1}{\log n} \frac{1}{(\log_2 n)^{(3+\varepsilon)/2}} \\
&= \frac{2(1+\varepsilon)}{\sqrt{\pi}} \frac{1}{\log n} \frac{1}{(\log_2 n)^{(2+\varepsilon)/2}}.
\end{aligned} \tag{5.4}$$

Let

$$n_k = \left[\exp\left(\frac{k}{\log k}\right)\right].$$

Then by the Borel–Cantelli lemma we get

$$M_{n_k} \leq (n_k(2\log_2 n_k + (3+\varepsilon)\log_3 n_k))^{1/2} \quad \text{a.s.}$$

for all but finitely many k. Let $n_k \leq n < n_{k+1}$. Then

$$\begin{aligned}
M_n \leq M_{n_{k+1}} &\leq (n_{k+1}(2\log_2 n_{k+1} + (3+\varepsilon)\log_3 n_{k+1}))^{1/2} \\
&\leq (n_k(2\log_2 n_k + (3+2\varepsilon)\log_3 n_k))^{1/2}
\end{aligned} \tag{5.5}$$

which implies (5.3).

Step 2. Introduce the following notations

$$n_l = \left[\exp\left(\frac{l}{\log l}\right)\right],$$
$$\varphi(n) = (2\log_2 n + 3\log_3 n)^{1/2},$$
$$\varphi^*(n) = (2\log_2 n + 6\log_3 n)^{1/2},$$
$$A_n = \{n^{-1/2}S_n \geq \varphi(n)\},$$
$$A_n^* = \{n^{-1/2}S_n \geq \varphi^*(n)\}.$$

Then clearly

$$\mathbf{P}\{A_{n_k}\} = O(k^{-1}(\log k)^{-1}),$$
$$\mathbf{P}\{A_{n_k}^*\} = O(k^{-1}(\log k)^{-5/2}),$$

and for any $j < k = j + m$ by (2.15) we have

$$\mathbf{P}\{A_{n_j}A_{n_k}\}$$
$$= \mathbf{P}\left\{\varphi^*(n_j) \geq n_j^{-1/2}S_{n_j} \geq \varphi(n_j),\ n_k^{-1/2}S_{n_k} \geq \varphi(n_k)\right\}$$
$$+ \mathbf{P}\left\{n_j^{-1/2}S_{n_j} \geq \varphi^*(n_j),\ n_k^{-1/2}S_{n_k} \geq \varphi(n_k)\right\}$$
$$\leq \mathbf{P}\left\{\left\{\varphi^*(n_j) \geq n_j^{-1/2}S_{n_j} \geq \varphi(n_j)\right\}\right.$$
$$\left.\cap \left\{\frac{S_{n_k} - S_{n_j}}{\sqrt{n_k - n_j}} \geq \sqrt{\frac{n_k}{n_k - n_j}}\varphi(n_k) - \sqrt{\frac{n_j}{n_k - n_j}}\varphi^*(n_j)\right\}\right\}$$
$$+ \mathbf{P}\left\{n_j^{-1/2}S_{n_j} \geq \varphi^*(n_j)\right\} = \mathbf{P}\left\{\varphi^*(n_j) \geq n_j^{-1/2}S_{n_j} \geq \varphi(n_j)\right\}$$
$$\times \mathbf{P}\left\{\frac{S_{n_k} - S_{n_j}}{\sqrt{n_k - n_j}} \geq \sqrt{\frac{n_k}{n_k - n_j}}\varphi(n_k) - \sqrt{\frac{n_j}{n_k - n_j}}\varphi^*(n_j)\right\}$$
$$+ O\left(j^{-1}(\log j)^{-5/2}\right)$$
$$= \left(O\left(j^{-1}(\log j)^{-1}\right) - O\left(j^{-1}(\log j)^{-5/2}\right)\right)\mathcal{P}_k$$
$$+ O\left(j^{-1}(\log j)^{-5/2}\right)$$

where

$$\mathcal{P}_k = \mathbf{P}\left\{\frac{S_{n_k} - S_{n_j}}{\sqrt{n_k - n_j}} \geq \mathcal{L}\right\}$$

and

$$\mathcal{L} = \sqrt{\frac{n_k}{n_k - n_j}}\varphi(n_k) - \sqrt{\frac{n_j}{n_k - n_j}}\varphi^*(n_j).$$

Observe that
$$\mathcal{L} = \varphi(n_k)\frac{\sqrt{n_k} - \sqrt{n_j}\frac{\varphi^*(n_j)}{\varphi(n_k)}}{\sqrt{n_k - n_j}}$$

and
$$\frac{\varphi^*(n_j)}{\varphi(n_k)} \leq \frac{\varphi^*(n_j)}{\varphi(n_j)} \leq 1 + \frac{3}{4}\frac{\log\log j}{\log j}.$$

Hence
$$\mathcal{L} \geq \varphi(n_k)\left(\frac{\sqrt{n_k} - \sqrt{n_j}}{\sqrt{n_k - n_j}} - \frac{\sqrt{n_j}}{\sqrt{n_k - n_j}}\frac{3}{4}\frac{\log\log j}{\log j}\right).$$

Since
$$\frac{\sqrt{x} - 1}{\sqrt{x - 1}} > \frac{1}{3}\sqrt{x - 1} \quad \text{if} \quad 1 < x < 4$$

and
$$\frac{\sqrt{x} - 1}{\sqrt{x - 1}} \geq 1 - \frac{1}{\sqrt{x}} \quad \text{if} \quad x > 1,$$

with
$$x = \frac{n_k}{n_j} = \exp\left(\frac{j + m}{\log(j + m)} - \frac{j}{\log j}\right)$$
$$= \exp\left(\frac{m}{\log(j + m)} - \frac{j}{\log j}\frac{\log\left(1 + \frac{m}{j}\right)}{\log(j + m)}\right)$$

for any j large enough, we obtain
$$\frac{\sqrt{n_k} - \sqrt{n_j}}{\sqrt{n_k - n_j}} \geq \frac{1}{3}\left(\frac{n_k}{n_j} - 1\right)^{1/2}$$
$$\geq \frac{1}{3}\left(\frac{m}{\log(j + m)} - \frac{j}{\log j}\frac{\log\left(1 + \frac{m}{j}\right)}{\log(j + m)}\right)^{1/2} \geq \frac{1}{4}\left(\frac{m}{\log j}\right)^{1/2}$$

if $1 \leq m \leq (\log 4)\log j$, and
$$\frac{\sqrt{n_k} - \sqrt{n_j}}{\sqrt{n_k - n_j}} \geq 1 - \frac{1}{\sqrt{n_k/n_j}} \geq 1 - \exp\left(-\frac{m}{2\log(j + m)}\right)$$

if $m > (\log 4)\log j$.

Similarly

$$\sqrt{\frac{n_j}{n_k - n_j}} \leq \left(\exp\left(\frac{m}{2\log(j+m)}\right) - 1\right)^{-1/2} \leq \left(\frac{2\log(j+m)}{m}\right)^{1/2}.$$

Hence
$$\mathcal{L} \geq O(m^{1/2}) \quad \text{if} \quad 2\log\log j \leq m \leq (\log 4)\log j \tag{5.6}$$

and
$$\mathcal{L} \geq \frac{4}{10}\phi(n_k) \quad \text{if} \quad (\log 4)\log j \leq m \leq (\log j)\log\log j.$$

In case $m > \log j(\log\log j)$ we obtain

$$\mathcal{P}_k = O\left(\frac{1}{k\log k}\right) \tag{5.7}$$

and
$$\sum_{k=j+1}^{j+\log j(\log\log j)} \mathbf{P}\{A_{n_j} A_{n_k}\} \leq O\left(j^{-1}(\log j)^{-1}\log\log j\right). \tag{5.8}$$

Having (5.7) and (5.8) a simple calculation gives

$$\sum_{1 \leq j,k \leq N} \mathbf{P}\{A_{n_j} A_{n_k}\} = O\left((\log\log N)^2\right)$$

and
$$\sum_{k=1}^{N} \mathbf{P}\{A_{n_k}\} = O(\log\log N).$$

Hence the Borel–Cantelli lemma 2* of Section 4.1 and the zero-one law of Section 3.2 imply the theorem. A simple consequence of EFKP LIL is

THEOREM 5.1 *The nonincreasing function* $-c(n) \in \text{LLC}(n^{-1/2}S_n)$ *if and only if* $I_1(c) < \infty$.

The Recurrence Theorem of Section 3.1 characterizes the monotone elements of $\text{LLC}(|S_n|)$. In fact

THEOREM 5.2 *A monotone function* $d(n) \in \text{LLC}(|S_n|)$ *if and only if* $d(n) < 0$ *for any n large enough.*

Remark 1. For the role of Kolmogorov in the proof of EFKP LIL see Bingham (1989).

5.3 The laws of Chung and Hirsch

The characterization of the lower classes of M_n and M_n^+ is not trivial at all. We present the following two results.

THEOREM OF CHUNG (1948). *The nonincreasing function* $\alpha(n) \in$ $\mathrm{LLC}(n^{-1/2} M_n)$ *if and only if*

$$I_2(\alpha) := \sum_{n=1}^{\infty} n^{-1}(\alpha(n))^{-2} \exp\left(-\frac{\pi^2}{8}\alpha^{-2}(n)\right) < \infty.$$

THEOREM OF HIRSCH (1965). *The nonincreasing function* $\beta(n) \in$ $\mathrm{LLC}(n^{-1/2} M_n^+)$ *if and only if*

$$I_3(\beta) := \sum_{n=1}^{\infty} n^{-1}\beta(n) < \infty.$$

Note that Theorem of Chung trivially implies

$$\liminf_{n \to \infty} \left(\frac{\log \log n}{n}\right)^{1/2} M_n = \frac{\pi}{\sqrt{8}} \quad \text{a.s.} \tag{5.9}$$

(5.9) is called the "Other LIL".

Remark 1. The proof of EFKP LIL is essentially the same as that of Consequence 1. However, it requires a lemma saying that if a monotone function $f(\cdot) \in \mathrm{UUC}(S_n)$ then $f(n) \geq b_n^{-1}/2$ and if $f(\cdot) \in \mathrm{ULC}(S_n)$ then $f(n) \leq 2b_n^{-1}$. The proofs of Theorems of Chung and Hirsch are also very similar to the above presented proof of Consequence 1 (Section 5.2). However, instead of (2.15) and (2.16) one should apply (2.17) and (2.18).

5.4 When will S_n be very large?

We say that S_n is very large if $S_n \geq b_n^{-1}$. EFKP LIL of Section 5.2 says that S_n is very large i.o. a.s. Define

$$\alpha(n) = \max\{k : 0 \leq k \leq n,\ S_k \geq b_k^{-1}\}, \tag{5.10}$$

i.e. $\alpha(n)$ is the last point before n where S_k is very large. The EFKP LIL also implies that $\alpha(n) = n$ i.o. a.s. Here we ask: how small can $\alpha(n)$ be? This question was studied by Erdős–Révész (1989). The result is:

THEOREM 5.3

$$\liminf_{n\to\infty} \frac{(\log\log n)^{1/2}}{(\log\log\log n)\log n} \log \frac{\alpha(n)}{n} = -C \quad a.s.$$

where C is a positive constant with

$$2^{-2} \leq C \leq 2^{14}.$$

Equivalently

$$\alpha(n) \geq n^{1-\delta_n} \quad a.s.$$

for all but finitely many n where

$$\delta_n = C \frac{\log\log\log n}{(\log\log n)^{1/2}}.$$

The exact value of C is unknown.

Clearly one could say that S_n is very large if

(i) $S_n \geq (1-\varepsilon)(2n\log\log n)^{1/2} \quad (0 < \varepsilon < 1)$ or

(ii) $S_n \geq (2n(\log\log n + 3\log\log\log n/2))^{1/2}$, etc.

These definitions of "very large" are producing different α's instead of the one defined by (5.10). It is natural to ask: what can be said about these new α's?

It is also interesting to investigate the time needed to arrive from a very large value to a very small one. Introduce the following notations: let

$$\alpha_1 = \min\{k: \ k \geq 3, \ S_k \geq b_k^{-1}\},$$
$$\beta_1 = \min\{k: \ k > \alpha_1, \ S_k \leq -b_k^{-1}\},$$
$$\alpha_2 = \min\{k: \ k > \beta_1, \ S_k \geq b_k^{-1}\},$$
$$\beta_2 = \min\{k: \ k > \alpha_2, \ S_k \leq -b_k^{-1}\},\ldots$$

Define a sequence of integers $\{n_k\}$ by

$$n_1 = 5, \quad n_k = \left[n_{k+1}^{1-\delta_{n_{k+1}}}\right] \quad (k = 2, 3, \ldots).$$

Then by Theorem 5.3 between n_k and n_{k+1} there exist integers j and l such that

$$S_j \geq b_j^{-1} \text{ and } S_l \leq -b_l^{-1}.$$

Hence we have

LÉVY CLASSES

THEOREM 5.4

$$\beta_k \leq n_k \quad a.s.$$

for all but finitely many k.

Very likely a lower estimate of β_k is also close to n_k but it is not proved.

Remark 1. The lim sup of the relative frequency of those i's ($1 \leq i \leq n$) for which $S_i \geq (1-\varepsilon)b_i^{-1}$ is investigated in Remark 2 of Section 8.1.

5.5 A theorem of Csáki

The Theorem of Chung (Section 5.3) implies that with probability 1 there are only finitely many n for which

$$M_n = \max(M_n^+, M_n^-) < (1-\varepsilon)\left(\frac{\pi^2 n}{8 \log \log n}\right)^{1/2},$$

or in other words there are only finitely many n for which simultaneously

$$M_n^+ < (1-\varepsilon)\left(\frac{\pi^2 n}{8 \log \log n}\right)^{1/2} \quad \text{and} \quad M_n^- < (1-\varepsilon)\left(\frac{\pi^2 n}{8 \log \log n}\right)^{1/2}$$

for any $0 < \varepsilon < 1$. At the same time Theorem of Hirsch (Section 5.3) implies that with probability 1 there are infinitely many n for which

$$M_n^+ < (1-\varepsilon)\left(\frac{\pi^2 n}{8 \log \log n}\right)^{1/2}.$$

In fact there are infinitely many n for which

$$M_n^+ < \frac{n^{1/2}}{\log n}.$$

Roughly speaking this means that if M_n^+ is small, smaller than

$$(1-\varepsilon)\left(\frac{\pi^2 n}{8 \log \log n}\right)^{1/2},$$

then M_n^- is not very small, it is larger than

$$(1-\varepsilon)\left(\frac{\pi^2 n}{8 \log \log n}\right)^{1/2}$$

provided that n is large enough. Csáki (1978) investigated the question of how big M_n^- must be if M_n^+ is very small. His result is

THEOREM 5.5 Let $a(n) > 0$, $b(n) > 0$ be nonincreasing functions. Then
$$\mathbf{P}\{M_n^+ \leq a(n)n^{1/2} \text{ and } M_n^- \leq b(n)n^{1/2} \text{ i.o.}\}$$
$$= \begin{cases} 1 & \text{if } I_4(a(n), b(n)) = \infty, \\ 0 & \text{otherwise} \end{cases}$$

where
$$I_4(a(n), b(n)) = \sum_{n=1}^{\infty} \frac{a(n)}{nc^3(n)} \exp\left(-\frac{\pi^2}{2c^2(n)}\right)$$

and $c(n) = a(n) + b(n)$.

The special case $a(n) = b(n)$ of this theorem also gives Chung's theorem. Formally this theorem does not contain Hirsch's theorem.

In order to illustrate what Csáki's theorem is all about we present here two examples.

Example 1. Put
$$a(n) = C(\log \log n)^{-1/2} \qquad (0 < C < \pi/\sqrt{8})$$

and
$$b(n) = D(\log \log n)^{-1/2} \qquad (D > 0).$$

Then
$$I_4(a(n), b(n)) < \infty \quad \text{if } D < \pi/\sqrt{2} - C$$

and
$$I_4(a(n), b(n)) = \infty \quad \text{if } D \geq \pi/\sqrt{2} - C.$$

Applying Csáki's theorem, this fact implies that the events
$$\left\{ M_n^+ < C\left(\frac{n}{\log \log n}\right)^{1/2} \text{ and } M_n^- < D\left(\frac{n}{\log \log n}\right)^{1/2} \right\}$$

occur infinitely often with probability 1 if $D \geq \pi/\sqrt{2} - C$. However, it is not so if $D < \pi/\sqrt{2} - C$. That is to say if n is large enough and
$$M_n^+ < C\left(\frac{n}{\log \log n}\right)^{1/2} \qquad (0 < C < \pi/\sqrt{8}),$$

then it follows that
$$M_n^- \geq D\left(\frac{n}{\log \log n}\right)^{1/2} \quad \text{for any } 0 < D < \pi/\sqrt{2} - C.$$

Example 2. Put
$$a(n) = (\log n)^{-\alpha} \quad (0 < \alpha < 1)$$
and
$$b(n) = E(\log \log n)^{-1/2} \quad (E > 0).$$
Then
$$I_4(a(n), b(n)) < \infty \quad \text{if} \quad 0 < \alpha < 1 \quad \text{and} \quad E < \pi(2(1-\alpha))^{-1/2}$$
and
$$I_4(a(n), b(n)) = \infty \quad \text{if} \quad 0 < \alpha < 1 \quad \text{and} \quad E \geq \pi(2(1-\alpha))^{-1/2}.$$
Observe also that
$$I_4((\log n)^{-1}, F(\log \log \log n)^{-1/2}) < \infty \quad \text{if} \quad F < \pi/\sqrt{2}$$
and
$$I_4((\log n)^{-1}, F(\log \log \log n)^{-1/2}) = \infty \quad \text{if} \quad F \geq \pi/\sqrt{2}.$$
Applying Csáki's theorem this fact implies that the events
$$\left\{ M_n^+ < n^{1/2}(\log n)^{-\alpha} \quad \text{and} \quad M_n^- < En^{1/2}(\log \log n)^{-1/2} \right\}$$
resp.
$$\left\{ M_n^+ < n^{1/2}(\log n)^{-1} \quad \text{and} \quad M_n^- < Fn^{1/2}(\log \log \log n)^{-1/2} \right\}$$
occur infinitely often with probability one if $E \geq \pi(2(1-\alpha))^{-1/2}$ (resp. $F \geq \pi/\sqrt{2}$). However, it is not so if $E < \pi(2(1-\alpha))^{-1/2}$ (resp. $F < \pi/\sqrt{2}$).

As we have mentioned already, Csáki's theorem states that if one of the r.v.'s M_n^+ and M_n^- is very small then the other cannot be very small. It is interesting to ask what happens if one of the r.v.'s M_n^+ and M_n^- is very large. In Section 8.1 we are going to prove Strassen's Theorem 1, which easily implies that for any $\varepsilon > 0$ the events
$$\left\{ M_n^+ \geq \frac{1-\varepsilon}{3}(2n \log \log n)^{1/2} \quad \text{and} \quad M_n^- \geq \frac{1-\varepsilon}{3}(2n \log \log n)^{1/2} \right\}$$
occur infinitely often with probability 1, but of the events
$$\left\{ M_n^+ \geq \frac{1+\varepsilon}{3}(2n \log \log n)^{1/2} \quad \text{and} \quad M_n^- \geq \frac{1+\varepsilon}{3}(2n \log \log n)^{1/2} \right\}$$
only finitely many occur with probability 1. In general one can say

THEOREM 5.6 *For any $\varepsilon > 0$ and $1/3 \leq q < 1$ the events*

$$\left\{ M_n^+ \geq (1-\varepsilon)qb_n^{-1} \quad and \quad M_n^- \geq (1-\varepsilon)\frac{1-q}{2}b_n^{-1} \right\}$$

occur infinitely often with probability 1, but of the events

$$\left\{ M_n^+ \geq (1+\varepsilon)qb_n^{-1} \quad and \quad M_n^- \geq (1+\varepsilon)\frac{1-q}{2}b_n^{-1} \right\}$$

only finitely many occur with probability 1.

As a trivial consequence of Theorems 5.5 and 5.6 we obtain

THEOREM 5.7 *Consider the range $M_n^* = M_n^+ + M_n^-$ of the random walk $\{S_n\}$. Then for any $\varepsilon > 0$ we have*

$$(1+\varepsilon)(2n \log\log n)^{1/2} \in \mathrm{UUC}(M_n^*),$$
$$(1-\varepsilon)(2n \log\log n)^{1/2} \in \mathrm{ULC}(M_n^*),$$
$$(1+\varepsilon)\left(\frac{\pi^2}{2}\frac{n}{\log\log n}\right)^{1/2} \in \mathrm{LUC}(M_n^*),$$
$$(1-\varepsilon)\left(\frac{\pi^2}{2}\frac{n}{\log\log n}\right)^{1/2} \in \mathrm{LLC}(M_n^*).$$

Theorems 5.5 and 5.6 describe the joint behaviour of M_n^+ and M_n^-. We also ask what can be said about the joint behaviour of S_n and M_n^- (say). In order to formulate the answer of this question we introduce the following notations.

Let $\gamma(n), \delta(n)$ be sequences of positive numbers satisfying the following conditions:
$$\gamma(n) \text{ monotone},$$
$$\delta(n) \downarrow 0,$$
$$n^{1/2}\gamma(n) \uparrow \infty,$$
$$n^{1/2}\delta(n) \uparrow \infty.$$

Further let $f(n) = n^{1/2}\psi(n) \in \mathrm{ULC}(S_n)$ with $\psi(n) \uparrow \infty$. Define the infinite random set of integers

$$\zeta = \zeta(f) = \{n : S_n \geq f(n)\}.$$

Then we have

THEOREM 5.8 (Csáki–Grill, 1988). *For any $f(n)=n^{1/2}\psi(n)\in\text{ULC}(S_n)$ the function*

$$g(n) = n^{1/2}\gamma(n) \in \text{UUC}(M_n^-, n \in \zeta)$$

if and only if

$$f(n) + 2g(n) \in \text{UUC}(S_n).$$

Further,

$$n^{1/2}\delta(n) \in \text{LLC}(M_n^-, n \in \zeta)$$

if and only if

$$\sum_{n=1}^{\infty} \frac{\delta(n)}{n}\psi^2(n)\exp\left(-\frac{\psi^2(n)}{2}\right) < \infty.$$

Remark 1. $n^{1/2}\gamma(n) \in \text{UUC}(M_n^-, n \in \zeta)$ means that $n^{1/2}\gamma(n) \geq M_n^-$ a.s. for all but finitely many such n for which $n \in \zeta$. In other words, the inequalities $S_n \geq f(n)$ and $M_n^- \geq n^{1/2}\gamma(n)$ simulteneously hold with probability 1 only for finitely many n.

Consequence 1. Let $V(n) = \min(M_n^+, M_n^-)$. Then $f(n) \in \text{UUC}(V)$ if and only if $3f(n) \in \text{UUC}(S_n)$.

Remark 2. Theorem 5.6 in case $q = 1/3$ follows from the above Consequence 1. For other q's ($1/3 < q < 1$) Theorem 5.8 implies Theorem 5.6.

Example 3. Let $f(n) = ((2-\varepsilon)n\log\log n)^{1/2}$ $(0 < \varepsilon < 2)$. Then we find the inequalities

$$S_n \geq \left(1 - \frac{\varepsilon}{2}\right)^{1/2} b_n^{-1} \quad \text{and} \quad M_n^- \geq \frac{1+\varepsilon}{2}\left(1 - \left(1 - \frac{\varepsilon}{2}\right)^{1/2}\right) b_n^{-1}$$

hold with probability 1 only for finitely many n. However,

$$S_n \geq \left(1 - \frac{\varepsilon}{2}\right)^{1/2} b_n^{-1} \quad \text{and} \quad M_n^- \geq \frac{1-\varepsilon}{2}\left(1 - \left(1 - \frac{\varepsilon}{2}\right)^{1/2}\right) b_n^{-1} \quad \text{i.o. a.s.}$$

The above two statements also follow from Strassen's theorem 1 (cf. Section 8.1). Further,

$$S_n \geq \left(1 - \frac{\varepsilon}{2}\right)^{1/2} b_n^{-1} \quad \text{and} \quad M_n^- \leq n^{1/2}(\log n)^{-\eta/2} \quad \text{i.o. a.s.} \tag{5.11}$$

if and only if $\eta \leq \varepsilon$.

Chapter 6

Wiener Process and Invariance Principle

6.1 Four lemmas

Clearly the r.v. $a^{-1}(S(k+a) - S(k))$ can be considered as the average speed of the particle in the interval $(k, k+a)$. Similarly the r.v.

$$a^{-1} I_1(n, a) = a^{-1} \max_{0 \leq k \leq n-a} (S(k+a) - S(k))$$

is the largest average speed of the particle in $(0, n)$ over the intervals of size a. We know (Theorem 2.9) that $a^{-1/2}(S(k+a) - S(k))$ is asymptotically $(a \to \infty)$ an $N(0,1)$ r.v. Hence $S(k+a) - S(k)$ behaves like $a^{1/2}$ or by the LIL of Khinchine

$$\limsup_{a \to \infty} \frac{S(k+a) - S(k)}{(2a \log \log a)^{1/2}} = 1 \quad \text{a.s.}$$

for any fixed k. We prove that even $I_1(n,a)$ cannot be much bigger than $a^{1/2}$. In fact we have

LEMMA 6.1 *Let $a = a_n \leq n^\alpha$ $(0 < \alpha < 1)$. Then*

$$\limsup_{n \to \infty} \frac{I_1(n, a)}{n^{\alpha/2} (\log n)^{1/2}} \leq C \quad \text{a.s.}$$

if $C > 2$.

Proof. By Theorem 2.10 for any k

$$\mathbf{P}\left\{ \frac{S(k+a) - S(k)}{n^{\alpha/2}(\log n)^{1/2}} \geq C \right\} \approx 1 - \Phi(C(\log n)^{1/2})$$

$$\leq \exp\left(-\frac{C^2}{2} \log n\right) = n^{-C^2/2}$$

as $n \to \infty$. Hence

$$\mathbf{P}\left\{ \frac{I_1(n,a)}{n^{\alpha/2}(\log n)^{1/2}} \geq C \right\} \leq n^{1 - C^2/2}$$

and the Borel–Cantelli lemma implies the statement.

Remark 1. Much stronger results than that of Lemma 6.1 can be found in Section 7.3.

LEMMA 6.2 *Let* $\{X_{ij};\ i = 1, 2, \ldots;\ j = 1, 2, \ldots\}$ *be a double array of i.i.d.r.v.'s with*

$$\mathbf{E} X_{ij} = 0, \quad \mathbf{E} X_{ij}^2 = 1, \quad \mathbf{E}(\exp(tX_{ij})) < \infty$$

for all t in some interval $\mid t \mid < t_0$. Then for any $K > 0$ there exists a positive constant $C = C(K)$ such that

$$(i \log i)^{-1/2} \sup_{n \leq Ki} \sum_{j=1}^{n} X_{ij} \leq C$$

for all but finitely many i.

Proof of Lemma 6.2 is essentially the same as that of Lemma 6.1 using Theorem 2.11 instead of Theorem 2.10.

LEMMA 6.3 *Let $X_1, X_2, \ldots$ be i.i.d.r.v.'s with*

$$\mathbf{P}\{X_i = 1\} = \mathbf{P}\{X_i = -1\} = \frac{1}{2}$$

and let

$$\nu = \min\{n : |X_1 + X_2 + \cdots + X_n| = 2\}.$$

Then

$$\mathbf{P}\{\nu = 2k\} = 2^{-k}, \quad (k = 1, 2, \ldots).$$

Consequently

$$\mathbf{E}\nu = 4,$$
$$\mathrm{Var}\ \nu = 8,$$
$$\mathbf{E} \exp(t(\nu - 4)) = \frac{1}{e^{2t}(2 - e^{2t})} \quad \text{if} \quad t < \frac{1}{2} \log 2, \qquad (6.1)$$
$$\mathbf{E} \exp(-t(\nu - 4)) = \frac{1}{2e^{2t} - 1} \quad \text{if} \quad t > 0. \qquad (6.2)$$

Proof is trivial.

WIENER PROCESS AND INVARIANCE PRINCIPLE

LEMMA 6.4 *Let $\nu_1, \nu_2, \ldots$ be i.i.d.r.v.'s with*

$$\mathbf{P}\{\nu = 2k\} = 2^{-k}, \quad (k = 1, 2, \ldots).$$

Then

$$\mathbf{P}\left\{\frac{|\nu_1 + \nu_2 + \cdots + \nu_n - 4n|}{2n^{1/2}} \geq K\right\} \leq \frac{1}{e^{K}-1}, \tag{6.3}$$

$$\mathbf{P}\left\{\max_{1 \leq n \leq N} \frac{|\nu_1 + \nu_2 + \cdots + \nu_n - 4n|}{2n^{1/2}} \geq N^{\varepsilon}\right\} \leq \frac{N}{e^{N^{\varepsilon}}-1}. \tag{6.4}$$

Proof. (6.3) follows from (6.1), (6.2) and the Markov inequality. (6.4) is a trivial consequence of (6.3).

6.2 Joining of independent random walks

Let $\{X(i,j,k),\ i,j,k = 1, 2, \ldots\}$ be an array of i.i.d.r.v.'s with

$$\mathbf{P}\{X_{ijk} = 1\} = \mathbf{P}\{X_{ijk} = -1\} = \frac{1}{2}$$

and let

$$S(i,j;n) = \sum_{k=1}^{n} X(i,j,k),$$

$$S(i,j;0) = 0,$$

$$\nu(i,j) = \min\{n:\ |S(i,j;n)| = 2\},$$

$$\mu(i,k) = \sum_{j=1}^{k} \nu(i,j),$$

$$T(1;n) = S(1,1;n).$$

Note that $\{T(1;n),\ n = 0, 1, 2, \ldots\}$ is a random walk. Now we define the sequence $\{T(2;n),\ n = 0, 1, 2, \ldots\}$ as follows:

$$T(2;n) = \frac{1}{2} S(2,1;n)\, \text{sign} S(1,1;n)\, \text{sign} S(2,1;\nu(2,1)) \quad \text{if} \quad 0 \leq n \leq \nu(2,1).$$

Note that $T(2; \nu(2,1)) = S(1,1;1)$.

Now we give the definition of $T(2;n)$ when $\mu(2,1) = \nu(2,1) \leq n \leq \mu(2,2)$. Let

$$T(2;n) = T(2;\nu(2,1)) + \frac{1}{2} S(2,2;n - \nu(2,1)) R(2,1)$$

where
$$R(2,1) = \operatorname{sign}(S(1,1;2) - S(1,1;1))\operatorname{sign}S(2,2;\nu(2,2))$$
and
$$\nu(2,1) = \mu(2,1) \leq n \leq \mu(2,2).$$

Similarly for $\mu(2,k) \leq n \leq \mu(2,k+1)$ $(k = 2,3,\ldots)$ let

$$T(2;n) = T(2;\mu(2,k)) + \frac{1}{2}S(2,k;n-\mu(2,k))R(2,k)$$

where
$$R(2,k) = \operatorname{sign}(S(1,1;k+1) - S(1,1;k))\operatorname{sign}S(2,k+1;\nu(2,k+1)).$$

On the properties of $T(2;n)$ note that

(i) $2T(2;n)$ is a random walk,

(ii) $T(2;\mu(2,k)) = S(1,1;k) = T(1;k)$ $(k = 1,2,\ldots)$.

Continuing this procedure we define $T(i;n)$ as follows:

$$T(i;n) = T(i;\mu(i,k)) + \frac{1}{2^{i-1}}S(i,k;n-\mu(i,k))R(i,k)$$

if
$$\mu(i,k) \leq n \leq \mu(i,k+1)$$

where
$$R(i,k) = \operatorname{sign}(T(i-1;k+1) - T(i-1;k))\operatorname{sign}S(i,k+1;\mu(i,k+1)).$$

On the properties of $T(\cdot;\cdot)$ note that

(i) $2^{i-1}T(i;n)$ $(n = 1,2,\ldots)$ is a random walk for any i fixed,

(ii) $T(i;\mu(i-1,k)) = T(i-1;k)$.

On the properties of $\mu(\cdot,\cdot)$ by (6.4) we have

$$\mathbf{P}\left\{\max_{1 \leq k \leq 4^i} \frac{|\mu(i,k) - 4k|}{2k^{1/2}} \geq 4^{\varepsilon i}\right\} \leq \frac{4^i}{\exp(4^{\varepsilon i} - 1)}. \tag{6.5}$$

Hence
$$\max_{1 \leq k \leq 4^i} |\mu(i,k) - 4k| \leq 2k^{1/2}4^{\varepsilon i} \quad \text{a.s.}$$

for all but finitely many i and

$$2^{i-1}T(i;\mu(i-1,k)) = 2^{i-1}T(i;4k+\vartheta(i;k)) \quad (k=1,2,\ldots,4^i)$$

where

$$|\vartheta(i;k)| \leq 2k^{1/2}4^{\varepsilon i}.$$

By Lemma 6.1

$$|2^{i-1}T(i;\mu(i-1,k)) - 2^{i-1}T(i;4k)| \leq (2k^{1/2}4^{\varepsilon i})^{1/2}C(\log(2k^{1/2}4^{\varepsilon i}))^{1/2}$$
$$\leq 4^{(1/4+2\varepsilon)i}.$$

Consequently

$$T(i-1;k) = T(i;4k) + 2^{-(1/2-2\varepsilon)i}. \tag{6.6}$$

6.3 Definition of the Wiener process

The random walk is not a very realistic model of the Brownian motion. In fact the assumption that the particle goes at least one unit in a direction before turning back, is hardly satisfied by the real Brownian motion. In a more realistic model of the Brownian motion the particle makes instantaneous steps to the right or to the left, that is a continuous time scale is used instead of a discrete one.

In order to build up such a model, assume that in a first experiment we can only observe the particle when it is located in integer points and further experiments describe the path of the particle between integers. Let

$$\{S(n) = S^{(0)}(n), \quad n = 0,1,2,\ldots\}$$

be the random walk which describes the location of the particle when it hits integer points.

In a next experiment we observe the particle when it hits the points $k/2$ ($k=0,\pm 1,\pm 2,\ldots$). Let $S^{(1)}_{(n)}$ be the obtained process, i.e. $S^{(1)}_{(n)}$ is a random walk with the properties:

(i) the particle moves $1/2$ to the right or to the left with probability $1/2$,

(ii) $S^{(1)}_{(n)}$ hits the integers in the same order as $S^{(0)}_{(n)}$ does.

Note that $(T(1;k), T(2;k))$ has the properties required from $(S^{(0)}(k), S^{(1)}(k))$.
Let

$$V_i(t) = T(i;[4^it]). \quad (0 \leq t \leq 1).$$

Then (6.6) easily implies that $\{V_i(t)\ 0 \leq t \leq 1\}$ converges a.s. uniformly to a continuous process $W(t)$. This limit process is called a *Wiener process*.

It is easy to see that this limit process has the following three properties:

(i) $W(t) - W(s) \in N(0, t-s)$ for all $0 \leq s < t < \infty$ and $W(0) = 0$,

(ii) $W(t)$ is an independent increment process that is $W(t_2) - W(t_1)$, $W(t_4) - W(t_3), \ldots, W(t_{2i}) - W(t_{2i-1})$ are independent r.v.'s for all $0 \leq t_1 < t_2 \leq t_3 < t_4 \leq \ldots \leq t_{2i-1} < t_{2i}$ $(i = 2, 3, \ldots)$,

(iii) the sample path function $W(t, \omega)$ is continuous a.s.

(i) and (ii) are simple consequences of the central limit theorem (Theorem 2.9). (iii) was proved above.

Remark 1. The above construction is closely related to the one of Knight (1981).

6.4 Invariance Principle

Define the sequence of r.v.'s $0 < \tau_1 < \tau_2 < \ldots$ as follows:

$$\tau_1 = \inf\{t : t > 0, |W(t)| = 1\},$$
$$\tau_2 = \inf\{t : t > \tau_1, |W(t) - W(\tau_1)| = 1\},$$
$$\ldots \quad \ldots$$
$$\tau_{i+1} = \inf\{t : t > \tau_i, |W(t) - W(\tau_i)| = 1\}, \ldots$$

Observe that

(i) $W(\tau_1), W(\tau_2) - W(\tau_1), W(\tau_3) - W(\tau_2), \ldots$ is a sequence of i.i.d.r.v.'s with distribution $\mathbf{P}\{W(\tau_1) = 1\} = \mathbf{P}\{W(\tau_1) = -1\} = 1/2$, i.e. $\{W(\tau_n)\}$ is a random walk,

(ii) $\tau_1, \tau_2 - \tau_1, \tau_3 - \tau_2, \ldots$ is a sequence of i.i.d.r.v.'s with distribution $\mathbf{P}\{\tau_1 > x\} = \mathbf{P}\{\sup_{t \leq x} |W(t)| < 1\}$.

Applying the reflection principle (formulated for S_n in Section 2.1, Proof 2 of (2.5)) for $W(\cdot)$ we obtain

$$\mathbf{P}\{\tau_1 > x\} = \frac{1}{\sqrt{2\pi}} \int_{-x^{-1/2}}^{x^{-1/2}} \sum_{k=-\infty}^{\infty} (-1)^k \exp\left(-\frac{(t - 2kx^{-1/2})^2}{2}\right) dt. \quad (6.7)$$

Evaluating the moments of τ_1 and applying the strong law of large numbers we obtain

$$\mathbf{E}\tau_1 = 1, \quad \mathbf{E}\tau_1^2 = 2, \quad \lim_{n \to \infty} n^{-1}\tau_n = 1 \quad \text{a.s.}$$

The above two observations are special cases of a theorem of Skorohod (1961). Because of (i) we say that a random walk can be embedded to a Wiener process (by the Skorohod embedding scheme).

Applying the LIL of Hartman–Wintner (1941) (Section 4.4) and some elementary properties of the random walk (formulated in Section 6.1) we obtain

$$\limsup_{n\to\infty} \frac{|W(\tau_n) - W(n)|}{(n\log\log n)^{1/4}(\log n)^{1/2}} < \infty \quad \text{a.s.}$$

This result can be formulated as follows:

THEOREM 6.1 *On a rich enough probability space $\{\Omega, \mathcal{F}, \mathbf{P}\}$ one can define a Wiener process $\{W(t), t \geq 0\}$ and a random walk $\{S_n, n = 0, 1, 2, \ldots\}$ such that*

$$\limsup_{n\to\infty} \frac{|S_n - W(n)|}{(n\log\log n)^{1/4}(\log n)^{1/2}} < \infty \quad a.s.$$

This result is a special case of a theorem of Strassen (1964).

A much stronger result was obtained by Komlós–Major–Tusnády (1975–76). A special case of their theorem runs as follows:

INVARIANCE PRINCIPLE 1 *On a rich enough probability space one can define a Wiener process $\{W(t), t \geq 0\}$ and a random walk $\{S_n, n = 0, 1, 2, \ldots\}$ such that*

$$|S_n - W(n)| = O(\log n) \quad a.s.$$

Remark 1. A theorem of Bártfai (1966) and Erdős–Rényi (1970) implies that the Invariance Principle 1 gives the best possible rate. In fact if $\{S_n, n = 0, 1, 2, \ldots\}$ and $\{W(t), t \geq 0\}$ are living on the same probability space $\{\Omega, \mathcal{F}, \mathbf{P}\}$ then $|S_n - W(n)| \geq O(\log n)$ a.s. except if S_n is the sum of i.i.d. $N(0,1)$ r.v.'s.

As a trivial consequence of Invariance Principle 1 we obtain

THEOREM 6.2 *Any of the EKFP LIL, the Theorems of Chung and Hirsch and Theorems 5.3, 5.4 and 5.5 remain valid replacing the random walk S_n by a Wiener process $W(t)$. As an example we mention: Let $Y(t)$ be any of the processes*

$$t^{-1/2}W(t),$$
$$t^{-1/2}|W(t)|,$$
$$t^{-1/2}\sup_{0\leq s\leq t}|W(s)|,$$
$$t^{-1/2}\sup_{0\leq s\leq t}W(s),$$
$$-t^{-1/2}\inf_{0\leq s\leq t}W(s).$$

Then a nondecreasing function $a(t) \in \mathrm{UUC}(Y(t))$ if and only if

$$\int_1^\infty \frac{a(t)}{t} \exp\left(-\frac{a^2(t)}{2}\right) dt < \infty. \tag{6.8}$$

Similarly a nonincreasing function $(\alpha(t))^{-1} \in \mathrm{LLC}(t^{-1/2}m(t))$ if and only if

$$\int_1^\infty t^{-1}(\alpha(t))^2 \exp\left(-\frac{\pi^2}{8}(\alpha(t))^2\right) dt < \infty. \tag{6.9}$$

Remark 2. In fact the EFKP LIL only implies that $a(n) \in \mathrm{UUC}(Y(n))$ ($n = 1, 2, \ldots$) if $a(\cdot)$ is nondecreasing and (6.8) is satisfied. In order to get our Theorem 6.2 completely we have to know something about the continuity of $W(\cdot)$, i.e. we have to see that the fluctuation $\sup_{k \le t} \sup_{k \le s \le k+1} |W(s) - W(k)|$ cannot be very big. For example, the complete result can be obtained by Theorem 7.13, especially Example 2 of Section 7.2, which says that the above fluctuation is asymptotically $(2 \log t)^{1/2}$ a.s.

Theorem 6.2 claimed that any of the strong theorems formulated up to now for S_n will be valid for $W(\cdot)$. The same is true for the limit distribution theorems of Section 2.2. In fact we have

THEOREM 6.3

$$\mathbf{P}\{t^{-1/2}m^+(t) > u\} = 2(1 - \Phi(u)) \quad (t > 0, \ u > 0),$$

$$\mathbf{P}\{t^{-1/2}m(t) < u\} = \frac{1}{\sqrt{2\pi}} \int_{-u}^u \sum_{k=-\infty}^{+\infty} (-1)^k \exp\left(-\frac{(x-2ku)^2}{2}\right) dx$$

$$= \frac{4}{\pi} \sum_{k=0}^\infty \frac{(-1)^k}{2k+1} \exp\left(\frac{-\pi^2(2k+1)^2}{8u^2}\right).$$

For later references we give a more general form of the above Invariance Principle.

INVARIANCE PRINCIPLE 2 (Komlós–Major–Tusnády, 1975–76). Let $F(x)$ be a distribution function with

$$\int_{-\infty}^\infty x\, dF(x) = 0, \quad \int_{-\infty}^\infty x^2\, dF(x) = 1,$$

$$\int_{-\infty}^\infty e^{tx} dF(x) < \infty \quad |t| \le t_0$$

with some $t_0 > 0$. Then, on a rich enough probability space $\{\Omega, \mathcal{F}, \mathbf{P}\}$, one can define a Wiener process $\{W(t),\ t \geq 0\}$ and a sequence of i.i.d.r.v.'s $Y_1, Y_2, \ldots$ with $\mathbf{P}\{Y_1 < x\} = F(x)$ such that

$$|T_n - W(n)| = O(\log n) \quad a.s.,$$

where $T_n = Y_1 + Y_2 + \cdots + Y_n$.

Chapter 7

Increments

7.1 Long head-runs

In connection with a teaching experiment in mathematics, T. Varga posed a problem. The experiment goes like this: his class of secondary school children is divided into two sections.

In one of the sections each child is given a coin which he then throws two hundred times, recording the resulting head and tail sequence on a piece of paper. In the other section the children do not receive coins but are told instead that they should try to write down a "random" head and tail sequence of length two hundred. Collecting these slips of paper, he then tries to subdivide them into their original groups. Most of the time he succeeds quite well. His secret is that he had observed that in a randomly produced sequence of length two hundred, there are, say, head-runs of length seven. On the other hand, he had also observed that most of those children who were to write down an imaginary random sequence are usually afraid of putting down head-runs of longer than four. Hence, in order to find the slips coming from the coin tossing group, he simply selects the ones which contain head-runs longer than five.

This experiment led T. Varga to ask: What is the length of the longest run of pure heads in n Bernoulli trials?

A trivial answer of this question is

THEOREM 7.1
$$\lim_{N \to \infty} \frac{Z_N}{\lg N} = 1 \quad a.s.$$

where Z_N is the length of longest head-run till N.

Proof.

Step 1. We prove that

$$\liminf_{N \to \infty} \frac{Z_N}{\lg N} \geq 1 \quad a.s. \tag{7.1}$$

Let $\varepsilon < 1$ be any positive number and introduce the notations:

$$t = [(1-\varepsilon)\lg N],$$
$$\bar{N} = \left[\frac{N}{t}\right] - 1,$$
$$U_k = S_{t(k+1)} - S_{tk} \quad (k = 0, 1, \ldots, \bar{N}).$$

Clearly $U_0, U_1, \ldots, U_{\bar{N}}$ are i.i.d.r.v.'s with

$$\mathbf{P}\{U_k = t\} = \frac{1}{2^t}.$$

Hence

$$\mathbf{P}\{U_0 < t, U_1 < t, \ldots, U_{\bar{N}} < t\} = \left(1 - \frac{1}{2^t}\right)^{\bar{N}}$$

and a simple calculation gives

$$\sum_{N=1}^{\infty}\left(1 - \frac{1}{2^t}\right)^{\bar{N}} < \infty$$

for any $\varepsilon > 0$. Now the Borel–Cantelli lemma implies (7.1).

Step 2. We prove that

$$\limsup_{N \to \infty} \frac{Z_N}{\lg N} \leq 1 \quad \text{a.s.} \tag{7.2}$$

Let ε be any positive number and introduce the following notations:

$$u = [(1+\varepsilon)\lg N],$$
$$V_k = S_{k+u} - S_k \quad (k = 0, 1, \ldots, N - u),$$
$$A_N = \bigcup_{k=0}^{N-u} \{V_k = u\}$$

and let T be any positive integer for which $T\varepsilon > 1$. Then

$$\mathbf{P}\{V_k = u\} = 2^{-u},$$

consequently

$$\mathbf{P}\{A_N\} \leq N 2^{-u} \quad \text{and} \quad \sum_{k=1}^{\infty} \mathbf{P}\{A_{k^T}\} < \infty.$$

Hence the Borel–Cantelli lemma implies

$$\limsup_{k \to \infty} \frac{Z_{k^T}}{\lg k^T} \leq 1 \quad \text{a.s.} \tag{7.3}$$

Let $k^T < n < (k+1)^T$ and observe that by (7.3)

$$Z_n \leq Z_{(k+1)^T} \leq (1+\varepsilon) \lg(k+1)^T \leq (1+2\varepsilon) \lg k^T \leq (1+2\varepsilon) \lg n$$

with probability 1 for all but finitely many n. Hence we have (7.2) as well as Theorem 7.1.

A much stronger statement is the following:

THEOREM 7.2 (Erdős–Révész, 1976). *Let $\{a_n\}$ be a sequence of positive numbers and let*

$$A(\{a_n\}) = \sum_{n=1}^{\infty} 2^{-a_n}.$$

Then

$$a_n \in \text{UUC}(Z_n) \quad \text{if} \quad A(\{a_n\}) < \infty, \tag{7.4}$$
$$a_n \in \text{ULC}(Z_n) \quad \text{if} \quad A(\{a_n\}) = \infty \tag{7.5}$$

and for any $\varepsilon > 0$

$$\mathcal{K}_n = [\lg n - \lg \lg \lg n + \lg \lg e - 1 + \varepsilon] \in \text{LUC}(Z_n), \tag{7.6}$$
$$\lambda_n = [\lg n - \lg \lg \lg n + \lg \lg e - 2 - \varepsilon] \in \text{LLC}(Z_n). \tag{7.7}$$

Example 1. If $\delta > 0$ and

$$a_n^* = \lg n + (1+\delta) \lg \lg n \quad \text{then} \quad A(\{a_n^*\}) < \infty.$$

Hence (7.4) and (7.7) together imply that

$$\mathbf{P}\{\lambda_n \leq Z_n \leq a_n^* \text{ for all but finitely many } n\} = 1.$$

Note that if $n = 2^{2^{20}} = 2^{1048576} \sim 10^{315621}$ and $\varepsilon = \delta = 0,1$ then $\lambda_n = 1048569$ and $a_n^* = 1048598$.

Remark 1. Clearly (7.4) and (7.5) are the best possible results while (7.6) and (7.7) are nearly the best possible ones.

A complete characterization of the lower classes was obtained by Guibas–Odlyzko (1980) and Samarova (1981). Their result is:

THEOREM 7.3 *Let*

$$\psi_n = \lg n - \lg \lg \lg n + \lg \lg e - 2.$$

Then

$$\liminf_{n\to\infty}[Z_n - \psi_n] = 0 \quad a.s.$$

It is also interesting to ask what is the length of the longest run containing at most one (or at most T, $T = 1, 2, \ldots$) (-1)'s. Let $Z_n(T)$ be the largest integer for which

$$I_1(n, Z_n(T)) \geq Z_n(T) - 2T$$

where

$$I_1(n, a) = \max_{0 \leq k \leq n-a}(S_{k+a} - S_k).$$

A generalization of Theorem 7.2 is the following:

THEOREM 7.4 *Let $\{a_n\}$ be a sequence of positive numbers and let*

$$A_T(\{a_n\}) = \sum_{n=1}^{\infty} a_n^T 2^{-a_n}.$$

Then

$$a_n \in \text{UUC}(Z_n(T)) \quad if \quad A_T(\{a_n\}) < \infty, \tag{7.8}$$
$$a_n \in \text{ULC}(Z_n(T)) \quad if \quad A_T(\{a_n\}) = \infty \tag{7.9}$$

and for any $\varepsilon > 0$

$$\mathcal{K}_n(T) = [\lg n + T \lg \lg n - \lg \lg \lg n - \lg T! + \lg \lg e - 1 + \varepsilon]$$
$$\in \text{LUC}(Z_n(T)), \tag{7.10}$$
$$\lambda_n(T) = [\lg n + T \lg \lg n - \lg \lg \lg n - \lg T! + \lg \lg e - 2 - \varepsilon]$$
$$\in \text{LLC}(Z_n(T)). \tag{7.11}$$

A very nice question connected to Theorem 7.2 was proposed by Benjamini, Häggström, Peres and Steif (2003). They considered the following model: let $\{\psi_n^{(j)}; j = 0, 1, 2, \ldots; n = 1, 2, \ldots\}$ be a sequence of independent Poisson processes with rate 1, also independent from the sequence $X_1, X_2, \ldots$. Let $\{X_n^{(j)}; j = 1, 2, \ldots; n = 1, 2, \ldots\}$ be an array of i.i.d.r.v.'s with

$$\mathbf{P}\{X_n^{(j)} = 1\} = \mathbf{P}\{X_n^{(j)} = -1\} = 1/2$$

and consider the processes
$$X_n(t) = X_n^{(j)} \quad \text{for} \quad \psi_n^{(j-1)} \le t < \psi_n^{(j)}.$$

Clearly for any t fixed the sequence $\{X_n(t)\ n = 1, 2, \ldots\}$ is a random walk. Hence any theorem formulated for random walks remains valid for $X_n(t)$ (for any fixed t). However, it can happen that for some t we get some new phenomenon.

For example for any $t > 0$, $Z_n(t)$ (the longest head-run of $X_n(t)$ up to n) obeys Theorem 7.2 i.e.
$$a_n \in \text{UUC}(Z_n(t)) \quad \text{if} \quad A(\{a_i\}) < \infty$$
and
$$a_n \in \text{ULC}(Z_n(t)) \quad \text{if} \quad A(\{a_n\}) = \infty.$$

Benjamini et al. proved that for some (random) t it is not the case. For example
$$\mathbf{P}\{\exists t \ge 0 \text{ such that } \{Z_n(t) \ge a_n \text{ i.o. }\}\}$$
$$= \begin{cases} 0 & \text{if} \quad \sum_{n=1}^{\infty} a_n 2^{-n} < \infty, \\ 1 & \text{if} \quad \sum_{n=1}^{\infty} a_n 2^{-n} = \infty. \end{cases}$$

Hence for fixed t the length of the longest run containing at most one tail only, might be as big as the length of the longest pure head-run for a randomly choosed t. This remark and Theorem 7.4 combined suggest the following

Conjecture.
$$\mathbf{P}\{\exists t \ge 0 \text{ such that } \{Z_n(T, t) \ge a_n \text{ i.o. }\}\}$$
$$= \begin{cases} 0 & \text{if} \quad \sum_{n=1}^{\infty} a_n^{T+1} 2^{-n} < \infty, \\ 1 & \text{if} \quad \sum_{n=1}^{\infty} a_n^{T+1} 2^{-n} = \infty. \end{cases}$$

A trivial reformulation of the question of T. Varga is: how many flips are needed in order to get a run of heads of size m. Formally speaking let $\tilde{Z}_m$ be the smallest integer for which
$$X_{\tilde{Z}_m - m + 1} = X_{\tilde{Z}_m - m + 2} = \cdots = X_{\tilde{Z}_m} = 1.$$

As a trivial consequence of Theorem 7.2 we obtain

THEOREM 7.5

$$\tilde{\lambda}_m \in \text{UUC}(\tilde{Z}_m),$$
$$\tilde{\mathcal{K}}_m \in \text{ULC}(\tilde{Z}_m),$$
$$\tilde{a}_m \in \text{LUC}(\tilde{Z}_m) \quad \text{if} \quad A(\{a_m\}) = \infty,$$
$$\tilde{a}_m \in \text{LLC}(\tilde{Z}_m) \quad \text{if} \quad A(\{a_m\}) < \infty$$

where $\tilde{\mathcal{K}}_m$ (resp. $\tilde{\lambda}_m$) are the inverse functions of $\mathcal{K}_m$ of (7.6) (resp. λ_m of (7.7)) and a_m is the inverse function of the positive increasing function $\tilde{a}_m$.

Instead of considering the pure head-runs of size m one can consider any given run of size m and investigate the waiting time till that given run would occur. This question was studied by Guibas–Odlyzko (1980).

Erdős asked about the waiting time V_m till all of the possible 2^m patterns of size m would occur at least once. An answer of this question was obtained by Erdős and Chen (1988). They proved

THEOREM 7.6 *For any* $\varepsilon > 0$

$$\frac{(1+\varepsilon)2^m m}{\lg e} \in \text{UUC}(V_m)$$

and

$$\frac{(1-\varepsilon)2^m m}{\lg e} \in \text{LLC}(V_m).$$

A much stronger version of this theorem was obtained by Móri (1989). He proved

THEOREM 7.7 *For any* $\varepsilon > 0$

$$(2^k k + (1+\varepsilon)2^k \lg k)(\lg e)^{-1} \in \text{UUC}(V_k),$$
$$(2^k k + (1-\varepsilon)2^k \lg k)(\lg e)^{-1} \in \text{ULC}(V_k),$$
$$(2^k k - (1-\varepsilon)2^k \lg \lg k)(\lg e)^{-1} \in \text{LUC}(V_k),$$
$$(2^k k - (1+\varepsilon)2^k \lg \lg k)(\lg e)^{-1} \in \text{LLC}(V_k).$$

We mention that the proof of Theorem 7.7 is based on the following limit distribution:

THEOREM 7.8 (Móri, 1989)

$$\lim_{k \to \infty} 2^{k/18} \sup_y \left| \mathbf{P}\left\{ 2^{-k} V_k - \frac{k}{\lg e} \leq y \right\} - e^{-e^{-y}} \right| = 0.$$

In order to compare Theorem 7.7 and Theorems 7.2 and 7.3 it is worthwhile to consider the inverse of V_k. Let

$$U_n = \max\{k : V_k \leq n\}.$$

Then Theorem 7.7 implies

Corollary 1. (Móri, 1989). *For any $\varepsilon > 0$ we have*

$$\left[\lg n - \lg\lg n - \varepsilon \lg e \frac{\lg\lg n}{\lg n}\right] \leq U_n \leq \left[\lg n - \lg\lg n + (1+\varepsilon)\lg e \frac{\lg\lg n}{\lg n}\right] \text{ a.s.}$$

for all but finitely many n. Consequently U_n is QAD.

Observe that U_n is "less random" than Z_n. In fact for some n's the lower and upper estimates of U_n are equal to each other and for the other n's they differ by 1. Clearly $U_n \leq Z_n$ but comparing Theorems 7.2, 7.3 and Corollary 1 it turns out that U_n is not much smaller than Z_n.

In Theorem 7.2 we have seen that for all n, big enough, there exists a block of size λ_n (of (7.7)) containing only heads but it is not true with $\mathcal{K}_n$ (of (7.6)). Now we ask what is the number of disjoint blocks of size λ_n containing only heads.

Let $\nu_n(k)$ be the number of disjoint blocks of size k (in the interval $[0, n]$) containing only heads, that is to say $\nu_n(k) = j$ if there exists a sequence $0 \leq t_1 < t_1 + k \leq t_2 < t_2 + k \leq \ldots \leq t_j < t_j + k \leq n$ such that

$$S_{t_i+k} - S_{t_i} = k \quad (i = 1, 2, \ldots, j)$$

but

$$S_{m+k} - S_m < k \quad \text{if} \quad t_i + k \leq m < t_{i+1} \quad (i = 1, 2, \ldots, j-1)$$

or $t_j + k \leq m \leq n - k$.

The proof of the following theorem is very simple.

THEOREM 7.9 (Révész, 1978). *For any $\varepsilon > 0$ there exist constants $0 < \alpha_1 = \alpha_1(\varepsilon) \leq \alpha_2 = \alpha_2(\varepsilon) < \infty$ such that*

$$\alpha_1 = \liminf_{n\to\infty} \frac{\nu_n(\lambda_n)}{\lg\lg n} \leq \limsup_{n\to\infty} \frac{\nu_n(\lambda_n)}{\lg\lg n} = \alpha_2 \quad a.s.$$

(for λ_n see (7.7)).

This theorem says that in the interval $[0, n]$ there are $O(\lg\lg n)$ blocks of size λ_n containing only heads. This fact is quite surprising knowing that it happens for infinitely many n that there is not any block of size $\lambda_n + 2 \geq \mathcal{K}_n$ containing only heads.

Deheuvels (1985) worked out a method to find some estimates of $\alpha_1(\varepsilon)$ and $\alpha_2(\varepsilon)$. In order to formulate his results let $Z_n = Z_n^{(1)}$ and let $Z_n^{(2)} \geq Z_n^{(3)} \geq \ldots$ be the length of the second, third, ... longest run of 1's observed in $X_1, X_2, \ldots, X_n$. Then

THEOREM 7.10 (Deheuvels, 1985). *For any integer $r \geq 3$ and $k \geq 1$ and for any $\varepsilon > 0$*

$$\lg n + \frac{1}{k}\left(\lg_2 n + \cdots + \lg_{r-1} n + (1+\varepsilon)\lg_r n\right) \in \mathrm{UUC}(Z_n^{(k)}), \quad (7.12)$$

$$\lg n + \frac{1}{k}\left(\lg_2 n + \cdots + \lg_{r-1} n + \lg_r n\right) \in \mathrm{ULC}(Z_n^{(k)}), \quad (7.13)$$

$$[\lg n - \lg_3 n + \lg \lg e - 1] \in \mathrm{LUC}(Z_n^{(k)}), \quad (7.14)$$

$$[\lg n - \lg_3 n + \lg \lg e - 2 - \varepsilon] \in \mathrm{LLC}(Z_n^{(k)}). \quad (7.15)$$

Remark 2. In case $k = 1$ (7.14) gives a stronger result than (7.6) but (7.14) and (7.15) together is not as strong as Theorem 7.3.

THEOREM 7.11 (Deheuvels, 1985). *Let $v \in (0, +\infty)$ be given, and let $0 < c'_v < 1 < c''_v < \infty$ be solutions of the equation*

$$c - 1 - \log c = \frac{1}{v}. \quad (7.16)$$

Then for any $\varepsilon > 0$ we have

$$[\lg n - \lg_3 n + \lg_2 e - \lg c''_v - 1 + \varepsilon] \in \mathrm{UUC}(Z_n^{[v \log_2 n]}), \quad (7.17)$$

$$[\lg n - \lg_3 n + \lg_2 e - \lg c''_v - 2 - \varepsilon] \in \mathrm{ULC}(Z_n^{[v \log_2 n]}), \quad (7.18)$$

$$[\lg n - \lg_3 n + \lg_2 e - \lg c'_v + \varepsilon] \in \mathrm{LUC}(Z_n^{[v \log_2 n]}), \quad (7.19)$$

$$[\lg n - \lg_3 n + \lg_2 e - \lg c'_v - 2 - \varepsilon] \in \mathrm{LLC}(Z_n^{[v \log_2 n]}). \quad (7.20)$$

Remark 3. This result is a modified version of the original form of the theorem. It is also due to Deheuvels (oral communication).

Theorem 7.2 also implies that

$$\liminf_{n \to \infty} \nu_n(l_n) = 0 \quad \text{a.s.}$$

if $l_n \geq \lambda_n$ but

$$\limsup_{n \to \infty} \nu_n([\lg n + (1+\delta)\lg \lg n]) \begin{cases} = 0 & \text{if } \delta > 0, \\ \geq 1 & \text{if } \delta \leq 0. \end{cases}$$

Now we are interested in $\limsup_{n\to\infty} \nu_n([\lg n + \lg\lg n])$ and formulate our simple

THEOREM 7.12 (Révész, 1978).

$$\limsup_{n\to\infty} \nu_n([\lg n + \lg\lg n]) \leq 2 \quad a.s. \tag{7.21}$$

Finally we mention a few unsolved problems (Erdős–Révész, 1987).

Problem 1. We ask about the properties of $Z_n - Z_n^{(2)} = Z_n^{(1)} - Z_n^{(2)}$. It is clear that $\mathbf{P}\{Z_n^{(1)} = Z_n^{(2)} \text{ i.o. }\} = 1$. The lim sup properties of $Z_n^{(1)} - Z_n^{(2)}$ look harder.

Problem 2. Let K_n be the largest integer for which

$$\mathbf{P}\left\{Z_n^{(1)} = Z_n^{(2)} = \cdots = Z_n^{(K_n)} \text{ i.o.}\right\} = 1.$$

Characterize the limit properties of K_n. Observe that Theorem 7.9 suggests

$$0 < \limsup_{n\to\infty} \frac{K_n}{\log\log n} < \infty.$$

Problem 3. Let Z_n^* be the length of the longest tail run, i.e. Z_n^* is the largest integer for which

$$I^*(n, Z_n^*) = -Z_n^*$$

where

$$I^*(n, k) = \min_{0 \leq j \leq n-k}(S_{j+k} - S_j).$$

How can we characterize the limit properties of $|Z_n - Z_n^*|$? Note that by Theorem 7.2

$$\limsup_{n\to\infty} \frac{|Z_n - Z_n^*|}{\log\log n} \leq 1 \quad a.s.$$

and clearly

$$\mathbf{P}\{Z_n = Z_n^* \text{ i.o.}\} = 1.$$

Problem 4. Let

$$U_n = \begin{cases} 0 & \text{if } Z_n \leq Z_n^*, \\ 1 & \text{if } Z_n > Z_n^* \end{cases}$$

and

$$\mathcal{L}_n = (\log n)^{-1} \sum_{k=1}^{n} k^{-1} U_k,$$

i.e. $U_n = 1$ if the longest head run up to n is longer than the longest tail run. We ask: does $\lim_{n\to\infty} \mathcal{L}_n$ exist with probability 1? In the case when $\lim_{n\to\infty} \mathcal{L}_n = \mathcal{L}$ a.s. then $\mathcal{L}$ is called the logarithmic density of the sequence $\{U_n\}$.

Problem 5. (Karlin–Ost, 1988). Consider two independent coin tossing sequences $X_1, X_2, \ldots, X_n$ and $X_1', X_2', \ldots, X_n'$. Let Y_n be the longest common "word" in these sequences, i.e. Y_n is the largest integer for which there exist a $1 \leq k_n < k_n + Y_n \leq n$ and a $1 \leq k_n' < k_n' + Y_n \leq n$ such that

$$X_{k_n+j} = X_{k_n'+j} \quad \text{if } j = 1, 2, \ldots, Y_n.$$

Karlin and Ost (1988) evaluated the limit distribution of Y_n. Its strong behaviour is unknown. Petrov (1965) and Nemetz and Kusolitsch (1982) investigated the length of the longest common word located in the same place, i.e. they defined Y_n assuming that $k_n = k_n'$. In this case they proved a strong law for Y_n.

7.2 The increments of a Wiener process

This paragraph is devoted to studying the limit properties of the processes $J_i(t, a_t)$ ($i = 1, 2, 3, 4, 5$) where a_t is a regular enough function (cf. Notations to the increments).

Note that the r.v. $a^{-1}(W(s + a) - W(s))$ can be considered as the average speed of the particle in the interval $(s, s + a)$. Similarly the r.v.

$$a^{-1} J_1(t, a) = a^{-1} \sup_{0 \leq s \leq t-a} (W(s + a) - W(s))$$

is the largest average speed of the particle in $(0, t)$ over the intervals of size a. The processes $J_i(t, a)$ ($i = 2, 3, 4, 5$, $t \geq a$) have similar meanings.

Note also that

$$J_1(t, a_t) \leq \min\{J_2(t, a_t), J_3(t, a_t)\},$$

$$\max\{J_2(t, a_t), J_3(t, a_t)\} \leq J_4(t, a_t).$$

To start with we present our

THEOREM 7.13 (Csörgő–Révész, 1979/A). *Let a_t ($t \geq 0$) be a nondecreasing function of t for which*

(i) $0 < a_t \leq t$,

(ii) t/a_t *is nondecreasing.*

Then for any $i = 1, 2, 3, 4$ we have

$$\limsup_{t \to \infty} \gamma_t J_i(t, a_t) = \limsup_{t \to \infty} \gamma_t |W(t + a_t) - W(t)|$$
$$= \limsup_{t \to \infty} \gamma_t (W(t + a_t) - W(t)) = 1 \quad a.s.$$

where

$$\gamma_t = \gamma(t, a_t) = \left(2a_t \left(\log \frac{t}{a_t} + \log\log t\right)\right)^{-1/2}.$$

If we also have

(iii)
$$\lim_{t \to \infty} \left(\log \frac{t}{a_t}\right) (\log\log t)^{-1} = \infty$$

then

$$\lim_{t \to \infty} \gamma_t J_i(t, a_t) = 1 \quad a.s.$$

In order to see the meaning of this theorem we present a few examples.

Example 1. For any $c > 0$ we have

$$\lim_{t \to \infty} \frac{J_i(t, c \log t)}{c \log t} = \left(\frac{2}{c}\right)^{1/2} \quad a.s. \quad (i = 1, 2, 3, 4). \tag{7.22}$$

This statement is also a consequence of the Erdős–Rényi (1970) law of large numbers.

Example 2.

$$\lim_{t \to \infty} \frac{J_i(t, 1)}{(2 \log t)^{1/2}} = 1 \quad a.s. \quad (i = 1, 2, 3, 4). \tag{7.23}$$

Example 3. For any $0 < c \leq 1$

$$\limsup_{t \to \infty} \frac{J_i(t, ct)}{(2ct \log\log t)^{1/2}} = 1 \quad a.s. \quad (i = 1, 2, 3, 4). \tag{7.24}$$

In case $c = 1$ we obtain the LIL for Wiener process (cf. Theorem 6.2). Note that (7.24) is also a consequence of Strassen's theorem of Section 8.1.

Having Theorem 7.13 it looks an interesting question to describe the Lévy-classes of the processes $J_i(t, a_t)$ $(i = 1, 2, 3, 4)$ in case of different a_t's. Unfortunately we do not have a complete description of the required Lévy-classes. We can only present the following results:

THEOREM 7.14 (Ortega–Wschebor, 1984). *Let $f(t)$ be a continuous nondecreasing function and assume that a_t satisfies conditions* (i) *and* (ii) *of Theorem* 7.13. *Then*

$$f(t) \in \text{UUC}\left(a_t^{-1/2} J_i(t, a_t)\right) \quad (i = 1, 2, 3, 4)$$

if

$$\int_1^\infty \frac{f^3(t)}{a_t} \exp\left(-\frac{f^2(t)}{2}\right) dt < \infty. \tag{7.25}$$

Further, if

$$\int_1^\infty \frac{f(t)}{a_t} \exp\left(-\frac{f^2(t)}{2}\right) dt = \infty \tag{7.26}$$

then

$$f(t) \in \text{UUC}\left(a_t^{-1/2} \sup_{0 \leq s \leq a_t} (W(t+s) - W(t))\right).$$

Remark 1. In case $a_t = t$ condition (7.26) is equivalent with the corresponding condition of the EFKP LIL of Section 5.2. However, condition (7.25) does not produce the correct UUC in case $a_t = t$. Hence it is natural to conjecture that, in general, the UUC can be characterized by the convergence of the integral of (7.26). It turns out that this conjecture is not exactly true. In fact Grill (1991) obtained the exact description of the upper classes under some weak regularity conditions on a_t. He proved

THEOREM 7.15 *Assume that*

$$a_t = C_0 \exp\left(\int_{C_1}^t \frac{g(y)}{y} dy\right) < \delta t$$

where $0 < \delta < 1$, $g(y)$ is a slowly varying function as $y \to \infty$, C_0, C_1 are positive constants.

Let $f(t) > 0$ $(t > 0)$ be a nondecreasing function. Then

$$f(t) \in \text{UUC}\left(a_t^{-1/2} J_i(t, a_t)\right) \quad (i = 1, 2, 3, 4)$$

if and only if

$$\int_1^\infty (1 + g(t) f^2(t)) \frac{f(t)}{a_t} \exp\left(-\frac{f^2(t)}{2}\right) dt < \infty.$$

In order to illustrate the meaning of this theorem we present a few examples.

INCREMENTS

Example 4. Let $a_t = (\log t)^\alpha$ ($\alpha > 0$). Then $g(t) = \alpha/\log t$ and

$$f_{p,\varepsilon}(t) = \left(2\log t + (3 - 2\alpha) \log_2 t + 2 \sum_{j=3}^{p-1} \log_j t + (2 + \varepsilon) \log_p t \right)^{1/2}$$

$\in \mathrm{UUC}\left(a_t^{-1/2} J_i\right)$ if and only if $\varepsilon > 0$ ($i = 1, 2, 3, 4$; $p = 3, 4, 5, \ldots$).

Example 5. Let $a_t = \exp((\log t)^\alpha)$ ($0 < \alpha < 1$). Then $g(t) = \alpha(\log t)^{\alpha-1}$ and

$$f_{p,\varepsilon}(t) = \left(2\log t - 2(\log t)^\alpha + (3 + 2\alpha) \log_2 t + 2 \sum_{j=3}^{p-1} \log_j t + (2 + \varepsilon) \log_p t \right)^{1/2}$$

$\in \mathrm{UUC}\left(a_t^{-1/2} J_i\right)$ if and only if $\varepsilon > 0$ ($i = 1, 2, 3, 4$; $p = 3, 4, 5, \ldots$).

Example 6. Let $a_t = t^\alpha$ ($0 < \alpha < 1$). Then $g(t) = \alpha$ and

$$f_{p,\varepsilon}(t) = \left(2(1-\alpha) \log t + 5 \log_2 t + 2 \sum_{j=3}^{p-1} \log_j t + (2 + \varepsilon) \log_p t \right)^{1/2}$$

$\in \mathrm{UUC}\left(a_t^{-1/2} J_i\right)$ if and only if $\varepsilon > 0$ ($i = 1, 2, 3, 4$; $p = 3, 4, 5, \ldots$).

Example 7. Let $a_t = \alpha t$ ($0 < \alpha < 1$). Then $g(t) = 1$ and

$$f_{p,\varepsilon}(t) = \left(2 \log_2 t + 5 \log_3 t + 2 \sum_{j=4}^{p-1} \log_j t + (2 + \varepsilon) \log_p t \right)^{1/2}$$

$\in \mathrm{UUC}\left(a_t^{-1/2} J_i\right)$ if and only if $\varepsilon > 0$ ($i = 1, 2, 3, 4; p = 4, 5, 6, \ldots$).

Theorem 7.15 does not cover the case $a_t/t \to 1$. As far as this case is concerned we present

THEOREM 7.16 (Grill, 1991). *Let $a_t = t(1 - \beta(t))$ where $\beta(t)$ is decreasing to 0 and slowly varying as $t \to \infty$ and $f(t) > 0$ be a nondecreasing continuous function. Then*

$$f(t) \in \mathrm{UUC}\left(a_t^{-1/2} J_i(t, a_t)\right) \quad (i = 1, 2, 3, 4)$$

if and only if

$$\int_1^\infty \left(1 + \beta(t) f^2(t)\right) \frac{f(t)}{t} \exp\left(-\frac{f^2(t)}{2}\right) dt < \infty.$$

The characterization of the lower classes is even harder. At first we present a theorem giving a nearly exact characterization of the lower classes when a_t is not very large.

THEOREM 7.17 (Grill, 1991). *Assume that*

$$a_t \uparrow \infty, \quad \Delta(t) = \frac{t}{a_t \log \log t} \uparrow \infty.$$

Then for any $i = 1, 2, 3, 4$ we have

$$(2\log \Delta(t) + \log\log \Delta(t) - K)^{1/2} \in \begin{cases} \mathrm{LLC}\left(a_t^{-1/2} J_i(t, a_t)\right) & \text{if } K > K_i, \\ \mathrm{LUC}\left(a_t^{-1/2} J_i(t, a_t)\right) & \text{if } K < K_i \end{cases}$$

where

$$\log \pi \leq K_1 \leq \log 4\pi,$$
$$\log \frac{\pi}{4} \leq K_2 \leq \log \pi,$$
$$\log \frac{\pi}{4} \leq K_3 \leq \log 4\pi,$$
$$\log \frac{\pi}{16} \leq K_4 \leq \log \pi.$$

If in addition either a_t is of the form

$$a_t = C_0 \exp\left(\int_{C_t}^t \frac{\eta(y)}{y} dy\right)$$

with

$$\limsup_{t \to \infty} \frac{\log \eta(t)}{\log \log t} \leq -t$$

or

$$\frac{\log \log t / a_t}{\log \log t} \to 0,$$

then

$$K_1 = \log \pi, \quad K_2 = \log \frac{\pi}{4},$$

$$\log \frac{\pi}{4} \leq K_3 \leq \log \pi,$$

$$\log \frac{\pi}{16} \leq K_4 \leq \log \frac{\pi}{4}.$$

Remark 2. A very similar result was obtained previously by Révész (1982). However, some of the constants given there are not correct.

Example 8. Let $a_t = te^{-r \log \log t}$ $(0 < r < \infty)$. Then

$$\Delta(t) = (\exp(r \log \log t))(\log \log t)^{-1} \uparrow \infty.$$

Hence

$$\liminf_{t \to \infty} \frac{J_i(t, a_t)}{(2 a_t r \log \log t)^{1/2}} = 1 \quad \text{a.s.}$$

This result was proved by Book and Shore (1978).

If a_t is so large that the condition $\Delta(t) \uparrow \infty$ does not hold, the situation is even more complicated. We have two special results (Theorems 7.18 and 7.19) only.

THEOREM 7.18 (Csáki–Révész, 1979, Grill, 1991). *If* $\Delta(t) = C > 0$, *i.e.* $a_t = Ct(\log \log t)^{-1}$ *then with probability 1 we have*

$$\liminf_{t \to \infty} J_1(t, a_t) = \begin{cases} +\infty & \text{if } C < \Gamma, \\ -\infty & \text{if } C > \Gamma, \end{cases}$$

where Γ *is an absolute, positive constant, its exact value is unknown.*

If $\Delta(t) \to 0$, *then*

$$\liminf_{t \to \infty} \frac{J_1(t, a_t)}{\sqrt{2t \log \log t}} \left(\beta \left(\frac{a_t}{t} \right) \right)^{-1} = -1 \quad \text{a.s.}$$

where

$$\beta(x) = \left(\frac{(2r+1)x - 1}{r(r+1)} \right)^{1/2}$$

and

$$r = \left[\frac{1}{x} \right].$$

Remark 3. Note that if $a_t = \alpha t$ and $1/\alpha$ is an integer, then $\beta(a_t/t) = \alpha$.

We return to the discussion of this theorem in Section 8.1 in the special case when $a_t = \alpha t$ ($0 < \alpha \leq 1$). The first part of Theorem 7.18 suggests the following question. Does there exist a function a_t for which $\liminf_{t\to\infty} J_1(t, a_t) = 0$ a.s.?

THEOREM 7.19 (Csáki–Révész, 1979). *If $\Delta(t) \to 0$, then*

$$1 \leq \liminf_{t\to\infty} \delta(t) J_4(t, a_t) \leq 2\sqrt{2} \quad a.s.$$

where

$$\delta(t) = \left(\frac{\pi^2}{8} a_t \Delta(t)\right)^{-1/2}.$$

Remark 4. In case $a_t = t$, Theorem of Chung (Section 5.3) implies that

$$\liminf_{t\to\infty} \delta(t) J_4(t, a_t) = 1 \quad \text{a.s.}$$

However, this relation does not follow from Theorem 7.19.

Remark 5. Ortega and Wschebor (1984) also investigated the upper classes of the "small increments" of $W(\cdot)$. These are defined as follows:

$$J_6(t, a_t) = \sup_{0 \leq s \leq t - a_t} (W(s + a_s) - W(s)),$$
$$J_7(t, a_t) = \sup_{0 \leq s \leq t - a_t} |W(s + a_s) - W(s)|,$$
$$J_8(t, a_t) = \sup_{0 \leq s \leq t - a_t} \sup_{0 \leq u \leq a_s} (W(s + u) - W(s)),$$
$$J_9(t, a_t) = \sup_{0 \leq s \leq t - a_t} \sup_{0 \leq u \leq a_s} |W(s + u) - W(s)|$$

where a_s is a function satisfying conditions (i) and (ii) of Theorem 7.13.

Remark 6. Hanson and Russo (1983/B) studied a strongly generalized version of the questions of the present paragraph. In fact they described the limit points of the sequence

$$\frac{W(\beta_k) - W(\alpha_k)}{(2(\beta_k - \alpha_k)(\log(\beta_k/(\beta_k - \alpha_k)) + \log\log\beta_k))^{1/2}}$$

for a large class of the sequences $0 \leq \alpha_k < \beta_k < \infty$.

Finally we present a result on the behaviour of $J_5(\cdot, \cdot)$.

THEOREM 7.20 (Csörgő–Révész, 1979/B). *Assume that a_t satisfies conditions* (i), (ii) *of Theorem 7.13. Then*

$$\liminf_{t\to\infty} \mathcal{K}_t J_5(t, a_t) = 1 \quad a.s.$$

where

$$J_5(t, a_t) = \inf_{0\leq s\leq t-a_t} \sup_{0\leq u\leq a_t} |W(s+u) - W(s)|$$

and

$$\mathcal{K}_t = \left(\frac{8(\log t a_t^{-1} + \log\log t)}{\pi^2 a_t}\right)^{1/2}.$$

If (iii) *of Theorem 7.13 is also satisfied then*

$$\lim_{t\to\infty} \mathcal{K}_t J_5(t, a_t) = 1 \quad a.s.$$

The following examples illustrate what this theorem is all about.

Example 9. Let $a_t = 8\log t/\pi^2$ hence $\mathcal{K}_t \to 1$ $(t \to \infty)$. Then Theorem 7.20 tells us that for all t large enough, for any $\varepsilon > 0$ and for almost all $\omega \in \Omega$ there exists a $0 \leq s = s(t,\varepsilon,\omega) \leq t - a_t$ such that

$$\sup_{0\leq u\leq \frac{8}{\pi^2}\log t} |W(s+u) - W(s)| \leq 1 + \varepsilon$$

but, for all $s \in [0, t-a_t]$ with probability 1

$$\sup_{0\leq u\leq \frac{8}{\pi^2}\log t} |W(s+u) - W(s)| \geq 1 - \varepsilon.$$

At the same time Theorem 7.13 stated the existence of an $s \in [0, t-a_t]$ for which, with probability 1,

$$\left|W\left(s + \frac{8}{\pi^2}\log t\right) - W(s)\right| \geq \left(\frac{4}{\pi} - \varepsilon\right)\log t$$

but for all $s \in [0, t-a_t]$

$$\sup_{0\leq u\leq \frac{8}{\pi^2}\log t} |W(s+u) - W(s)| \leq \left(\frac{4}{\pi} + \varepsilon\right)\log t.$$

Example 10. Let $a_t = t$. Then Theorem 7.20 implies

$$\liminf_{t\to\infty} \left(\frac{8\log\log t}{\pi^2 t}\right)^{1/2} \sup_{0\leq s\leq t} |W(s)| = 1 \quad a.s.$$

Hence we have the Other LIL (cf. (5.9)).

Example 11. Let $a_t = (\log t)^{1/2}$ hence $\mathcal{K}_t \approx 2\sqrt{2}(\log t)^{1/4}/\pi$. Then Theorem 7.20 claims that for all t large enough, for any $\varepsilon > 0$ and for almost all $\omega \in \Omega$ there exists an $s = s(t, \varepsilon, \omega) \in [0, t - a_t]$ such that

$$\sup_{0 \leq u \leq (\log t)^{1/2}} |W(s+u) - W(s)| \leq (1+\varepsilon)\frac{\pi}{\sqrt{8}}(\log t)^{-1/4}.$$

That is to say the interval $[0, t - a_t]$ has a subinterval of length $(\log t)^{1/2}$ where the sample function of the Wiener process is nearly constant. In fact, the fluctuation away from a constant is as small as $(1+\varepsilon)\pi 8^{-1/2}(\log t)^{-1/4}$.

This result is sharp in the sense that for all t large enough and all $s \in [0, t - a_t]$ we have with probability 1

$$\sup_{0 \leq u \leq (\log t)^{1/2}} |W(s+u) - W(s)| \geq (1-\varepsilon)\frac{\pi}{\sqrt{8}}(\log t)^{-1/4}.$$

Clearly, replacing the condition $a_t = (\log t)^{1/2}$ in Example 11 by $a_t = o(\log t)$, we also find that there exists a subinterval of $[0, t - a_t]$ of size a_t where the sample function is nearly constant. Csáki and Földes (1984/A) were interested in the analogue problem when the term "nearly constant" is replaced by "nearly zero". They proved

THEOREM 7.21 *Assume that a_t satisfies conditions* (i), (ii) *of Theorem 7.13. Then*

$$\liminf_{t \to \infty} h_t \inf_{0 \leq s \leq t-a_t} \sup_{0 \leq u \leq a_t} |W(s+u)| = 1 \quad a.s.$$

where

$$h_t = \left(\frac{4\log(ta_t^{-1}) + 8\log\log t}{\pi^2 a_t}\right)^{1/2}.$$

If (iii) *of Theorem 7.13 is also satisfied then*

$$\lim_{t \to \infty} h_t \inf_{0 \leq s \leq t-a_t} \sup_{0 \leq u \leq a_t} |W(s+u)| = 1 \quad a.s.$$

Example 12. Letting $a_t = t$ we obtain the Other LIL (cf. (5.9)).

Example 13. If $a_t = o(\log t)$ then $h_t \to \infty$ and

$$\lim_{t \to \infty} \inf_{0 \leq s \leq t-a_t} \sup_{0 \leq u \leq a_t} |W(s+u)| = 0$$

while in case $a_t = 4c^2\pi^{-2}\log t$ we have $h_t \to c^{-1}$ as $t \to \infty$ and

$$\lim_{t\to\infty} \inf_{0\leq s\leq t-a_t} \sup_{0\leq u\leq a_t} |W(s+u)| = c.$$

(Compare Example 13 in case $c = 1$ and $c = \sqrt{2}$ with the first part of Example 9.)

Theorems 7.13–7.19 gave a more or less complete description of the strong behaviour of $J_i(t,a_t)$ ($i = 1, 2, 3, 4$). To complete this Section we give the following weak law:

THEOREM 7.22 (Deheuvels–Révész, 1987). *Let $t/a_t = d_t$. Assume that*

$$\lim_{t\to\infty} d_t = \infty. \tag{7.27}$$

Then for any $i = 1, 2, 3, 4$ in probability

$$\lim_{t\to\infty} \frac{(2\log d_t)^{1/2}(a_t^{-1/2} J_i(t,a_t) - (2\log d_t)^{1/2})}{\log\log d_t} = 1/2. \tag{7.28}$$

We also mention that the proof of Theorem 7.22 is based on the following:

THEOREM 7.23 (Deheuvels–Révész, 1987). *Assume (7.27). Then for any $-\infty < y < \infty$ we have*

$$\mathbf{P}\left\{(2\log d_t)^{1/2}\left(\frac{J_i(t,a_t)}{a_t^{1/2}} - (2\log d_t)^{1/2} - \frac{\log_2 d_t + \log(4/\pi) + 2y}{2(2\log d_t)^{1/2}}\right) \leq 0\right\}$$

$$\to \exp(-e^{-y}) \quad (t\to\infty)$$

if $i = 2, 3, 4$ and

$$\mathbf{P}\left\{(2\log d_t)^{1/2}\left(\frac{J_1(t,a_t)}{a_t^{1/2}} - (2\log d_t)^{1/2} - \frac{\log_2 d_t - \log\pi + 2y}{2(2\log d_t)^{1/2}}\right) \leq 0\right\}$$

$$\to \exp(-e^{-y}) \quad (t\to\infty).$$

Note that in the above two theorems we have no regularity conditions on a_t except (7.27).

In order to understand the meaning of (7.28) consider the case $i = 2$ and assume

$$\frac{\log d_t}{\log\log t} = r \quad (0 < r < \infty). \tag{7.29}$$

In this case Theorem 7.13 implies that $J_2(t,a_t)$ can be as large as

$$\gamma_t^{-1} = (2a_t)^{1/2}((r+1)\log\log t)^{1/2}.$$

In the same case Theorem 7.17 implies that $J_2(t, a_t)$ can be as small as

$$(2a_t)^{1/2}(r \log \log t)^{1/2}.$$

(7.28) describes the "typical" behaviour of $J_2(t, a_t)$ under the condition of (7.29). Namely it behaves like

$$a_t^{1/2}\left\{(2r \log_2 t)^{1/2} + \frac{\log_3 t}{2(2r \log_2 t)^{1/2}}\right\} \approx (2a_t)^{1/2}(r \log_2 t)^{1/2}.$$

It is worthwhile to mention an equivalent but simpler form of Theorem 7.23.

THEOREM 7.24

$$\lim_{t \to \infty} \mathbf{P}\left\{\sup_{0 \leq s \leq t}(W(s+1) - W(s)) \leq f(y,t)\right\} = \exp(-e^{-y})$$

and

$$\lim_{t \to \infty} \mathbf{P}\left\{\sup_{0 \leq s \leq t}|W(s+1) - W(s)| \leq f(y,t)\right\} = \exp(-2e^{-y})$$

where

$$f(y,t) = (2\log t)^{-1/2}\left(y + 2\log t - \frac{1}{2}\log \log t - \frac{1}{2}\log \pi\right).$$

Let us give a summary of the results of this section. To study the properties of the processes $J_i(t, a_t)$ ($i = 1, 2, 3, 4, 5$) we have to assume different conditions on a_t. For the sake of simplicity from now on we always assume that a_t is nondecreasing and satisfies conditions (i) and (ii) of Theorem 7.13.

Then the limit distributions of $J_i(\cdot, \cdot)$ for $i = 1, 2, 3, 4$ are given in Theorem 7.23. Observe that the limit distributions in case $i = 1$ and in case $i = 2, 3, 4$ are different. The limit distribution of $J_5(\cdot, \cdot)$ is unknown. The exact distribution is not known in any case.

A description of the upper classes of $J_i(\cdot, \cdot)$ ($i = 1, 2, 3, 4$) is given in Theorem 7.14 but there is a big gap in this theorem between the description of UUC(J_i) and ULC(J_i), i.e. there is a big class of very regular functions for which Theorem 7.14 does not tell us whether they belong to the UUC(J_i) or to the ULC(J_i). This gap is filled in by Theorem 7.15 if a_t satisfies a weak regularity condition. However, this regularity condition excludes the case $a_t/t \to 1$. This case is studied in Theorem 7.16. The above-mentioned results do not tell us anything about $J_5(\cdot, \cdot)$. In case if a_t is not very large (condition (iii) of Theorem 7.13 is satisfied) a very weak result is given in Theorem 7.20.

The lower classes of $J_i(\cdot,\cdot)$ ($i=1,2,3,4$) are "almost" completely described if $a_t \ll t/\log\log t$ by Theorem 7.17. If a_t does not satisfy this condition then Theorems 7.18, 7.19 (resp. 7.20) tell something about the lower classes of $J_1(\cdot,\cdot)$, $J_4(\cdot,\cdot)$ (resp. $J_5(\cdot,\cdot)$) but we do not have a complete characterization and we do not have any results (except trivial ones) about the lower classes of $J_2(\cdot,\cdot)$ and $J_3(\cdot,\cdot)$.

7.3 The increments of S_N

By the Invariance Principle 1 (cf. Section 6.3) we obtain that any theorem of Section 7.2 will remain true, replacing the Wiener process by a random walk (i.e. replacing J_i by I_i ($i=1,2,3,4,5$)) provided that γ_n^{-1} is big enough or equivalently a_n is big enough. In fact Theorems 7.13, 7.18–7.21 (resp. 7.14–7.17) remain true if $a_n \gg \log n$ (resp. $a_n \gg (\log n)^3$) while Theorems 7.23 and 7.24 remain true as they are. Hence we only study the increments of S_n in the case when $\lim_{n\to\infty} a_n (\log n)^{-3-\varepsilon} = 0$ for any $\varepsilon > 0$.

A trivial consequence of Theorem 7.1 (resp. Theorem 7.13) (cf. also Example 1 of Section 7.2) is

THEOREM 7.25

$$\lim_{n\to\infty} \frac{I_1(n, \lg n)}{\lg n} = 1 \quad a.s. \tag{7.30}$$

$$\lim_{n\to\infty} \frac{J_1(n, \lg n)}{\lg n} = \lim_{n\to\infty} \frac{J_1\left(n, \dfrac{\log n}{\log 2}\right)}{\dfrac{\log n}{\log 2}} = (2\log 2)^{1/2} \quad a.s. \tag{7.31}$$

Remark 1. Comparing (7.30) and (7.31) we can see that the behaviours of I_1 and J_1 are different indeed if $a_n = c\log n$ ($0 < c < \infty$). As a consequence we also obtain that the rate $O(\log n)$ of Invariance Principle 1 (cf. Section 6.3) is the best possible one. This observation is due to Bártfai (1966) and Erdős–Rényi (1970).

Theorems 7.2, 7.3, 7.4 imply much stronger results on the behaviour of $I_i(\cdot,\cdot)$ than (7.30) of Theorem 7.25. In fact we obtain

THEOREM 7.26 *Assuming different growing conditions on $\{a_n\}$ we get*

(i) *If for some $\varepsilon > 0$*

$$a_n \leq [\lg n - \lg\lg\lg n + \lg\lg e - 2 - \varepsilon] = \lambda_n.$$

Then
$$I_i(n, a_n) = a_n \quad a.s. \quad (i = 1, 2, 3, 4)$$
for all but finitely many n, i.e. $I_i(n, a_n)$ *is* AD.

(ii) *If for some* $\varepsilon > 0$
$$\lambda_n \leq a_n \leq [\lg n + \lg\lg n - \lg\lg\lg n + \lg\lg e - 2 - \varepsilon] = \lambda_n(1).$$
Then
$$I_i(n, a_n) = a_n \text{ or } a_n - 2 \quad a.s \quad (i = 1, 2, 3, 4)$$
for all but finitely many n, i.e. $I_i(n, a_n)$ *is* QAD.

(iii) *If for some* $\varepsilon > 0$
$$\lambda_n(1) \leq a_n \leq [\lg n + \lg\lg n + (1+\varepsilon)\lg\lg\lg n] = d_n(1).$$
Then
$$I_i(n, a_n) = a_n \text{ or } a_n - 2 \text{ or } a_n - 4 \quad a.s. \quad (i = 1, 2, 3, 4)$$
for all but finitely many n, i.e. $I_i(n, a_n)$ *is* QAD.

(iv) *In general, if*
$$d_n(T) = [\lg n + T\lg\lg n + (1+\varepsilon)\lg\lg\lg n] < a_n \leq \lambda_n(T+1)$$
$$= [\lg n + (T+1)\lg\lg n - \lg\lg\lg n - \lg((T+1)!) + \lg\lg e - 2 - \varepsilon].$$
Then
$$I_i(n, a_n) = a_n - 2T - 2 \text{ or } a_n - 2T \quad a.s.$$
for all but finitely many n, i.e. $I_i(n, a_n)$ *is* QAD,
and if
$$\lambda_n(T+1) < a_n \leq d_n(T+1)$$
then
$$I_i(n, a_n) = a_n - 2T - 4 \text{ or } a_n - 2T - 2 \text{ or } a_n - 2T,$$
for all but finitely many n, i.e. $I_i(n, a_n)$ *is* QAD.

The above theorem essentially tells us that $I_i(n, a_n)$ is QAD with not more than three possible values if $a_n \leq \lg n + T\lg\lg n$ for some $T > 0$. The next theorem applies for somewhat larger a_n.

INCREMENTS

THEOREM 7.27 (Deheuvels–Erdős–Grill–Révész, 1987). *Let $a_n = O(\lg n)$ and $0 < T = T_{a_n} < a_n/2$ be nondecreasing sequences of integers. Then*

$$a_n - 2T \in \mathrm{LLC}(I_i(n, a_n)) \quad if \quad \sum_{n=1}^{\infty} \exp(-2^n p(2^n)) < \infty,$$

$$a_n - 2T \in \mathrm{LUC}(I_i(n, a_n)) \quad if \quad \sum_{n=1}^{\infty} \exp(-2^n p(2^n)) = \infty,$$

$$a_n - 2T \in \mathrm{ULC}(I_i(n, a_n)) \quad if \quad \sum_{n=1}^{\infty} 2^n p(2^n) = \infty,$$

$$a_n - 2T \in \mathrm{UUC}(I_i(n, a_n)) \quad if \quad \sum_{n=1}^{\infty} 2^n p(2^n) < \infty,$$

where

$$p(n) = \left(1 - \frac{2T_{a_n}}{a_n - 1}\right) 2^{-a_n - 1} \binom{a_n - 1}{T_{a_n}}.$$

Here we present a few consequences.

Consequence 1. Let

$$a_n = \lg n + f(n)$$

be a nondecreasing sequence of positive integers with $f(n) = o(\lg n)$.

(i) Assume that

$$\lim_{n \to \infty} \frac{f(n)}{(\lg n)^\varepsilon} = 0 \text{ for any } \varepsilon > 0.$$

Then $I_i(n, a_n)$ is QAD and there exist an $\alpha_1(n) \in \mathrm{UUC}(I_i(n, a_n))$ and an $\alpha_4(n) \in \mathrm{LLC}(I_i(n, a_n))$ such that $\alpha_1(n) - \alpha_4(n) \leq 3$.

(ii) Assume that

$$f(n) = O\left((\lg n)^\Theta\right) \quad (0 < \Theta < 1).$$

Then $I_i(n, a_n)$ is QAD and there exist an $\alpha_1(n) \in \mathrm{UUC}(I_i(n, a_n))$ and an $\alpha_4(n) \in \mathrm{LLC}(I_i(n, a_n))$ such that $\alpha_1(n) - \alpha_4(n) \leq 2/(1 - \Theta) + 1$.

(iii) Assume that

$$\lim_{n \to \infty} \frac{f(n)}{(\lg n)^{1-\varepsilon}} = \infty \text{ for any } \varepsilon > 0.$$

Then $I_i(n, a_n)$ is not QAD.

Consequence 2. Let $a_n = [C \lg n]$ with $C > 1$. Then

$$C(1 - 2\beta)\lg n + (1 + \varepsilon)2\rho \lg \lg n \in \text{UUC}(I_i(n, a_n)),$$
$$C(1 - 2\beta)\lg n + (1 - \varepsilon)2\rho \lg \lg n \in \text{ULC}(I_i(n, a_n)),$$
$$C(1 - 2\beta)\lg n - 2\rho \lg \lg n - 4\rho \lg \lg \lg n$$
$$+4\rho \lg(1 - 2\beta) + 4\rho \lg \lg e + 2\rho \lg \pi + 6\rho + 1 + \varepsilon \in \text{LUC}(I_i(n, a_n)),$$
$$C(1 - 2\beta)\lg n - 2\rho \lg \lg n - 4\rho \lg \lg \lg n$$
$$+4\rho \lg(1 - 2\beta) + 4\rho \lg \lg e + 2\rho \lg \pi + 6\rho + 1 - \varepsilon \in \text{LLC}(I_i(n, a_n)),$$

where β is the solution of the equation

$$\left(2\beta^\beta (1-\beta)^{1-\beta}\right)^C = 2,$$

$$\rho = \left(2 \lg \frac{1-\beta}{\beta}\right)^{-1}$$

and ε is an arbitrary positive number.

Remark 2. Consequence 2 above is a stronger version of an earlier result of Deheuvels–Devroye–Lynch (1986).

In the case $a_n \gg \lg n$ we present the following:

THEOREM 7.28 (Deheuvels–Steinebach, 1987). *Let a_n be a sequence of positive integers with $a_n = [\tilde{a}_n]$ where $\tilde{a}_n (\log n)^{-p}$ is decreasing for some $p > 0$ and $\tilde{a}_n / \log n$ is increasing for some $p > 0$. Then for any $\varepsilon > 0$ we have*

$$\alpha_n a_n - t_n^{-1} \log a_n + (3/2 + \varepsilon) t_n^{-1} \log \log n \in \text{UUC}(I_i(n, a_n)),$$
$$\alpha_n a_n - t_n^{-1} \log a_n + (3/2 - \varepsilon) t_n^{-1} \log \log n \in \text{ULC}(I_i(n, a_n)),$$
$$\alpha_n a_n - t_n^{-1} \log a_n + (1/2 + \varepsilon) t_n^{-1} \log \log n \in \text{LUC}(I_i(n, a_n)),$$
$$\alpha_n a_n - t_n^{-1} \log a_n + (1/2 - \varepsilon) t_n^{-1} \log \log n \in \text{LLC}(I_i(n, a_n))$$

where α_n is the unique positive solution of the equation

$$\exp(-(\log n)/a_n) = (1 + \alpha_n)^{(1+\alpha_n)/2}(1 - \alpha_n)^{(1-\alpha_n)/2}$$

and

$$t_n = \frac{1}{2} \log \frac{1 + \alpha_n}{1 - \alpha_n}.$$

Note that $\alpha_n \approx (2 a_n^{-1} \log n)^{1/2}$.

In order to study the properties of I_5 resp.

$$I_5^*(n, a_n) = \min_{0 \le j \le n - a_n} \max_{0 \le i \le a_n} |S_{j+i}|,$$

first we mention that by the Invariance Principle the properties of J_5 resp.

$$J_5^*(t, a_t) = \inf_{0 \le s \le t - a_t} \sup_{0 \le u \le a_t} |W(s + u)|$$

will be inherited if $a_n \ge (\log n)^{3+\varepsilon}$ ($\varepsilon > 0$). In fact Theorems 7.20 and 7.21 will remain true if J_5 (resp. J_5^*) are replaced by I_5 (resp. I_5^*) and $a_n \ge (\log n)^{3+\varepsilon}$ ($\varepsilon > 0$). Hence we have to study the properties of I_5 (resp. I_5^*) only when $a_n (\log n)^{-3-\varepsilon} \to 0$ ($n \to \infty$) for any $\varepsilon > 0$. It turns out that Theorem 7.21 remains true if $a_n / \log n \to \infty$ ($n \to \infty$). In fact we have

THEOREM 7.29 (Csáki–Földes, 1984/B). *Assume that a_n satisfies conditions* (i) *and* (ii) *of Theorem 7.13 and*

$$\lim_{n \to \infty} a_n (\log n)^{-1} = \infty.$$

Then

$$\liminf_{n \to \infty} h_n I_5^*(n, a_n) = 1 \quad a.s.$$

where h_n is defined in Theorem 7.21. If condition (iii) *of Theorem 7.13 is also satisfied, then*

$$\lim_{n \to \infty} h_n I_5^*(n, a_n) = 1 \quad a.s.$$

If $a_n = [c \log n]$ then we have

THEOREM 7.30 (Csáki–Földes, 1984/B). *Let $a_n = [c \log n]$ ($c > 0$) and define $\alpha^* = \alpha^*(c) > 1$ as the solution of the equation*

$$\cos \frac{\pi}{2\alpha^*} = \exp\left(-\frac{1}{2c}\right),$$

if $\alpha^*(c)$ is not an integer then

$$I_5^*(n, a_n) = [\alpha^*(c)] \quad a.s.$$

*for all but finitely many n, i.e. I_5^* is AD,
if $\alpha^*(c)$ is an integer then*

$$\alpha^*(c) - 1 \le I_5^*(n, a_n) \le \alpha^*(c) \quad a.s.$$

for all but finitely many n, i.e. I_5^* is QAD. Moreover

$$I_5^*(n, a_n) = \alpha^*(c) - 1 \quad i.o. \ a.s.$$

and

$$I_5^*(n, a_n) = \alpha^*(c) \quad i.o. \ a.s.$$

The properties of I_5 are unknown when $\log n \ll a_n \leq (\log n)^3$. However, we have

THEOREM 7.31 (Csáki–Földes, 1984/B). *Let $a_n = [c \log n]$ $(c > 0)$ and define $\alpha = \alpha(c) > 1$ as the solution of the equation*

$$\cos \frac{\pi}{2\alpha} = \exp\left(-\frac{1}{c}\right),$$

if $\alpha(c)$ is not an integer then

$$I_5(n, a_n) = [\alpha(c)] \quad a.s.$$

*for all but finitely many n, i.e. I_5 is AD,
if $\alpha(c)$ is an integer then*

$$\alpha(c) - 1 \leq I_5(n, a_n) \leq \alpha(c) \quad a.s.$$

for all but finitely many n, i.e. I_5 is QAD. Moreover

$$I_5 = \alpha(c) - 1 \quad i.o. \ a.s.$$

and

$$I_5 = \alpha(c) \quad i.o. \ a.s.$$

Chapter 8

Strassen Type Theorems

8.1 The theorem of Strassen

The Law of Iterated Logarithm of Khinchine (Section 4.4) implies that *for any $\varepsilon > 0$ and for almost all $\omega \in \Omega$ there exists a random sequence of integers $0 < n_1 = n_1(\varepsilon, \omega) < n_2 = n_2(\varepsilon, \omega) < \ldots$ such that*

$$S(n_k) \geq (1-\varepsilon)(2n_k \log\log n_k)^{1/2} = (1-\varepsilon)(b(n_k))^{-1}. \tag{8.1}$$

We ask what can be said about the sequence $\{S(j); j = 1, 2, \ldots, n_k\}$ (provided that (8.1) holds). In order to illuminate the meaning of this question we prove

THEOREM 8.1 *Assume that $n_k = n_k(\varepsilon, \omega)$ satisfies (8.1). Then*

$$S([n_k/2]) \geq (1-\varepsilon)\frac{1}{2}S(n_k) \geq \frac{1-2\varepsilon}{2}(b(n_k))^{-1} \quad a.s. \tag{8.2}$$

for all but finitely many k.

Proof. Let $0 \leq \alpha < 1 - 2\varepsilon$ and assume that

$$\alpha(b(n_k))^{-1} \leq S([n_k/2]) \leq (\alpha+\varepsilon)(b(n_k))^{-1}. \tag{8.3}$$

Then by (8.1)

$$S(n_k) - S([n_k/2]) \geq (1 - \alpha - 2\varepsilon)(b(n_k))^{-1}. \tag{8.4}$$

By Theorem 2.10 the probability that the inequalities (8.3) and (8.4) simultaneously hold is equal to $O((\log n_k)^{-2(\alpha^2 + (1-\alpha-2\varepsilon)^2)})$.

Note that if $\alpha \neq 1/2$ and ε is small enough then $2(\alpha^2 + (1-\alpha-2\varepsilon)^2) > 1$. Hence by the method used in the proof of Khinchine's theorem (Step 1) we obtain that the inequalities (8.3) and (8.4) will be satisfied only for finitely many k with probability 1. This fact easily implies Theorem 8.1.

Similarly one can prove that *for any $0 < x < 1$*

$$S([xn_k]) \geq (1-\varepsilon)xS(n_k) \quad a.s. \tag{8.5}$$

83

for all but finitely many k.

(8.5) suggests that if n_k satisfies (8.1) and k is big enough then the process $\{S([xn_k]); 0 \leq x \leq 1\}$ will be close to the process $\{xS(n_k); 0 \leq x \leq 1\}$. It is really so and it is a trivial consequence of

STRASSEN'S THEOREM 1 (1964). *The sequence*

$$s_n(x) = b_n \left(S_{[nx]} + \left(x - \frac{[nx]}{n} \right) X_{[nx]+1} \right) \quad (0 \leq x \leq 1; \ n = 1, 2, \ldots)$$

is relatively compact in $C(0,1)$ with probability 1 and the set of its limit points is $\mathcal{S}$ (see notations to Strassen type theorems).

The meaning of this statement is that there exists an event $\Omega_0 \subset \Omega$ of probability zero with the following two properties:

Property 1. For any $\omega \notin \Omega_0$ and any sequence of integers $0 < n_1 < n_2 < \ldots$ there exist a random subsequence $n_{k_j} = n_{k_j}(\omega)$ and a function $f \in \mathcal{S}$ such that

$$s_{n_{k_j}}(x, \omega) \to f(x) \text{ uniformly in } x \in [0,1].$$

Property 2. For any $f \in \mathcal{S}$ and $\omega \notin \Omega_0$ there exists a sequence of integers $n_k = n_k(\omega, f)$ such that

$$s_{n_k}(x, \omega) \to f(x) \text{ uniformly in } x \in [0,1].$$

The Invariance Principle 1 of Section 6.3 implies that the above theorem is equivalent to

STRASSEN'S THEOREM 2 (1964). *The sequence*

$$w_n(x) = b_n W(nx) \quad (0 \leq x \leq 1; \ n = 1, 2, \ldots)$$

is relatively compact in $C(0,1)$ with probability 1 and the set of its limit points is $\mathcal{S}$.

Remark 1. Since $|f(1)| \leq 1$ for any function $f \in \mathcal{S}$ and $f(x) = x \in \mathcal{S}$, Strassen's theorem 1 implies Khinchine's LIL.

Consequence 1. For any $\varepsilon > 0$ and for almost all $\omega \in \Omega$ there exists a $T_0 = T_0(\varepsilon, \omega)$ such that if

$$W(T) \geq (1-\varepsilon)(b(T))^{-1} \text{ for some } T \geq T_0$$

then

$$\sup_{0 \leq t \leq T} \left| W(t) - t \left(\frac{2 \log \log T}{T} \right)^{1/2} \right| \leq 2\varepsilon (b(T))^{-1}.$$

STRASSEN TYPE THEOREMS

Consequence 1 tells us that if $W(t)$ "wants" to be as big in point T as it can be at all then it has to increase in $(0, T)$ nearly linearly (that is to say it has to minimize the used energy).

The proof of Strassen's theorem 2 will be based on the following three lemmas.

LEMMA 8.1 *Let d be a positive integer and $\alpha_1, \alpha_2, \ldots, \alpha_d$ be a sequence of real numbers for which*

$$\sum_{i=1}^{d} \alpha_i^2 = 1.$$

Further, let

$$W^*(n) = \alpha_1 W(n) + \alpha_2 (W(2n) - W(n)) + \cdots + \alpha_d (W(dn) - W((d-1)n)).$$

Then

$$\limsup_{n \to \infty} b_n W^*(n) = 1 \quad a.s. \tag{8.6}$$

and

$$\liminf_{n \to \infty} b_n W^*(n) = -1 \quad a.s. \tag{8.7}$$

Proof of this lemma is essentially the same as that of the Khinchine's LIL. The details will be omitted.

The next lemma gives a characterization of S.

LEMMA 8.2 (Riesz–Sz.–Nagy, 1955, p. 75). *Let f be a real valued function on $[0, 1]$. The following two conditions are equivalent:*

(i) *f is absolutely continuous and $\int_0^1 (f')^2 dx \leq 1$,*

(ii)

$$\sum_{i=1}^{r} \frac{\left(f\left(\frac{i}{r}\right) - f\left(\frac{i-1}{r}\right)\right)^2}{1/r} \leq 1 \text{ for any } r = 1, 2, \ldots$$

and f is continuous on $[0, 1]$.

In order to formulate our next lemma we introduce some notations. For any real valued function $f \in C(0, 1)$ and positive integer d, let $f^{(d)}$ be the linear interpolation of f over the points i/d, that is

$$f^{(d)}(x) = f\left(\frac{i}{d}\right) + d\left(f\left(\frac{i+1}{d}\right) - f\left(\frac{i}{d}\right)\right)\left(x - \frac{i}{d}\right)$$

if $i/d \leq x \leq (i+1)/d$, $i = 0, 1, \ldots, d-1$.
Let
$$C_d = \{f^{(d)} : f \in C(0,1)\} \subset C(0,1),$$
$$\mathcal{S}_d = \{f^{(d)} : f \in \mathcal{S}\},$$

where $\mathcal{S}_d \subset \mathcal{S}$ by Lemma 8.2.

LEMMA 8.3 *The sequence $\{w_n^{(d)}(x); \ 0 \leq x \leq 1\}$ is relatively compact in C_d with probability 1 and the set of its limit points is $\mathcal{S}_d$.*

Proof. By Khinchine's LIL and continuity of Wiener process our statement holds when $d = 1$. We prove it for $d = 2$. For larger d the proof is similar and immediate. Let $V_n = (W(n), W(2n) - W(n))$ $(n = 1, 2, \ldots)$ and α, β be real numbers such that $\alpha^2 + \beta^2 = 1$. Then by Lemma 8.1 and continuity of W the set of limit points of the sequence

$$\left\{ \frac{V_n\binom{\alpha}{\beta}}{\sqrt{2n \log \log n}} \right\}_{n=1}^{\infty} = \left\{ \frac{\alpha W(n) + \beta(W(2n) - W(n))}{\sqrt{2n \log \log n}} \right\}_{n=1}^{\infty}$$

is the interval $[-1, +1]$. This implies that the set of limit points of the sequence $\{b_n V_n\}$ is a subset of the unit disc and the boundary of the unit circle belongs to this limit set.

Now let $V_n^* = (W(n), W(2n) - W(n), W(3n) - W(2n))$. In the same way as above one can prove that the set of limit points of $\{b_n V_n^*\}$ is a subset of the unit ball of $\mathbb{R}^3$ which contains the boundary of the unit sphere. This fact in itself already implies that the set of limit points of $\{b_n V_n\}$ is the unit disc of $\mathbb{R}^2$ and this, in turn, is equivalent to our statement.

Proof of Strassen's Theorem 2. For each $\omega \in \Omega$ we have

$$\sup_{0 \leq x \leq 1} \left| w_n(x) - w_n^{(d)}(x) \right| \leq \sup_{0 \leq x \leq 1} \sup_{0 \leq s \leq 1/d} |w_n(x+s) - w_n(x)|,$$

hence by Theorem 7.13 (cf. also Example 3 of Section 7.2)

$$\limsup_{n \to \infty} \sup_{0 \leq x \leq 1} \left| w_n(x) - w_n^{(d)}(x) \right| = d^{-1/2} \quad \text{a.s.}$$

Consequently we have the Theorem by Lemmas 8.2 and 8.3 where we also use the fact that Lemma 8.2 guarantees that $\mathcal{S}$ is closed.

The discreteness of n is inessential in this Theorem. In fact if we define

$$w_t(x) = b_t W(tx) \quad (x \in [0,1], \ t > 0)$$

then we have

STRASSEN'S THEOREM 3 (1964). *The net $w_t(x)$ is relatively compact in $C(0,1)$ with probability 1 and the set of its limit points is S.*

As an application of Strassen's theorem we sketch the proof of Theorem 7.18 in the special case $a_t = \alpha t$.

At first we mention that Strassen's theorem implies that

$$\limsup_{t\to\infty} b_t J_1(t,\alpha t) = \alpha^{1/2} \quad \text{a.s.}$$

which can be obtained by considering the function

$$f(x) = \begin{cases} x\alpha^{-1/2} & \text{if } 0 \leq x \leq \alpha, \\ \alpha^{1/2} & \text{if } \alpha < x \leq 1 \end{cases}$$

which belongs to the Strassen's class S (cf. also Theorem 7.13 and Example 3 of Section 7.2). The fact that b_t is the right normalizing factor for the lim inf also follows from Strassen's theorem. Let

$$C_\alpha = -\liminf_{t\to\infty} b_t J_1(t,\alpha t).$$

In case $\alpha = 1$ it is well known that $C_1 = 1$ and this can be obtained by considering the function $f(s) = -s$ ($0 \leq s \leq 1$) in S. Considering the function $f(s) = -s$ it is also immediate that $C_\alpha \geq \alpha$. Theorem 7.18 claims, however, that equality holds (i.e. $C_\alpha = \alpha$) if and only if $1/\alpha$ is an integer; in other cases $C_\alpha > \alpha$.

Now we show that the Strassen's theorem implies that

$$\liminf_{t\to\infty} b(t) J_1(t,\alpha t) \leq -C_\alpha = -\left(\frac{(2r+1)\alpha - 1}{r(r+1)}\right)^{1/2}. \tag{8.8}$$

Define the function $x(s)$ as follows: if $1/\alpha = r$ (an integer), then let $x(s) = -s$. If $1/\alpha = r + \tau$, where r is an integer and $0 < \tau < 1$, then split the interval $[0,1]$ into $2r+1$ parts with the points

$$u_{2i} = i\alpha, \quad (i = 0,1,2,\ldots,r),$$
$$u_{2i+1} = (i+\tau)\alpha. \quad (i = 0,1,2,\ldots,r).$$

Let $x(s)$ be a continuous piecewise linear function starting from 0 (i.e. $x(0) = 0$) and having slopes

$$x'(s) = \begin{cases} -\dfrac{1}{r+1}\left(\dfrac{1}{\alpha}\dfrac{r(r+1)}{r+1-\tau}\right)^{1/2} & \text{if } u_{2i} < s < u_{2i+1}, \\ \\ -\dfrac{1}{r}\left(\dfrac{1}{\alpha}\dfrac{r(r+1)}{r+1-\tau}\right)^{1/2} & \text{if } u_{2i-1} < s < u_{2i}. \end{cases}$$

It is easily seen that $x(s)$ so defined is in Strassen's class, i.e. $x(0) = 0$, $x(s)$ is absolutely continuous for $0 < s < 1$ and $\int_0^1 x'^2(s)ds = 1$. Since

$$x(s+\alpha) - x(s) = -C_\alpha, \quad 0 \leq s \leq 1 - \alpha$$

we have (8.8). Unfortunately we cannot accomplish the proof of Theorem 7.18 by showing that $x(s)$ defined above is extremal within $\mathcal{S}$. In Csáki–Révész (1979) the proof was completed by some direct probabilistical ideas. The details are omitted here.

Here we mention a few further applications of Strassen's theorem given by Strassen (1964). At first we present the following:

Consequence of Strassen's Theorems 1 and 2. If φ is a continuous functional from $C(0,1)$ to $\mathbb{R}^1$ then with probability 1 the sequences $\varphi(w_n(t))$ and $\varphi(s_n(t))$ are relatively compact and the sets of limit points coincide with $\varphi(\mathcal{S})$. Consequently

$$\limsup_{n \to \infty} \varphi(w_n(t)) = \limsup_{n \to \infty} \varphi(s_n(t)) = \sup_{x \in \mathcal{S}} \varphi(x) \quad \text{a.s.}$$

Applying this consequence to the functional

$$\varphi(x) = \int_0^1 x(t)f(t)dt \quad (x \in C(0,1))$$

where $f(t)$ ($0 \leq t \leq 1$) is a Riemann integrable function with

$$F(t) = \int_t^1 f(s)ds \in L^2[0,1],$$

we obtain

$$\limsup_{n \to \infty} \int_0^1 w_n(t)f(t)dt = \limsup_{n \to \infty} \int_0^1 s_n(t)f(t)dt$$

$$= \limsup_{n \to \infty} \frac{1}{n} \sum_{i=1}^n f\left(\frac{i}{n}\right) b(n) S_i = \sup_{x \in \mathcal{S}} \varphi(x), \quad (8.9)$$

and by integration by parts we get

$$\sup_{x \in \mathcal{S}} \varphi(x) = \left(\int_0^1 (F(t))^2 dt \right)^{1/2}. \quad (8.10)$$

The above consequence also implies:

For any $a \geq 1$ we have

$$\limsup_{n\to\infty} n^{-1}(b(n))^a \sum_{i=1}^n |S_i|^a = \frac{2(a+2)^{a/2-1}}{\left(\int_0^1 \frac{dt}{(1-t^a)^{1/2}}\right)^a a^{a/2}} \quad \text{a.s.}, \quad (8.11)$$

in particular

$$\limsup_{n\to\infty} n^{-1} b(n) \sum_{i=1}^n |S_i| = 3^{-1/2} \quad \text{a.s.},$$

$$\limsup_{n\to\infty} n^{-1}(b(n))^2 \sum_{i=1}^n S_i^2 = 4\pi^{-2} \quad \text{a.s.}$$

Remark 2. In order to prove (8.11) we have to prove only that

$$\sup_{x\in S} \int_0^1 |x(t)|^a dt = \frac{2(a+2)^{a/2-1}}{\left(\int_0^1 \frac{dt}{(1-t^a)^{1/2}}\right)^a a^{a/2}}.$$

This can be done by an elementary but hard calculation.

Similarly we obtain

$$\limsup_{n\to\infty} b(n) \frac{\sum_{i=1}^n S_i^2}{\sum_{i=1}^n |S_i|} = 2p \quad \text{a.s.}$$

where p is the largest solution of the equation

$$(1-p)^{1/2} \sin\left(p^{-1}(1-p)^{1/2}\right) + \cos\left(p^{-1}(1-p)^{1/2}\right) = 0.$$

A further application given by Strassen is the following. Let $0 \leq c \leq 1$ and

$$c_i = \begin{cases} 1 & \text{if } S_i > c(b(i))^{-1}, \\ 0 & \text{otherwise.} \end{cases}$$

Consider the relative frequency $\gamma_n = n^{-1} \sum_{i=3}^n c_i$. We have

$$\limsup_{n\to\infty} \gamma_n = 1 - \exp\left(-4\left(\frac{1}{c^2} - 1\right)\right) \quad \text{a.s.}$$

Strassen also notes:

"For $c = 1/2$ as an example we get the somewhat surprising result that with probability 1 for infinitely many n the percentage of times $i \leq n$ when $S_i \geq 1/2(2i \log \log i)^{1/2}$ exceeds 99.999 but only for finitely many n exceeds 99.9999."

Finally we mention a very trivial consequence of Strassen's theorem.

THEOREM 8.2 *The set*

$$\{b_t m^+(xt); \ 0 \leq x \leq 1\} \quad (t \to \infty)$$

and the sequence

$$\left\{ b_n \left(M^+_{[nx]} + \left(x - \frac{[nx]}{n} \right) (M^+_{[nx]+1} - M^+_{[nx]}) \right); \ 0 \leq x \leq 1 \right\} \quad (n \to \infty)$$

are relatively compact in $C(0,1)$ with probability 1 and the set of their limit points is the set of the nondecreasing elements of S. The analogous statements for $m(t)$ and $M(n)$ are also valid.

8.2 Strassen theorems for increments

As we have already mentioned, Khinchine's LIL is a simple consequence of Strassen's theorem 1. Here we are interested in getting such a Strassen type generalization of Theorem 7.13. At first we mention a trivial consequence of Theorem 7.13.

Consequence 1. For almost all $\omega \in \Omega$ and for all $\varepsilon > 0$ there exists a $T_0 = T_0(\varepsilon, \omega)$ such that for all $T \geq T_0$ there is a corresponding $0 \leq t = t(\omega, \varepsilon, T) \leq T - a_T$ such that

$$W(t + a_T) - W(t) \geq (1 - \varepsilon)(\gamma(T, a_T))^{-1} \approx (1 - \varepsilon)(2a_T \log T a_T^{-1})^{1/2} \quad (8.12)$$

provided that a_T satisfies conditions (i), (ii), (iii) of Theorem 7.13.

Knowing Consequence 1 of Section 8.1 we might pose the following question: does inequality (8.12) imply that $W(x)$ is increasing nearly linearly in $(t, t + a_T)$? The answer to this question is positive in the same sense as in the case of Consequence 1 of Section 8.1.

In order to formulate our more general result introduce the following notations:

(i) $\Gamma_{t,T}(x) = \gamma(T, a_T)(W(t + xa_T) - W(t))$ $(0 \leq x \leq 1)$,

(ii) for all $\omega \in \Omega$ define the set $V_T = V_T(\omega) \subset C(0,1)$ as follows:

$$V_T = \{\Gamma_{t,T}(x): \ 0 \leq t \leq T - a_T\},$$

(iii) for any $A \subset C(0,1)$ and $\varepsilon > 0$ denote $U(A, \varepsilon)$ the ε-neighbourhood of A, that is a continuous function $\alpha(x)$ is an element of $U(A, \varepsilon)$ if there exists an $a(x) \in A$ such that $\sup_{0 \le x \le 1} |\alpha(x) - a(x)| \le \varepsilon$.

Now we present

THEOREM 8.3 (Révész, 1979). *For almost all $\omega \in \Omega$ and for all $\varepsilon > 0$ there exists a $T_0 = T_0(\omega, \varepsilon)$ such that*

$$U(V_T(\omega), \varepsilon) \supset \mathcal{S} \tag{8.13}$$

and

$$U(\mathcal{S}, \varepsilon) \supset V_T(\omega) \tag{8.14}$$

if $T \ge T_0$ provided that a_T satisfies conditions (i), (ii), (iii) *of Theorem* 7.13.

To grasp the meaning of this Theorem let us mention that it says that:

(a) for all T large enough and for all $s(x) \in \mathcal{S}$ there exists a $0 < t < T - a_T$ such that $\Gamma_{t,T}(x)$ $(0 \le x \le 1)$ will approximate the given $s(x)$,

(b) for all T large enough and for all $0 < t < T - a_T$ the function $\Gamma_{t,T}(x)$ $(0 \le x \le 1)$ can be approximated by a suitable element $s(x) \in \mathcal{S}$.

We have to emphasize that in Theorem 8.3 we assumed all the conditions (i), (ii), (iii) of Theorem 7.13. If we only assume conditions (i) and (ii) then we get a weaker result which contains Strassen's theorem 3 in case $a_t = T$.

THEOREM 8.4 (Révész, 1979). *Assume that a_T satisfies conditions* (i) *and* (ii) *of Theorem* 7.13. *Then for almost all $\omega \in \Omega$ and for all $\varepsilon > 0$ there exists a $T_0 = T_0(\varepsilon, \omega)$ such that*

$$V_T(\omega) \subset U(\mathcal{S}, \varepsilon)$$

if $T \ge T_0$.

Further, for any $s = s(x) \in \mathcal{S}$, $\varepsilon > 0$ and for almost all $\omega \in \Omega$ there exist a $T = T(\varepsilon, \omega, s)$ and a $0 \le t = t(\varepsilon, \omega, s) \le T - a_T$ such that

$$\sup_{0 \le x \le 1} |\Gamma_{t,T}(x) - s(x)| \le \varepsilon.$$

Remark 1. The important difference between Theorems 8.3 and 8.4 is the fact that in Theorem 8.3 we stated that for *every* T big enough and for *every* $s(x) \in \mathcal{S}$ there exists a $0 \le t \le T - a_T$ such that $\Gamma_{t,T}(x)$ approximates the given $s(x)$; while in Theorem 8.4 we only stated that for every $s(x) \in \mathcal{S}$

there exists a T (in fact there exist infinitely many T but not all T are suitable as in Theorem 8.3) and a $0 \leq t \leq T - a_T$ such that $\Gamma_{t,T}(x)$ approximates the given $s(x)$.

In other words if a_T is small (condition (iii) holds true), then for every T (big enough) the random functions $\Gamma_{t,T}(x)$ will approximate every element of $\mathcal{S}$ as t runs over the interval $[0, T - a_T]$. However, if a_T is large then for any fixed T the random functions $\Gamma_{t,T}(x)$ ($0 \leq t \leq T - a_T$) will approximate some elements of $\mathcal{S}$ but not all of them; all of them will be approximated when T is also allowed to vary.

8.3 The rate of convergence in Strassen's theorems

Strassen's theorems 1 and 2 imply: *for any $\varepsilon > 0$ and for almost all $\omega \in \Omega$ there exists an integer $n_0 = n_0(\varepsilon, \omega) > 0$ such that*

$$s_n(x, \omega) \in U(\mathcal{S}, \varepsilon) \quad and \quad w_n(x, \omega) \in U(\mathcal{S}, \varepsilon)$$

if $n \geq n_0$, equivalently there exists a sequence $\varepsilon_n = \varepsilon_n \searrow 0$ such that

$$s_n(x, \omega) \in U(\mathcal{S}, \varepsilon_n) \quad and \quad w_n(x, \omega) \in U(\mathcal{S}, \varepsilon_n) \quad a.s.$$

for all but finitely many n.

It is natural to ask how can we characterize the possible ε_n sequences in the above statement. This question was proposed and firstly investigated by Bolthausen (1978). A better result was given by Grill (1987/A), who proved

THEOREM 8.5 *Let*

$$\psi_\delta(n) = (\log \log n)^{-\delta}.$$

Then

$$s_n(x) \in U(\mathcal{S}, \psi_\delta(n)) \quad and \quad w_n(x) \in U(\mathcal{S}, \psi_\delta(n)) \quad a.s.$$

for all but finitely many n if $\delta < 2/3$; while for $\delta > 2/3$

$$s_n(x) \notin U(\mathcal{S}, \psi_\delta(n)) \quad and \quad w_n(x) \notin U(\mathcal{S}, \psi_\delta(n)) \quad i.o. \ a.s.$$

Clearly Theorem 8.5 implies Property 1 of Section 8.1 but it does not contain Property 2 of Section 8.1. As far as Property 2 of Section 8.1 is concerned one can ask the following question.

Let $f(x)$ be an arbitrary element of $\mathcal{S}$. We know that for all $\varepsilon > 0$ and for almost all $\omega \in \Omega$ there exists an integer $n = n(\varepsilon, \omega)$ resp. $\overline{n} = \overline{n}(\varepsilon, \omega)$ such that

$$\sup_{0 \leq x \leq 1} |s_n(x) - f(x)| \leq \varepsilon \quad \text{resp.} \quad \sup_{0 \leq x \leq 1} |w_{\overline{n}}(x) - f(x)| \leq \varepsilon.$$

Replacing ε by ε_n in the above inequalities, they remain true if $\varepsilon_n \downarrow 0$ slowly enough. We ask how such an ε_n can be chosen. This question was raised and studied by Csáki. He proved

THEOREM 8.6 (Csáki, 1980). *For any $f(x) \in \mathcal{S}$ and $0 < c < 1$ we have*

$$\sup_{0 \leq x \leq 1} |w_n(x) - f(x)| < c(\log\log n)^{-1/2} \quad i.o. \; a.s.$$

and

$$\sup_{0 \leq x \leq 1} |w_n(x) - f(x)| \geq \frac{\pi}{4}(1 - c)(\log\log n)^{-1} \quad a.s.$$

for all but finitely many n.

If $\int_0^1 (f'(x))^2 dx = \alpha < 1$ then a stronger result can be obtained:

THEOREM 8.7 (Csáki, 1980, de Acosta, 1983). *If $f(x) \in \mathcal{S}$, $0 < c < 1$ and $\int_0^1 (f'(x))^2 dx = \alpha < 1$ then*

$$\sup_{0 \leq x \leq 1} |w_n(x) - f(x)| < \frac{\pi(1+c)}{4(1-\alpha)^{1/2} \log\log n} \quad i.o. \; a.s.$$

and

$$\sup_{0 \leq x \leq 1} |w_n(x) - f(x)| > \frac{\pi(1-c)}{4(1-\alpha)^{1/2} \log\log n} \quad a.s.$$

for all but finitely many n.

In case $\int_0^1 (f'(x))^2 dx = 1$ the best possible rate is available only for piecewise linear and quadratic functions. Let $f(x)$ be a continuous piecewise linear function with $f(0) = 0$ and

$$f'(x) = \beta_i \quad \text{if} \quad a_{i-1} < x < a_i \quad (i = 1, 2, \ldots, k)$$

where $a_0 = 0 < a_1 < \ldots < a_{k-1} < a_k = 1$. Then we have

THEOREM 8.8 (Csáki, 1980). *If $f(x) \in \mathcal{S}$ and $\int_0^1 (f'(x))^2 dx = 1$ then for any $\varepsilon > 0$*

$$\sup_{0 \leq x \leq 1} |w_n(x) - f(x)| < \pi^{2/3} 2^{-5/3} B^{-1/3} (1+\varepsilon)(\log\log n)^{-2/3} \quad i.o. \; a.s.$$

and

$$\sup_{0 \le x \le 1} |w_n(x) - f(x)| > \pi^{2/3} 2^{-5/3} B^{-1/3} (1-\varepsilon)(\log \log n)^{-2/3} \quad a.s.$$

for all but finitely many n where

$$B = |\beta_2 - \beta_1| + \cdots + |\beta_k - \beta_{k-1}| + |\beta_k|.$$

THEOREM 8.9 (Csáki, 1990). *Let*

$$f(x) = \frac{a}{2}x^2 + bx \quad (a \ge 0)$$

and

$$\int_0^1 (f'(x))^2 dx = 1.$$

Then

$$\liminf_{t \to \infty} (\log \log t)^{2/3} \sup_{0 \le x \le 1} |b_n W(nx) - f(x)| = \left(\frac{\mu_0}{4}\right)^{1/3} \quad a.s.$$

where μ_0 is the smallest eigenvalue of the differential equation

$$\frac{1}{2} y'' + \mu(ax = |a+b|)y = 0$$

with boundary condition $y(-1) = y(1) = 0$.

The rate of convergence for a larger class of functions $f(\cdot)$ is given by Gorn and Lifschitz (1998).

Remark 1. Theorems 8.5 (resp. 8.8) imply that for any $\beta < 2/3$

$$W(t) \le \left(t \left(2 \log \log t + (\log \log t)^{1-\beta}\right)\right)^{1/2} \quad a.s., \qquad (8.15)$$

if t is big enough resp.

$$W(t) \ge \left(t \left(2 \log \log t - (1+\varepsilon)(\log \log t)^{1/3} \pi^{2/3} 2^{1/3} B^{-1/3}\right)\right)^{1/2} \quad \text{i.o. a.s.}$$
$$(8.16)$$

(8.15) and (8.16) clearly imply the Khinchine's LIL but they are much weaker than EFKP LIL of Section 5.2 (cf. Consequence 1 of Section 5.2).

Remark 2. Applying Theorem 8.8 for $f(x) = 0$ ($0 \le x \le 1$) we obtain (8.16) as a special case.

Remark 3. It looks an interesting question to find a common generalization of the results of Section 8.2 and those of Section 8.3, i.e. to investigate the rate of convergence in Theorems 8.3 and 8.4. This question was studied by Goodman and Kuelbs (1988).

8.4 A theorem of Wichura

We have seen that Strassen's theorem is a natural generalization of Khinchine's LIL. Wichura proposed to find a similar (Strassen type) generalization of the Other LIL (cf. (5.9)). He proved

WICHURA'S THEOREM (Wichura, 1977, Mueller, 1983). *Consider the sequence*

$$\hat{s}_n(u) = \left\{ \left(\frac{\log\log n}{n}\right)^{1/2} \sup_{x \le u} \left| S_{[nx]} + \left(x - \frac{[nx]}{n}\right) X_{[nx]+1} \right|; \ 0 \le u \le 1 \right\}$$

$(n = 1, 2, \ldots)$ *and the net*

$$\hat{w}_t(u) = \left\{ \left(\frac{\log\log t}{t}\right)^{1/2} \sup_{x \le u} |W(tx)|; \ 0 \le u \le 1 \right\} \ (t \to \infty).$$

Let $\mathcal{G}$ be the set of nondecreasing, nonnegative functions g on $[0, 1]$ satisfying

$$\int_0^1 \frac{dx}{(g(x))^2} \le \frac{8}{\pi^2}.$$

Then with probability 1, the set of limit points of $\hat{s}_n(u)$ (resp. $\hat{w}_t(u)$) in the weak topology, as $n \nearrow \infty$ (resp. $t \nearrow \infty$) is $\mathcal{G}$.

Remark 1. In order to see that this theorem implies the Other LIL we only have to prove that $g(1) \ge \sqrt{8/\pi}$ for any $g(\cdot) \in \mathcal{G}$.

Chapter 9

Distribution of the Local Time

9.1 Exact distributions

Let
$$\rho_1(k) = \min\{n : S_n = k\} \quad (k = 1, 2, \ldots).$$
Then
$$\mathbf{P}\{\rho_1(k) > n\} = \mathbf{P}\{M_n^+ < k\}.$$
Hence by Theorem 2.4 we obtain

THEOREM 9.1
$$\mathbf{P}\{\rho_1(k) > n\} = 2^{-n} \sum_{j=0}^{k-1} \binom{n}{[(n-j)/2]} \quad (k = 1, 2, \ldots, \ n = 1, 2, \ldots).$$

Especially
$$\mathbf{P}\{\rho_1(1) > n\} = 2^{-n} \binom{n}{[n/2]}.$$

Consequently
$$\mathbf{P}\{\rho_1(1) = 2m+1\} = 2^{-2m-1}(m+1)^{-1}\binom{2m}{m} \tag{9.1}$$

and
$$\mathbf{P}\{\rho_1(k) = n\} = \frac{k}{n}\binom{n}{(n+k)/2}2^{-n}.$$

Note that $\rho_1(2k+1)$ takes only odd, $\rho_1(2k)$ takes only even numbers.

THEOREM 9.2 *Let $\rho_0 = 0$ and $\rho_k = \min\{j : j > \rho_{k-1}, S_j = 0\}$ $(k = 1, 2, \ldots)$. Then $\rho_1, \rho_2 - \rho_1, \rho_3 - \rho_2, \ldots$ is a sequence of i.i.d.r.v's with*

$$\mathbf{P}\{\rho_1 = 2k\} = 2^{-2k+1}k^{-1}\binom{2k-2}{k-1} \quad (k = 1, 2, \ldots) \tag{9.2}$$

and
$$\mathbf{P}\{\rho_1 > 2n\} = 2^{-2n}\binom{2n}{n} = \mathbf{P}\{S_{2n} = 0\}.$$

Proof. The statement that $\rho_1, \rho_2 - \rho_1, \rho_3 - \rho_2, \ldots$ are i.i.d.r.v.'s taking only even values is trivial. Hence we prove (9.2) only.

$$\mathbf{P}\{\rho_1 = 2k\} = \frac{1}{2}\mathbf{P}\{\rho_1 = 2k \mid X_1 = +1\} + \frac{1}{2}\mathbf{P}\{\rho_1 = 2k \mid X_1 = -1\}.$$

Clearly

$$\mathbf{P}\{\rho_1 = 2k \mid X_1 = +1\} = \mathbf{P}\{\rho_1 = 2k \mid X_1 = -1\}$$
$$= \mathbf{P}\{\rho_1(1) = 2k - 1\}. \tag{9.3}$$

Hence by (9.1) we have (9.2). The second statement of Theorem 9.2 is a simple consequence of (9.2).

Remark 1. A simple calculation gives

$$\mathbf{P}\{\rho_1 < \infty\} = \sum_{k=1}^{\infty} \mathbf{P}\{\rho_1 = 2k\} = \sum_{k=1}^{\infty} 2^{-2k+1} k^{-1} \binom{2k-2}{k-1} = 1.$$

Hence the particle returns to the origin with probability 1, i.e. we obtained a new proof of Pólya Recurrence Theorem of Section 3.1. However, observe that

$$\mathbf{E}\rho_1 = \sum_{k=1}^{\infty} 2^{-2k+2} \binom{2k-2}{k-1} = \infty,$$

i.e. the expectation of the waiting time of the recurrence is infinite.

Consider a random walk $\{S_k; \ k = 0, 1, 2, \ldots\}$ and observe how many times a given $x (x = 0, \pm 1, \pm 2, \ldots)$ was visited during the first n steps, i.e. observe

$$\xi(x, n) = \#\{k: \ 0 < k \leq n, \ S_k = x\}.$$

The process $\xi(x, n)$ (of two variables) is the *local time* of the random walk $\{S_k\}$.

THEOREM 9.3 *For any* $k = 0, 1, 2, \ldots, n; \ n = 1, 2, \ldots$ *we have*

$$\mathbf{P}\{\xi(0, 2n) = k\} = \mathbf{P}\{\xi(0, 2n+1) = k\} = 2^{-2n+k}\binom{2n-k}{n}.$$

Equivalently

$$\mathbf{P}\{\xi(0, n) = k\} = 2^{-2[n/2]+k}\binom{2[n/2]-k}{[n/2]},$$

$$\mathbf{P}\{\rho_k > 2n\} = \mathbf{P}\{\xi(0, 2n) < k\} = 2^{-2n}\sum_{j=0}^{k-1} 2^j \binom{2n-j}{n}.$$

DISTRIBUTION OF THE LOCAL TIME

Proof. (9.3) implies that the distribution of ρ_1 is identical with that of $\rho_1(1) + 1$. It follows that the distribution of $\rho_k - k$ is identical to that of $\rho_1(k)$, i.e.

$$\mathbf{P}\{\rho_k - k > n\} = \mathbf{P}\{\rho_1(k) > n\} = \mathbf{P}\{M_n^+ < k\}.$$

Further, we have

$$\begin{aligned}\mathbf{P}\{\xi(0,2n) = k\} &= \mathbf{P}\{\rho_k \leq 2n, \rho_{k+1} > 2n\} \\ &= \mathbf{P}\{M_{2n-k}^+ \geq k, M_{2n-k-1}^+ < k+1\} \\ &= \mathbf{P}\{M_{2n-k}^+ = k\} = 2^{-2n+k}\binom{2n-k}{n}\end{aligned}$$

which implies the Theorem.

Applying Theorems 9.1 and 9.3 we can get the distribution of $\xi(x,n)$. In fact we have

THEOREM 9.4 *Let $x > 0$, $k > 0$. Then for any $k = 1, 2, \ldots$*

$$\mathbf{P}\{\xi(x,n) = k\}$$
$$= \sum_{j=x}^{n-2k} \mathbf{P}\{\xi(x,n) = k \mid \rho_1(x) = j\}\mathbf{P}\{\rho_1(x) = j\}$$
$$= \sum_{j=x}^{n-2k} \mathbf{P}\{\xi(0, n-j) = k-1\}\mathbf{P}\{\rho_1(x) = j\}$$
$$= \sum_{j=x}^{n-2k} 2^{-2[(n-j)/2]-j+k-1}\binom{2[(n-j)/2] - k + 1}{[(n-j)/2]}\binom{j}{(j+x)/2}\frac{x}{j}$$
$$= \begin{cases}\dfrac{1}{2^{n-k+1}}\dbinom{n-k+1}{(n+x)/2} & \text{if } n+x \text{ is even,} \\ \dfrac{1}{2^{n-k}}\dbinom{n-k}{(n+x-1)/2} & \text{if } n+x \text{ is odd.}\end{cases}$$

For $k = 0$, $x > 0$ by (2.9) we have

$$\mathbf{P}\{\xi(x,n) = 0\} = \mathbf{P}\{\rho_1(x) > n\} = \mathbf{P}\{M_n^+ < x\}$$
$$= 1 - \mathbf{P}\{M_n^+ \geq x\} = 1 - 2^{-n}\sum_{j=x}^{n}\binom{n}{[(n-j)/2]}.$$

Theorem 9.3 gave the distribution of the number of zeros in the path $S_0, S_1, \ldots, S_{2n+1}$. We ask also about the distribution of the number of those zeros where the path changes its sign. Let

$$\Theta(n) = \#\{k : 1 \leq k < n, \; S_{k-1}S_{k+1} < 0\}$$

be the number of crossings (sign changes). Then as a trivial consequence of Theorem 9.3 we obtain

THEOREM 9.5

$$\mathbf{P}\{\Theta(2n+1) = k\}$$
$$= \sum_{j=k}^{n} \mathbf{P}\{\Theta(2n+1) = k \mid \xi(0, 2n) = j\}\mathbf{P}\{\xi(0, 2n) = j\}$$
$$= \sum_{j=k}^{n} \binom{j}{k} \frac{1}{2^j} \binom{2n-j}{n} \frac{1}{2^{2n-j}} = \frac{1}{2^{2n}} \binom{2n+1}{n+k+1}$$
$$= 2\mathbf{P}\{S_{2n+1} = 2k+1\}.$$

Proof. Observe that $\mathbf{P}\{S_{k-1}S_{k+1} < 0 \mid S_k = 0\} = 1/2$.

THEOREM 9.6

$$\mathbf{P}\{\max_{0 \leq i \leq \rho_1} S_i \geq n\} = (2n)^{-1} \quad (n = 1, 2, \ldots).$$

Proof. It is a trivial consequence of Lemma 3.1.

THEOREM 9.7 *For any $k = \pm 1, \pm 2, \ldots$ and $l = 0, 1, 2, \ldots$ we have*

$$\mathbf{P}\{\xi(k, \rho_1) = 0\} = \frac{2|k| - 1}{2|k|}, \tag{9.4}$$

$$\mathbf{P}\{\xi(k, \rho_1) = l + 1\} = \left(\frac{1}{2|k|}\right)^2 \left(\frac{2|k| - 1}{2|k|}\right)^l, \tag{9.5}$$

$$\mathbf{E}\xi(k, \rho_1) = 1, \quad \mathbf{E}(\xi(k, \rho_1) - 1)^2 = 4|k| - 2. \tag{9.6}$$

Proof. Without loss of generality we may assume that $k > 0$. Then

$$\{\xi(k, \rho_1) = 0\} = \{X_1 = -1\} \cup \{X_1 = 1, S_2 \neq k, S_3 \neq k, \ldots, S_{\rho_1 - 1} \neq k\}.$$

Hence by Lemma 3.1

$$\mathbf{P}\{\xi(k, \rho_1) = 0\} = \frac{1}{2} + \frac{1}{2}\frac{k-1}{k} = \frac{2k-1}{2k}$$

and (9.4) is proved.

For any $x = \pm 1, \pm 2, \ldots$ define

$$\rho_0(x) = 0,$$
$$\rho_1(x) = \inf\{l : l > 0, \ S_l = x\},$$
$$\ldots \ldots \ldots \ldots$$
$$\rho_{i+1}(x) = \inf\{l : l > \rho_i(x), \ S_l = x\} \ (i = 0, 1, 2, \ldots).$$

Then in case $m > 0$ we have

$$\{\xi(k, \rho_1) = m\} = \{0 < \rho_1(k) < \rho_2(k) < \ldots < \rho_m(k) < \rho_1 < \rho_{m+1}(k)\}.$$

Hence, again by Lemma 3.1

$$\mathbf{P}\{\xi(k,\rho_1) = m\} = \frac{1}{2}\frac{1}{k}\left(\sum_{j=0}^{m-1}\binom{m-1}{j}\left(\frac{1}{2}\right)^j\left(\frac{1}{2}\frac{k-1}{k}\right)^{m-1-j}\right)\frac{1}{2}\frac{1}{k}$$

$$= \left(\frac{1}{2k}\right)^2\left(\frac{2k-1}{2k}\right)^{m-1}$$

and (9.5) is also proved.

(9.6) is a simple consequence of (9.4) and (9.5).

Define the r.v.'s ζ_{2n} as follows: ζ_{2n} $(n = 1, 2, \ldots)$ is the number of those terms of the sequence $S_1, S_2, \ldots, S_{2n}$ which are positive or which are equal to 0 but the preceding term of which is positive. Then ζ_{2n} takes on only even numbers and its distribution is described by

THEOREM 9.8

$$\mathbf{P}\{\zeta_{2n} = 2k\} = \binom{2k}{k}\binom{2n-2k}{n-k}2^{-2n} \quad (k = 0, 1, \ldots, n). \tag{9.7}$$

Proof. (Rényi, 1970/A). Clearly

$$\{\zeta_{2n} = 0\} = \{M_{2n}^+ = 0\}.$$

Hence by Theorem 2.4

$$\mathbf{P}\{\zeta_{2n} = 0\} = \mathbf{P}\{M_{2n}^+ = 0\} = \binom{2n}{n}2^{-2n},$$

i.e. (9.7) holds for $k = 0$. It is also easy to see that (9.7) holds for $n = 1$, $k = 0, 1$. Now we use induction on n. Suppose that (9.7) is true for $n \leq N - 1$ and consider $\mathbf{P}\{\zeta_{2N} = 2k\}$ for $1 \leq k \leq N - 1$. If $\zeta_{2N} = 2k$

and $1 \leq k \leq N-1$ then the sequence $S_1, S_2, \ldots, S_{2N}$ has to contain both positive and negative terms, and thus it contains at least one term equal to 0. Let $\rho_1 = 2l$. Then either $S_n > 0$ for $n < 2l$ and $S_{2l} = 0$ or $S_n < 0$ for $n < 2l$ and $S_{2l} = 0$. Both possibilities have the probability

$$\frac{1}{2}\mathbf{P}\{\rho_1 = 2l\} = \frac{1}{2^{2l}l}\binom{2l-2}{l-1}$$

(cf. (9.2)).

Now if $S_n > 0$ for $n < 2l$ and $S_{2l} = 0$, further if $\zeta_{2N} = 2k$, then among the numbers $S_{2l+1}, \ldots, S_{2N}$ there are $2k-2l$ positive ones or zeros preceded by a positive term, while in case $S_n < 0$ for $n < 2l$, $S_{2l} = 0$ and $\zeta_{2N} = 2k$, the number of such terms is $2k$. Hence

$$\mathbf{P}\{\zeta_{2N} = 2k\} = \frac{1}{2}\sum_{l=1}^{k}\mathbf{P}\{\rho_1 = 2l\}\mathbf{P}\{\zeta_{2N-2l} = 2k - 2l\}$$
$$+ \frac{1}{2}\sum_{l=1}^{N-k}\mathbf{P}\{\rho_1 = 2l\}\mathbf{P}\{\zeta_{2N-2l} = 2k\}$$

and we obtain (9.7) by an elementary calculation.

It is worthwhile to mention that the distribution of the location of the last zero up to $2n$, i.e. the distribution of

$$\Psi(2n) = \max\{k:\ 0 \leq k \leq n,\ S_{2k} = 0\}$$

agrees with the distribution of ζ_{2n}. In fact we have

THEOREM 9.9

$$\mathbf{P}\{\Psi(2n) = 2k\} = \binom{2k}{k}\binom{2n-2k}{n-k}2^{-2n} \qquad (k = 0, 1, \ldots, n).$$

Proof. Clearly by Theorem 9.2

$$\mathbf{P}\{\Psi(2n) = 2k\} = \mathbf{P}\{S_{2k} = 0\}\mathbf{P}\{\rho_1 > 2n - 2k\}$$
$$= \mathbf{P}\{S_{2k} = 0\}\mathbf{P}\{S_{2n-2k} = 0\}.$$

Hence the Theorem.

The distribution of the location of the maximum also agrees with those of ζ_n and $\Psi(n)$. In fact we have

THEOREM 9.10 *Let*

$$\mu^+(n) = \inf\{k: 0 \leq k \leq n \quad for \quad which \quad S(k) = M^+(n)\}.$$

DISTRIBUTION OF THE LOCAL TIME

Then

$$\mathbf{P}\{\mu^+(2n) = k\} = \begin{cases} \binom{2[k/2]}{[k/2]}\binom{2n - 2[k/2]}{n - [k/2]}2^{-2n-1} & \text{if } k = 1, 2, \ldots, 2n, \\ \binom{2n}{n}2^{-2n} & \text{if } k = 0. \end{cases}$$

Proof. Clearly the number of paths for which

$$\{S_0 < S_k, S_1 < S_k, \ldots, S_{k-1} < S_k\}$$

is equal to the number of paths for which

$$\{S_1 > 0, S_2 > 0, \ldots, S_k > 0\}.$$

Hence

$$\mathbf{P}\{\mu^+(2n) = k\}$$
$$= \mathbf{P}\{\max_{1 \leq i \leq k-1} S_i < S_k, \max_{k+1 \leq i \leq 2n} S_i \leq S_k\}$$
$$= \mathbf{P}\{S_0 < S_k, S_1 < S_k, \ldots, S_{k-1} < S_k\}\mathbf{P}\{S_1 \geq 0, S_2 \geq 0, \ldots, S_{2n-k} \geq 0\}$$
$$= \mathbf{P}\{S_1 > 0, S_2 > 0, \ldots, S_k > 0\}\mathbf{P}\{S_1 \geq 0, \ldots, S_{2n-k} \geq 0\}.$$

Then we obtain Theorem 9.10 by Theorem 2.4.

9.2 Limit distributions

Applying the above given exact distributions and the Stirling formula we obtain

THEOREM 9.11

$$\lim_{k \to \infty} \mathbf{P}\left\{\frac{\rho_1(k)}{k^2} < x\right\} = \lim_{k \to \infty} \mathbf{P}\left\{\frac{\rho_k}{k^2} < x\right\}$$
$$= \frac{1}{\sqrt{2\pi}} \int_0^x v^{-3/2} e^{-1/2v} dv, \qquad (9.8)$$

$$\mathbf{P}\{\rho_1 = 2k\} = \frac{1}{2\sqrt{\pi}} k^{-3/2} \exp(\vartheta_k/k) \text{ where } |\vartheta_k| \leq 1, \qquad (9.9)$$

$$\mathbf{P}\{\rho_1 \geq x\} = \sqrt{\frac{2}{\pi x}}\left(1 + O\left(\frac{1}{x}\right)\right) \quad (x \to \infty), \qquad (9.10)$$

$$\lim_{n\to\infty} \mathbf{P}\left\{n^{-1/2}\xi(0,n) < x\right\} = \lim_{n\to\infty} \mathbf{P}\left\{n^{-1/2}\xi(z,n) < x\right\}$$
$$= \sqrt{\frac{2}{\pi}} \int_0^x e^{-u^2/2} du \quad (x = \pm 1, \pm 2, \ldots), \tag{9.11}$$

$$\lim_{n\to\infty} \mathbf{P}\left\{\frac{\mu^+(n)}{n} < x\right\} = \lim_{n\to\infty} \mathbf{P}\left\{\frac{\Psi(n)}{n} < x\right\}$$
$$= \lim_{n\to\infty} \mathbf{P}\left\{\frac{\zeta_n}{n} < x\right\} = \frac{2}{\pi} \arcsin\sqrt{x} \quad (0 < x < 1). \tag{9.12}$$

Remark 1. (9.12) is called *arcsine law*. It is worthwhile to mention that by (9.12) we obtain

$$\lim_{n\to\infty} \mathbf{P}\left\{0,45 < \frac{\zeta_n}{n} < 0,55\right\} = 0,063769\ldots$$

and

$$\lim_{n\to\infty} \mathbf{P}\left\{\frac{\zeta_n}{n} < 0,1\right\} = \lim_{n\to\infty} \mathbf{P}\left\{\frac{\zeta_n}{n} > 0,9\right\} = 0,204833\ldots$$

The exact distribution (9.7) of ζ_{2n} also implies that the most improbable value of ζ_{2n} is n and the most probable values are 0 and $2n$. In other words, with a big probability the particle spends a long time on the left-hand side of the line and only a short time on the right-hand side or conversely but it is very unlikely that it spends the same (or nearly the same) time on the positive and on the negative side.

9.3 Definition and distribution of the local time of a Wiener process

It is easy to see that the number of the time points before any given T, where a Wiener process W is equal to a given x, is 0 or ∞ a.s., i.e. for any $T > 0$ and any real x

$$\#\{t: \ 0 \le t < T, \ W(t) = x\} = 0 \text{ or } \infty \quad \text{a.s.}$$

Hence if we want to characterize the amount of time till T which the Wiener process spends in x (or nearby) then we have to find a more adequate definition than the number of time points. P. Lévy proposed the following idea.

DISTRIBUTION OF THE LOCAL TIME

Let $H(A,t)$ be the *occupation time* of the Borel set $A \subset \mathbb{R}^1$ by $W(\cdot)$ in the interval $(0,t)$, formally

$$H(A,t) = \lambda\{s: \ s \leq t, \ W(s) \in A\}$$

where λ is the Lebesgue measure.

For any fixed $t > 0$ and for almost all $\omega \in \Omega$ the occupation time $H(A,t)$ is a measure on the Borel sets of the real line. Trotter (1958) proved that this measure is absolutely continuous with respect to the Lebesgue measure and its Radon–Nikodym derivate $\eta(x,t)$ is continuous in (x,t). The stochastic process $\eta(x,t)$ is called the *local time* of W. (It characterizes the amount of time that the Wiener process W spends till t "near" to the point x.) Our first aim is to evaluate the distribution of the r.v. $\eta(0,t)$.

In fact we prove

THEOREM 9.12 *For any $x > 0$ and $t > 0$*

$$\mathbf{P}\left\{t^{-1/2}\eta(0,t) < x\right\} = \sqrt{\frac{2}{\pi}} \int_0^x e^{-u^2/2} du. \tag{9.13}$$

Proof. For any $N = 1, 2, \ldots$ define the sequence $0 < \tau_1 = \tau_1^{(N)} < \tau_2 = \tau_2^{(N)} < \ldots$ as follows:

$$\tau_1 = \inf\{t: \ t > 0, \ |W(t)| = N^{-1}\},$$
$$\tau_2 = \inf\{t: \ t > \tau_1, \ |W(t) - W(\tau_1)| = N^{-1}\},$$
$$\ldots \quad \ldots$$
$$\tau_{i+1} = \inf\{t: \ t > \tau_i, \ |W(t) - W(\tau_i)| = N^{-1}\},$$
$$\ldots \quad \ldots$$

(cf. Skorohod embedding scheme, Section 6.4) and let

$$S_k^{(N)} = W(\tau_k) \quad (k = 1, 2, \ldots),$$
$$\xi^{(N)}(x,n) = \#\{k: 0 < k \leq n, S_k^{(N)} = x\},$$
$$\nu = \nu^{(N)} = \max\{i: \tau_i \leq 1\}.$$

Note that $\tau_1, \tau_2 - \tau_1, \ldots$ is a sequence of i.i.d.r.v.'s with

$$\mathbf{E}\tau_1 = N^{-2} \text{ and } \mathbf{E}\tau_1^4 < \infty \tag{9.14}$$

(cf. (6.7)).

The interval (τ_i, τ_{i+1}) will be called type (a,b) $(a = jN^{-1}, \ |b-a| = N^{-1}, j = 0,1,2,\ldots)$ if $|W(\tau_i)| = a$ and $|W(\tau_{i+1})| = b$. The infinite random

set of those i's for which (τ_i, τ_{i+1}) is an interval of type (a,b) will be denoted by $I^{(N)}(a,b) = I(a,b)$. It is clear that $|W(t)|$ can be smaller than N^{-1} if t is an element of an interval of type $(0, N^{-1})$, $(N^{-1}, 0)$ or $(N^{-1}, 2N^{-1})$. Let
$$A = A^{(N)} = \{i:\ 0 \leq i \leq \nu,\ i \in I(0, N^{-1}) \cup I(N^{-1}, 0)\}.$$
Then by the law of large numbers and (9.14)
$$\frac{N^2 \sum_{i \in A}(\tau_{i+1} - \tau_i)}{2\xi^{(N)}(0,\nu)} \to 1 \quad \text{a.s.} \quad (N \to \infty). \tag{9.15}$$

(In fact (9.15) can be obtained using the "Method of high moments" of Section 4.2, to obtain it by "Gap method" seems to be hard.)

Studying the local time of $W(\cdot)$ in intervals of type $(N^{-1}, 2N^{-1})$, we obtain
$$\sum_{i \in B}(\eta(0, \tau_{i+1}) - \eta(0, \tau_i)) = 0 \quad \text{a.s.} \tag{9.16}$$
for any $N = 1, 2, \ldots$ where
$$B = B^N = \{i:\ 0 \leq i \leq \nu,\ i \in I(N^{-1}, 2N^{-1})\}. \tag{9.17}$$

((9.16) follows from the simple fact that for almost all $\omega \in \Omega$ there exists an $\varepsilon_0 = \varepsilon_0(\omega, N)$ such that $|W(t)| \geq \varepsilon_0$ if $t \in \cup_{i \in B}(\tau_i, \tau_{i+1})$.)

Hence
$$\eta(0, 1) = \lim_{N \to \infty} \frac{N}{2} \sum_{i \in A}(\tau_{i+1} - \tau_i) \quad \text{a.s.} \tag{9.18}$$

Then, taking into account that $\lim_{N \to \infty} N^{-2}\nu_N = 1$ a.s., (9.11), (9.15) and (9.18) combined imply that
$$\mathbf{P}\{\eta(0,1) < x\} = \sqrt{\frac{2}{\pi}} \int_0^x e^{-u^2/2} du \tag{9.19}$$

and Theorem 9.12 follows from (9.19) and from the simple transformation: for any $T > 0$
$$\{\eta(x, tT), x \in \mathbb{R}^1, 0 \leq t \leq 1\} \stackrel{\mathcal{D}}{=} \{T^{1/2}\eta(xT^{-1/2}, t), x \in \mathbb{R}^1, 0 \leq t \leq 1\}. \tag{9.20}$$

Theorem 9.12 clearly implies that for any $x \in \mathbb{R}^1$ we have
$$\lim_{T \to \infty} \mathbf{P}\left\{\frac{\eta(x,T)}{\sqrt{T}} < u\right\} = \sqrt{\frac{2}{\pi}} \int_0^u e^{-v^2/2} dv.$$

Lévy (1948) also proved

THEOREM 9.13 *For any $x \in \mathbb{R}^1$, $T > 0$ and $u > 0$ we have*

$$\mathbf{P}\{\eta(x,T) < u\} = 2\Phi\left(\frac{u + |x|}{T^{1/2}}\right) - 1.$$

To evaluate the distribution of $\eta(t) = \sup_{-\infty < x < \infty} \eta(x,t)$ is much harder. This was done by Csáki and Földes (1986). They proved

THEOREM 9.14

$$\mathbf{P}\left\{t^{-1/2}\eta(t) < z\right\} = \sum_{k=1}^{\infty} a_k \exp\left(-\frac{2j_k^2}{z^2}\right) + \sum_{k=1}^{\infty}\left(b_k + \frac{c_k}{z^2}\right)\exp\left(-\frac{2k^2\pi^2}{z^2}\right)$$

where $0 < j_1 < j_2 < \ldots$ are the positive zeros (roots) of the Bessel function $J_0(x) = I_0(ix)$ and for any $k = 1, 2, \ldots$

$$a_k = \frac{4}{\sin^2 j_k},$$

$$b_k = 4\left(-1 + \frac{J_1(k\pi)}{k\pi J_0(k\pi)} - \left(\frac{J_1(k\pi)}{J_0(k\pi)}\right)^2\right),$$

$$c_k = 16k\pi \frac{J_1(k\pi)}{J_0(k\pi)},$$

$$J_1(x) = \frac{1}{i} I_1(ix).$$

Furthermore

$$\mathbf{P}\left\{t^{-1/2}\eta(t) < z\right\} \approx a_1 \exp\left(-\frac{2j_1^2}{z^2}\right) \quad \text{as} \quad z \to 0.$$

Remark 1. The proof of Theorem 9.14 is based on a result of Borodin (1982), who evaluated the Laplace transform of the distribution of $t^{-1/2}\eta(t)$.

As we mentioned above the occupation time $H(A,t)$ is absolutely continuous with respect to the Lebesgue measure. Consequently

$$\lim_{\varepsilon \to 0} \frac{1}{\varepsilon} H([x, x+\varepsilon), t) = \eta(x,t) \quad \text{a.s.} \tag{9.21}$$

Hence (9.21) can be considered as a possible definition of the local time. A number of results are known on the rate of convergence in (9.21). The strongest one is

THEOREM 9.15 (Khoshnevisan, 1994). *For all $t > 0$ with probability 1,*

$$\limsup_{\substack{\varepsilon \to 0 \\ x \in \mathbb{R}^1}} \frac{|\varepsilon^{-1} H([x, x+\varepsilon), t) - \eta(x,t)|}{(\varepsilon \log \varepsilon^{-1})^{1/2}} = \left(\frac{4}{3}\eta(t)\right)^{1/2}$$

where $\eta(t) = \sup_{x \in \mathbb{R}^1} \eta(x,t)$.

The local version of the above theorem is

THEOREM 9.16 (Khoshnevisan, 1994). *For every $t > 0$ and $x \in \mathbb{R}^1$ we have*
$$\limsup_{\varepsilon \to 0} \frac{|\varepsilon^{-1} H([x, x+\varepsilon), t) - \eta(x,t)|}{(\varepsilon \log \log \varepsilon^{-1})^{1/2}} = \left(\frac{4}{3}\eta(x,t)\right)^{1/2}.$$

Chapter 10

Local Time and Invariance Principle

10.1 An invariance principle

The main result of this Section claims that the local time $\xi(x,n)$ of a random walk can be approximated by the local time $\eta(x,n)$ of a Wiener process uniformly in x as $n \to \infty$. In fact we have

THEOREM 10.1 (Révész, 1981). *Let $\{W(t), t \geq 0\}$ be a Wiener process defined on a probability space $\{\Omega, \mathcal{F}, \mathbf{P}\}$. Then on the same probability space $\{\Omega, \mathcal{F}, \mathbf{P}\}$ one can define a sequence $X_1, X_2, \ldots$ of i.i.d.r.v.'s with $\mathbf{P}\{X_i = 1\} = \mathbf{P}\{X_i = -1\} = 1/2$ such that*

$$\lim_{n \to \infty} n^{-1/4-\varepsilon} \sup_x |\xi(x,n) - \eta(x,n)| = 0 \quad a.s. \tag{10.1}$$

for any $\varepsilon > 0$ where the sup is taken over all integers, η is the local time of W and ξ is the local time of $S_n = X_1 + X_2 + \cdots + X_n$.

For the sake of simplicity, instead of (10.1) we prove only

$$\lim_{n \to \infty} n^{-1/4-\varepsilon} |\xi(0,n) - \eta(0,n)| = 0 \quad a.s. \tag{10.2}$$

for any $\varepsilon > 0$. The proof of (10.1) does not require any new idea. Only a more tiresome calculation is needed.

Proof of (10.2). Define the r.v.'s $\tau_0 = 0 < \tau_1 < \tau_2 < \ldots$ just like in Section 6.3. Further let $1 < \mu_1 < \mu_2 < \ldots$ be the time-points where the random walk $\{S_k\} = \{W(\tau_k)\}$ visits 0, i.e. let

$$\mu_1 = \min\{k : k > 0, \ W(\tau_k) = S_k = 0\},$$
$$\mu_2 = \min\{k : k > \mu_1, \ W(\tau_k) = S_k = 0\},$$
$$\ldots \quad \ldots$$
$$\mu_n = \min\{k : k > \mu_{n-1}, \ W(\tau_k) = S_k = 0\},$$
$$\ldots \quad \ldots$$

Then

$$\xi(0,n) = \max\{k : \mu_k \leq n\}$$

and
$$\eta(0,\tau_k) = \sum_{j=1}^{\xi(0,k)} (\eta(0,\tau_{\mu_j+1}) - \eta(0,\tau_{\mu_j})).$$

The proof of Theorem 9.12 implies
$$\mathbf{E}(\eta(0,\tau_{\mu_j+1}) - \eta(0,\tau_{\mu_j})) = 1.$$

Hence
$$\eta(0,\tau_k) = \xi(0,k) + o\left((\xi(0,k))^{1/2+\varepsilon}\right) \quad \text{a.s.} \quad (k \to \infty).$$

(6.1) easily implies that
$$\tau_k = k + o\left(k^{1/2+\varepsilon}\right) \quad \text{a.s.}$$

Then (10.2) easily follows from
$$\xi(0,k) = o\left(k^{1/2+\varepsilon}\right) \quad \text{a.s.} \quad (k \to \infty) \tag{10.3}$$

and
$$\sup_{j \leq k} \left(\eta\left(0, j + k^{1/2+\varepsilon}\right) - \eta(0,j)\right) = o\left(k^{1/4+\varepsilon}\right) \quad \text{a.s.} \quad (k \to \infty). \tag{10.4}$$

(10.3) and (10.4) can be easily proved. Their proofs are omitted here because more general results will be given in Chapter 11 (Theorems 11.1 and 11.7).

Remark 1. Borodin (1986/A,B) proved that n^ε in (10.1) can be replaced by $\log n$, more than that M. Csörgő–Horváth (1989) proved that n^ε in (10.1) can be replaced by $(\log n)^{1/2}(\log \log n)^{1/4+\varepsilon}$.

Remark 2. It turns out that the rate of convergence in Theorem 10.1 is nearly the best possible. In fact Remark 3 of Section 11.5 implies that *if a Wiener process $W(\cdot)$ and a random walk $\{S_n\}$ are defined on the same probability space then*
$$\limsup_{n \to \infty} n^{-1/4} \sup_x |\xi(x,n) - \eta(x,n)| > 0 \quad a.s. \tag{10.5}$$

However, the answer to the following question is unknown. Assume that a Wiener process and a random walk are defined on the same probability space and
$$\lim_{n \to \infty} n^{-\alpha}|\xi(0,n) - \eta(0,n)| = 0 \quad \text{a.s.}$$

What can be said about α?

Remark 3. It can also be proved that in Theorem 10.1 the random walk S_n and the Wiener process $W(t)$ can be constructed so that besides (10.1)

$$|S_n - W(n)| = O(\log n) \quad \text{a.s.}$$

Remark 4. A trivial consequence of Theorem 10.1 is

$$\lim_{n \to \infty} \mathbf{P}\left\{n^{-1/2}\xi(n) < z\right\} = \mathbf{P}\{\eta(1) < z\} \tag{10.6}$$

for any $z > 0$ where $\xi(n) = \max_x \xi(x, n)$ (cf. Theorem 9.14).

10.2 A theorem of Lévy

Theorem 10.1 tells us that the properties of the process $\xi(x, n)$ are the same (or more or less the same) as those of $\eta(x, n)$. In other words, studying the behaviour of one of the processes $\xi(x, n)$, $\eta(x, n)$ we can automatically claim that the behaviour of the other process is the same. The main results of the present Section tell us that the properties of $\xi(0, n)$ (resp. $\eta(0, n)$) are the same as those of $M^+(n)$ (resp. $m^+(n)$). Hence the theorems proved for $M^+(n)$ (resp. $m^+(n)$) will be inherited by $\xi(0, n)$ (resp. $\eta(0, n)$).

Let
$$y(t) = m^+(t) - W(t) \; (t \geq 0)$$

and
$$Y(n) = M^+(n) - S(n) \; (n = 0, 1, 2, \ldots).$$

Then a celebrated result of P. Lévy reads as follows (see for example, Knight (1981), Theorem 5.3.7):

THEOREM 10.2 *We have*

$$\{y(t), m^+(t); \; t \geq 0\} \stackrel{\mathcal{D}}{=} \{|W(t)|, \eta(0, t); \; t \geq 0\},$$

in other words, the finite dimensional distributions of the vector valued process $\{y(t), m^+(t); \; t \geq 0\}$ are equal to the corresponding distributions of $\{|W(t)|, \eta(0, t); \; t \geq 0\}$.

In order to see the importance of this theorem, we mention that applying the LIL of Khinchine (Section 4.4, see also Theorem 6.2) for $m^+(t)$ as a trivial consequence of Theorem 10.2 we obtain

Consequence 1.

$$\limsup_{t\to\infty} \frac{\eta(0,t)}{(2t\log\log t)^{1/2}} = 1 \quad \text{a.s.} \tag{10.7}$$

In fact the Lévy classes can also be obtained for $\eta(0,t)$.

Applying Theorem 10.1, Consequence 1 in turn implies

Consequence 2.

$$\limsup_{n\to\infty} \frac{\xi(0,n)}{(2n\log\log n)^{1/2}} = 1 \quad \text{a.s.} \tag{10.8}$$

Remark 1. (10.7) was proved (directly) by Kesten (1965). (10.8) is due to Chung and Hunt (1949).

A natural question arises: what is the analogue of Theorem 10.2 in the case of a random walk? In fact we ask: does Theorem 10.2 remain true if we replace $W(t), y(t), m^+(t), \eta(0,t)$ by $S(n), Y(n), M^+(n), \xi(0,n)$ respectively? The answer to this question is negative, which can be seen by comparing the distributions of $\xi(0, 2n)$ and $M^+(2n)$ (cf. Theorems 2.4 and 9.3).

In spite of this disappointing fact we prove that Theorem 10.2 is "nearly true" for random walks. In fact we have

THEOREM 10.3 (Csáki–Révész, 1983). *Let $X_1, X_2, \ldots$ be a sequence of i.i.d.r.v.'s with $\mathbf{P}\{X_1 = 1\} = \mathbf{P}\{X_1 = -1\} = 1/2$ defined on a probability space $\{\Omega, \mathcal{F}, \mathbf{P}\}$. Then one can define a sequence $\hat{X}_1, \hat{X}_2, \ldots$ of i.i.d.r.v.'s on the same probability space such that $\mathbf{P}\{\hat{X}_1 = 1\} = \mathbf{P}\{\hat{X}_1 = -1\} = 1/2$ and for any $\varepsilon > 0$*

$$n^{-\varepsilon}|\hat{Y}(n) - |S(n)|| \to 0 \quad \text{a.s.}$$

and

$$n^{-1/4-\varepsilon}|\hat{M}^+(n) - \xi(0,n)| \to 0 \quad \text{a.s.},$$

where

$$\hat{M}^+(n) = \max_{0\leq k\leq n} \hat{S}(k), \quad \hat{S}(0) = 0, \quad \hat{S}(n) = \sum_{k=1}^{n} \hat{X}_k \quad (n = 1, 2, \ldots),$$

$$\hat{Y}(n) = \hat{M}^+(n) - \hat{S}(n).$$

Remark 2. This theorem is a bit stronger than that of Csáki–Révész (1983). The proof is presented below.

Remark 3. Consequence 2 can also be obtained by applying the LIL of Khinchine (cf. Section 4.4) for M_n^+ and Theorem 10.3.

Theorem 10.3 tells us that the vector $(|S(n)|, \xi(0,n))$ can be approximated by the vector $(\hat{Y}(n), \hat{M}^+(n))$ in order $n^{1/4+\varepsilon}$. Unfortunately we do not know what the best possible rate here is. However, we can show that by considering the number of crossings $\Theta(n)$ instead of the number of roots $\xi(0,n)$, better rates can be achieved than that of Theorem 10.3. Let

$$\Theta(n) = \#\{k: \ 1 \le k \le n, \ S(k-1)S(k+1) < 0\}$$

be the number of crossings. Then we have

THEOREM 10.4 (Csáki–Révész, 1983 and Simons, 1983). *Let $X_1, X_2, \ldots$ be a sequence of i.i.d.r.v.'s with $\mathbf{P}\{X_1 = 1\} = \mathbf{P}\{X_1 = -1\} = 1/2$ defined on a probability space $\{\Omega, \mathcal{F}, \mathbf{P}\}$. Then one can define a sequence $\bar{X}_1, \bar{X}_2, \ldots$ of i.i.d.r.v.'s on the same probability space $\{\Omega, \mathcal{F}, \mathbf{P}\}$ such that $\mathbf{P}\{\bar{X}_1 = 1\} = \mathbf{P}\{\bar{X}_1 = -1\} = 1/2$ and*

$$|\bar{M}^+(n) - 2\Theta(n)| \le 1, \tag{10.9}$$
$$|\bar{Y}(n) - |S(n)|| \le 2, \tag{10.10}$$

for any $n = 1, 2, \ldots$ where

$$\bar{M}^+(n) = \max_{0 \le k \le n} \bar{S}(k), \quad \bar{S}(0) = 0, \quad \bar{S}(n) = \sum_{k=1}^{n} \bar{X}_k \quad (n = 1, 2, \ldots),$$

$$\bar{Y}(n) = \bar{M}^+(n) - \bar{S}(n).$$

Proof. Let

$$\tau_1 = \min\{i: \ i > 0, \ S(i-1)S(i+1) < 0\},$$
$$\tau_2 = \min\{i: \ i > \tau_1, \ S(i-1)S(i+1) < 0\},$$
$$\ldots \ldots$$
$$\tau_{l+1} = \min\{i: \ i > \tau_l, \ S(i-1)S(i+1) < 0\}, \ldots$$

and

$$\bar{X}_j = \begin{cases} -X_1 X_{j+1} & \text{if } 1 \le j \le \tau_1, \\ X_1 X_{j+1} & \text{if } \tau_1 + 1 \le j \le \tau_2, \\ \ldots \ldots \\ (-1)^{l+1} X_1 X_{j+1} & \text{if } \tau_l + 1 \le j \le \tau_{l+1} \\ \ldots \ldots \end{cases}$$

This transformation was given by Csáki and Vincze (1961). The following lemma is clearly true.

LEMMA 10.1

(i) $\bar{X}_1, \bar{X}_2, \ldots$ *is a sequence of i.i.d.r.v.'s with*

$$\mathbf{P}\{\bar{X}_1 = +1\} = \mathbf{P}\{\bar{X}_1 = -1\} = 1/2.$$

(ii)
$$\bar{S}(k) - \bar{S}(\tau_l) = \sum_{j=\tau_l+1}^{k} \bar{X}_j = (-1)^{l+1} X_1 \sum_{j=\tau_l+1}^{k} X_{j+1} =$$

$$(-1)^{l+1} X_1 (S(k+1) - S(\tau_l+1)) \begin{cases} \leq 1 & \text{if } \tau_l + 1 \leq k \leq \tau_{l+1} - 2, \\ = 1 & \text{if } k = \tau_{l+1} - 1, \\ = 2 & \text{if } k = \tau_{l+1}. \end{cases}$$

(iii) $2\Theta(\tau_l) = 2l = \bar{S}(\tau_l) = \bar{M}^+(\tau_l),\ l = 1, 2, \ldots$.

(iv) *For any* $\tau_l \leq n < \tau_{l+1}$ *we have* $\Theta(n) = l,\ 2l \leq \bar{M}^+(n) \leq 2l + 1$, *consequently* $0 \leq \bar{M}^+(n) - 2\Theta(n) \leq 1$.

(v)
$$\bar{S}(k) = \begin{cases} 2l + 1 - |S(k+1)| & \text{if } \tau_l + 1 \leq k \leq \tau_{l+1} - 1, \\ 2l + 2 - |S(k)| & \text{if } k = \tau_{l+1}, \end{cases}$$

therefore

$$\bar{Y}(k) = \bar{M}^+(k) - \bar{S}(k) \leq |S(k+1)| \leq |S(k)| + 1$$

and

$$\bar{Y}(k) = \bar{M}^+(k) - \bar{S}(k) \geq |S(k+1)| - 1 \geq |S(k)| - 2.$$

This proves Theorem 10.4.

Proof of Theorem 10.3. Clearly we have

$$n^{-1/4-\varepsilon}|\xi(0,n) - 2\Theta(n)| \to 0 \quad \text{a.s.} \tag{10.11}$$

Hence we obtain Theorem 10.3 as a trivial consequence of Theorem 10.4.

Applying the Invariance Principle 1 (cf. Section 6.3), Theorems 10.2 and 10.4 as well as (10.11) we easily obtain

Consequence 3. (Csáki–Révész, 1983). On a rich enough probability space $\{\Omega, \mathcal{F}, \mathbf{P}\}$ one can define a Wiener process $\{W(t); t \geq 0\}$ and a sequence $X_1, X_2, \ldots$ of i.i.d.r.v.'s with $\mathbf{P}\{X_1 = 1\} = \mathbf{P}\{X_1 = -1\} = 1/2$ such that

$$|S(n) - W(n)| = O(\log n) \quad \text{a.s.}, \tag{10.12}$$

$$|2\Theta(n) - \eta(0,n)| = O(\log n) \quad \text{a.s.} \tag{10.13}$$

and for any $\varepsilon > 0$
$$|\xi(0,n) - \eta(0,n)| = o(n^{1/4+\varepsilon}). \tag{10.14}$$

Remark 4. Hence we obtain a new proof of Theorem 10.1 when only a fixed x is considered.

Remark 5. Having Theorem 10.3, Theorem 10.2 can be easily deduced (cf. Csáki–Révész 1983, Simons 1983).

Question. Is it possible to define two random walks $\{S_n^{(1)}\}$ and $\{S_n^{(2)}\}$ on a probability space such that
$$n^{-\alpha}|\xi^{(1)}(0,n) - 2\Theta^{(2)}(n)| \to 0 \quad a.s.$$
for some $0 < \alpha \leq 1/4$ (cf. (10.14)) where $\xi^{(1)}(0,n)$ is the local time of $S_n^{(1)}$ and $\Theta^{(2)}$ is the number of crossings of $S_n^{(2)}$? If a positive answer can be obtained, then in (10.11) a better rate can also be obtained. However, if the answer is negative, then (10.11) also gives the best possible rate (except that perhaps n^ε can be replaced by some $\log n$ power). Hence this question is equivalent to the question of Remark 1 of Section 10.1.

Now we formulate another trivial consequence of Theorem 10.2.

Consequence 4. Let
$$\mathcal{F}(T) = \max_{0 \leq u \leq v \leq T}(W(u) - W(v))$$
be the maximal fall of $W(\cdot)$ in $[0, T]$. Then
$$(1+\varepsilon)(b(T))^{-1} \in \mathrm{UUC}(\mathcal{F}(T)),$$
$$(1-\varepsilon)(b(T))^{-1} \in \mathrm{ULC}(\mathcal{F}(T)),$$
$$(1+\varepsilon)\left(\frac{\pi^2}{8}\frac{n}{\log\log n}\right)^{1/2} \in \mathrm{LUC}(\mathcal{F}(T)),$$
$$(1-\varepsilon)\left(\frac{\pi^2}{8}\frac{n}{\log\log n}\right)^{1/2} \in \mathrm{LLC}(\mathcal{F}(T)),$$
$$\mathbf{P}\left\{T^{-1/2}\mathcal{F}(T) < x\right\} = G(x) = H(x)$$
where $G(\cdot)$ and $H(\cdot)$ are defined in Theorem 2.13.

Proof of Consequence 4. Observe that $\mathcal{F}(T) = \max_{0 \leq t \leq T} y(t)$ and apply Theorems 10.2, 2.13, the LIL of Khinchine (Section 4.4) and the Other LIL (5.9).

It is also interesting to study the properties of $\mathcal{F}(T)$ for those T's for which $W(T)$ is very large. In fact we prove

THEOREM 10.5 *Let C_1 and C_2 be two positive constants. Then there exists a sequence $0 < t_1 = t_1(\omega; C_1, C_2) < t_2 = t_2(\omega; C_1, C_2) < \ldots$ such that*

$$\mathcal{F}(t_n) \leq C_2 \left(\frac{t_n}{\log \log t_n}\right)^{1/2} \quad \text{and} \quad W(t_n) \geq C_1(b(t_n))^{-1}$$

if

$$C_1^2 + \frac{\pi^2}{2C_2^2} < 1.$$

The proof of the above theorem is based on the following:

THEOREM 10.6 (Mogul'skii, 1979).

$$\lim_{u \to 0} u^2 \log \mathbf{P}\left\{\sup_{0 \leq t \leq T}\left|W(t) - \frac{t}{T}W(T)\right| \leq uT^{1/2}\right\} = -\frac{\pi^2}{8}.$$

Proof of Theorem 10.5. Observe that the conditional distribution of

$$\{W(t),\ 0 \leq t \leq T\} \quad \text{given} \quad W(T) = C(b(T))^{-1}$$

is equal to the distribution of

$$\{B_T(t) + Ct(Tb(T))^{-1},\ 0 \leq t \leq T\}.$$

Further, the maximal fall of $B_T(t) + Ct(Tb(T))^{-1}$ is less than or equal to $2\max_{0 \leq t \leq T}|B_T(t)|$. Hence we obtain

$$\mathbf{P}\left\{\mathcal{F}(T) \leq C_2\left(\frac{T}{\log \log T}\right)^{1/2},\ W(T) \geq C_1(b(T))^{-1}\right\}$$

$$\geq \mathbf{P}\left\{\max_{0 \leq t \leq T}|B_T(t)| \leq \frac{C_2}{2}\left(\frac{T}{\log \log T}\right)^{1/2}\right\}\mathbf{P}\left\{W(T) \geq C_1(b(T))^{-1}\right\}$$

$$\geq \exp\left(-\frac{\pi^2}{8}\frac{4+\varepsilon}{C_2^2}\log \log T\right)\exp\left(-C_1^2 \log \log T\right)$$

$$\geq (\log T)^{-(C_1^2 + \pi^2(1+\varepsilon)/2C_2^2)}$$

where $B_T(t) = W(t) - tW(T)$ ($0 \leq t \leq T$) and the proof follows by the usual way (cf. e.g. the proof of the LIL of Khinchine, Section 4.4).

Chapter 11

Strong Theorems of the Local Time

11.1 Strong theorems for $\xi(x,n)$ and $\xi(n)$

The Recurrence Theorem (cf. Section 3.1) clearly implies that for any $x = 0, \pm 1, \pm 2, \ldots$

$$\lim_{n\to\infty} \xi(x,n) = \infty \quad \text{a.s.} \tag{11.1}$$

In order to get the rate of convergence in (11.1) it is enough to observe that by Theorem 10.3 the limit behaviour of $\xi(0,n)$ (and consequently that of $\xi(x,n)$) is the same as that of M_n^+. Hence by the EFKP LIL (cf. Section 5.2) and by the Theorem of Hirsch (cf. Section 5.3) we obtain

THEOREM 11.1 *The nondecreasing function*

$$f(n) \in \text{UUC}\left(n^{-1/2}\xi(x,n)\right)$$

if and only if

$$I_1(f) = \sum_{n=1}^{\infty} \frac{f(n)}{n} \exp\left(-\frac{f^2(n)}{2}\right) < \infty.$$

The nonincreasing function

$$g(n) \in \text{LLC}\left(n^{-1/2}\xi(x,n)\right)$$

if and only if

$$I_3(g) = \sum_{n=1}^{\infty} \frac{g(n)}{n} < \infty$$

where x is an arbitrary fixed integer.

Having Theorem 10.2 (instead of Theorem 10.3) and Theorem 6.2 or applying Theorem 11.1 and Theorem 10.1 we obtain

THEOREM 11.2 *Theorem 11.1 remains true if we replace $\xi(\cdot,\cdot)$ by $\eta(\cdot,\cdot)$.*

Remark 1. Theorem 11.1 was proved originally by Chung and Hunt (1949). Theorem 11.2 is due to Kesten (1965).

The study of $\xi(n) = \max_x \xi(x,n)$ is much harder than that of $\xi(x,n)$. However, having Theorem 9.14 (cf. also (10.6)) one can prove

THEOREM 11.3 (Kesten, 1965, Csáki–Földes, 1986).

$$\limsup_{n\to\infty} b(n)\xi(n) = \limsup_{t\to\infty} b(t)\eta(t) = 1 \quad a.s. \tag{11.2}$$

$$\liminf_{n\to\infty} n^{-1/2}(\log\log n)^{1/2}\xi(n) = \liminf_{t\to\infty} t^{-1/2}(\log\log t)^{1/2}\eta(t)$$
$$= \gamma = 2^{1/2} j_1 \quad a.s. \tag{11.3}$$

where j_1 is the first positive root of the Bessel function $J_0(x)$.

Remark 2. (11.2) is due to Kesten (1965). (11.3) is also due to Kesten without obtaining the exact value of γ.

The result of Csáki and Földes (1986) is much stronger than (11.3). In fact they proved:

THEOREM 11.4 Let $u(t) > 0$ be a nonincreasing function such that $\lim_{t\to\infty} u(t) = 0$, $u(t)t^{1/2}$ is nondecreasing and $\lim_{t\to\infty} u(t)t^{1/2} = \infty$. Then

$$u(t) \in \text{LLC}\left(t^{-1/2}\eta(t)\right) \quad \text{and} \quad u(n) \in \text{LLC}\left(n^{-1/2}\xi(n)\right)$$

if and only if

$$\int_1^\infty t^{-1} u^{-2}(t) \exp\left(-\frac{2j_1^2}{u^2(t)}\right) dt < \infty.$$

Remark 3. The proof of Theorem 11.4 is based on Theorem 9.14.

The upper classes of $\eta(t)$ and those of $\xi(n)$ were also described by Csáki (1989). He proved

THEOREM 11.5 Let $a(t) > 0$ $(t \geq 1)$ be a nondecreasing function. Then

$$a(t) \in \text{UUC}\left(t^{-1/2}\eta(t)\right) \quad \text{and} \quad a(n) \in \text{UUC}\left(n^{-1/2}\xi(n)\right)$$

if and only if

$$\int_1^\infty \frac{a^3(t)}{t} \exp\left(-\frac{a^2(t)}{2}\right) dt < \infty.$$

Since $\xi(0, \rho_n) = n$, i.e. ρ_n is the inverse function of $\xi(0,n)$, by Theorem 11.1 we can also obtain the Lévy classes of ρ_n. Here we present only the simplest consequence.

THEOREM 11.6 *For any $\varepsilon > 0$ we have*

$$n^2(\log n)^{2+\varepsilon} \in \mathrm{UUC}(\rho_n),$$
$$n^2(\log n)^{2-\varepsilon} \in \mathrm{ULC}(\rho_n),$$
$$(1+\varepsilon)\frac{n^2}{2\log\log n} \in \mathrm{LUC}(\rho_n),$$
$$(1-\varepsilon)\frac{n^2}{2\log\log n} \in \mathrm{LLC}(\rho_n).$$

Perkins (1981/B) proposed to study the limit behaviour of

$$\eta^-(t, h(t)) = \inf_{|x| \leq h(t)} \eta(x, t)$$

and he proved

THEOREM 11.7 *There is a nonincreasing function $\Theta(\alpha)$ such that*

$$\limsup_{t \to \infty} \frac{\eta^-\left(t, \alpha\left(\frac{t}{2\log\log t}\right)^2\right)}{(2t\log\log t)^{1/2}} = \Theta(\alpha) \quad a.s.$$

for all $\alpha > 0$.

It also looks interesting to study the properties of $\eta(x, \rho_r^*)$ where $\rho_r^* = \inf\{t: t \geq 0, \eta(0, t) \geq r\}$. An analogue of Theorem 11.7 for stopping time ρ_r^* is

THEOREM 11.8 (Földes, 1989). *Let $f(x)$ be a nondecreasing function with $\lim_{x\to\infty} f(x) = \infty$. Then*

$$\mathbf{P}\left\{\limsup_{r \to \infty} \inf_{|x| \leq rf(r)} \frac{\eta(x, \rho_r^*)}{r} = 1\right\} = 1$$

if and only if

$$\int_1^\infty \frac{dx}{xf^2(x)} = \infty.$$

11.2 Increments of $\eta(x, t)$

First we give the analogue of Theorem 7.13.

THEOREM 11.9 (Csáki–Csörgő–Földes–Révész, 1983). *Let $a_t (t \geq 0)$ be a nondecreasing function of t for which*

(i) $0 < a_t \le t$,

(ii) t/a_t is nondecreasing.

Then

$$\limsup_{t\to\infty} \delta_t \sup_{0\le s\le t-a_t} (\eta(x, s + a_t) - \eta(x, s))$$
$$= \limsup_{t\to\infty} \delta_t(\eta(x, t) - \eta(x, t - a_t)) = 1 \quad a.s.$$

If we also have

(iii)
$$\lim_{t\to\infty} (\log(t/a_t))(\log\log t)^{-1} = \infty$$

then

$$\lim_{t\to\infty} \delta_t \sup_{0\le s\le t-a_t} (\eta(x, s + a_t) - \eta(x, s)) = 1 \quad a.s.$$

for any fixed $x \in \mathbb{R}^1$ *where* $\delta_t = a_t^{-1/2}(\log(t/a_t) + 2\log\log t)^{-1/2}$.

By Theorem 10.2 as a trivial consequence of the above Theorem we obtain

THEOREM 11.10 *Theorem* 11.9 *remains true replacing* $\eta(x,t)$ *by* $m^+(t)$.

Remark 1. Clearly

$$\sup_{0\le s\le t-a_t} (m^+(s + a_t) - m^+(s)) \le \sup_{0\le s\le t-a_t} (W(s + a_t) - W(s)). \quad (11.4)$$

Comparing δ_t and γ_t of Theorem 7.13 we obtain in (11.4) that for a sequence $t = t_n \uparrow \infty$ we may have strict inequality whenever (iii) does not hold true.

The investigation of the largest possible increment in t when x is also varying seems to be also interesting. We obtained

THEOREM 11.11 (Csáki–Csörgő–Földes–Révész, 1983). *Let* a_t $(t \ge 0)$ *be a nondecreasing function of* t *satisfying conditions* (i) *and* (ii) *of Theorem* 11.9. *Then*

$$\limsup_{t\to\infty} \gamma_t \sup_{x\in\mathbb{R}^1} \sup_{0\le s\le t-a_t} (\eta(x, s + a_t) - \eta(x, s)) = 1 \quad a.s.$$

If we also assume that (iii) *of Theorem* 11.9 *holds then*

$$\lim_{t\to\infty} \gamma_t \sup_{x\in\mathbb{R}^1} \sup_{0\le s\le t-a_t} (\eta(x, s + a_t) - \eta(x, s)) = 1 \quad a.s.$$

To find the analogue of Theorem 7.20 seems to be much more delicate. At first we ask about the length of the longest zero-free interval. Let

$$r(t) = \sup\{a : \text{for which } \exists 0 < s < t - a \text{ such that } \eta(0, s+a) - \eta(0, s) = 0\}$$

be the length of the longest zero-free interval. Then we have

THEOREM 11.12 (Chung–Erdős, 1952). *Let $f(x)$ be a nondecreasing function for which $f(x) \nearrow \infty$ and $x/f(x) \nearrow \infty$. Then*

$$t\left(1 - \frac{1}{f(t)}\right) \in \text{UUC}(r(t))$$

if and only if

$$L(f) = \int_1^\infty \frac{dx}{x(f(x))^{1/2}} < \infty.$$

Remark 2. Originally this theorem was formulated for random walk instead of Wiener process.

Example 1. Since $L(f) < \infty$ if $f(x) = (\log x)^{2+\varepsilon} (\varepsilon > 0)$ and $L(f) = \infty$ if $f(x) = (\log x)^2$, we obtain

$$t\left(1 - \frac{1}{(\log t)^{2+\varepsilon}}\right) \in \text{UUC}(r(t))$$

and

$$t\left(1 - \frac{1}{(\log t)^2}\right) \in \text{ULC}(r(t)),$$

or equivalently

$$\liminf_{t \to \infty} \inf_{0 \leq s \leq t - a_t} (\eta(0, s + a_t) - \eta(0, s)) = \liminf_{t \to \infty} (\eta(0, t) - \eta(0, t - a_t)) > 0$$

$$\text{if} \quad a_t = t\left(1 - \frac{1}{(\log t)^{2+\varepsilon}}\right) \quad (\varepsilon > 0),$$

and

$$\liminf_{t \to \infty} \inf_{0 \leq s \leq t - a_t} (\eta(0, s + a_t) - \eta(0, s)) = \liminf_{t \to \infty} (\eta(0, t) - \eta(0, t - a_t)) = 0$$

$$\text{if} \quad a_t = t\left(1 - \frac{1}{(\log t)^2}\right).$$

This example shows that the study of the properties of the above lim inf (i.e. the analogue of Theorem 7.20) is interesting only if $a_t > t(1 - (\log t)^{-2})$. This question was studied by Csáki and Földes (1986). They proved

THEOREM 11.13 *Let $f_1(t) = t(t - a_t)^{-1}$ be a nondecreasing function for which $t/f_1(t) \nearrow \infty$ and $\lim_{t\to\infty} f_1(t) = \infty$. Further, let $f_2(t)$ be a nonincreasing function for which $\lim_{t\to\infty} f_2(t) = 0$, $t^{1/2} f_2(t)$ is nondecreasing and $\lim_{t\to\infty} t^{1/2} f_2(t) = \infty$. Then*

$$t^{1/2} f_2(t) \in \mathrm{LUC} \left\{ \inf_{0 \leq s \leq t - a_t} (\eta(0, s + a_t) - \eta(0, s)) \right\}$$

if

$$L(f_1) = \infty \quad or \quad L^*(f_2) = \infty,$$

and

$$t^{1/2} f_2(t) \in \mathrm{LLC} \left\{ \inf_{0 \leq s \leq t - a_t} (\eta(0, s + a_t) - \eta(0, s)) \right\}$$

if

$$L(f_1) < \infty \quad and \quad L^*(f_2) < \infty,$$

where

$$L(f) = \int_1^\infty \frac{dx}{x(f(x))^{1/2}} \quad and \quad L^*(f) = \int_1^\infty \frac{f(x)}{x} dx.$$

Example 2. Let $a_t = t(1 - (\log t)^{-2-\varepsilon})(\varepsilon > 0)$. Then $f_1(t) = (\log t)^{2+\varepsilon}$ and $L(f_1) < \infty$. Since

$$L^*((\log t)^{-1-\delta}) \begin{cases} = \infty & \text{if } \delta = 0, \\ < \infty & \text{if } \delta > 0, \end{cases}$$

we obtain

$$\liminf_{t\to\infty} (\log t)^{1+\delta} t^{-1/2} \inf_{0 \leq s \leq t - a_t} (\eta(0, s + a_t) - \eta(0, s)) = \begin{cases} \infty & \text{if } \delta > 0 \quad \text{a.s.} \\ 0 & \text{if } \delta = 0 \quad \text{a.s.} \end{cases}$$

Remark 3. By Theorems 10.1, 10.2 and 10.3, we find that the statement of Example 2 remains true replacing $\eta(0, t)$ by $m^+(t)$ or $M^+(n)$. (Compare this result with the Theorem of Hirsch of Section 5.3.)

Finally we mention the following analogue of Theorem 11.11 (cf. also Theorem 11.3).

THEOREM 11.14 (Csáki–Földes, 1986). *Let $a_t(t \geq 0)$ be a nondecreasing function of t satisfying conditions* (i) *and* (ii) *of Theorem 11.9. Then*

$$\liminf_{t\to\infty} \vartheta_t Q(t) = 1 \quad a.s.$$

where

$$Q(t) = \inf_{0 \leq s \leq t - a_t} \sup_{x \in \mathbb{R}^1} (\eta(x, s + a_t) - \eta(x, s))$$

and
$$\vartheta_t = \left(\frac{\log(t/a_t) + \log\log t}{2j_1^2 a_t}\right)^{1/2}.$$

If we also assume that (iii) of Theorem 11.9 holds then
$$\lim_{t\to\infty} \vartheta_t Q(t) = 1 \quad a.s.$$

Remark 4. In case $a_t = t$ we obtain (11.3) as a special case of Theorem 11.14.

Remark 5. The study of the increments of $\eta(x,t)$ in x or in both variables looks a challenging question.

11.3 Increments of $\xi(x, n)$

In Section 10.1 we have seen that the strong theorems valid for $\eta(x,t)$ (resp. $\eta(t)$) remain valid for $\xi(x,n)$ (resp. $\xi(n)$) due to the Invariance Principle (Theorem 10.1). In Section 7.3 we have seen that the strong theorems proved for the increments of a Wiener process remain valid for those of a random walk if $a_n \gg \log n$ (resp. $a_n \gg (\log n)^3$) depending on what kind of theorems we are talking about. This latter fact is due to the Invariance Principle 1 (Section 6.3) and especially the rate $O(\log n)$ in it. Since the rate in Theorem 10.1 is much worse (it is $o(n^{1/4+\varepsilon})$ only) we can only claim (as a consequence of the Invariance Principle) that the results of Section 11.2 remain valid for $\xi(x,n)$ (instead of $\eta(x,t)$) if $a_n \geq n^{1/2+\varepsilon}$. The case $a_n < n^{1/2+\varepsilon}$ requires a separate study. This was done by Csáki and Földes (1984/C). They proved that Theorem 11.9 remain valid for $\xi(x,n)$ if $a_n \gg \log n$. In fact they proved the following two theorems:

THEOREM 11.15 Let $0 < a_n \leq n$ $(n = 1, 2, \ldots)$ be an integer valued nondecreasing sequence. Assume that a_n/n is nonincreasing and
$$\lim_{n\to\infty} \frac{a_n}{\log n} = \infty.$$

Then
$$\limsup_{n\to\infty} \delta_n \sup_{0\leq k\leq n-a_n} (\xi(x, k+a_n) - \xi(x,k)) = 1 \quad a.s.$$

If we also have
$$\lim_{n\to\infty} \frac{\log(n/a_n)}{\log\log n} = \infty$$

then
$$\lim_{n\to\infty} \delta_n \sup_{0\leq k\leq n-a_n} (\xi(x, k+a_n) - \xi(x,k)) = 1 \quad a.s.$$
for any fixed $x \in \mathbb{Z}^1$ where
$$\delta_n = a_n^{-1/2} \left(\log(na_n^{-1}) + 2\log\log n\right)^{-1/2}.$$

THEOREM 11.16 *Let $c > 0$. Then for any fixed $x \in \mathbb{Z}^1$*
$$\lim_{n\to\infty} \max_{0\leq k\leq n-[c\log n]} \frac{\xi(x, k+[c\log n]) - \xi(x,k)}{[c\log n]} = \alpha(c) \quad a.s.$$
where $\alpha(c) = 1/2$ if $c \leq (\log 2)^{-1}$ and it is the only solution of the equation
$$\frac{1}{c} = (1-2\alpha)\log(1-2\alpha) - 2(1-\alpha)\log(1-\alpha)$$
if $c > (\log 2)^{-1}$.

Remark 1. The above theorem suggests the conjecture:
$$\max_{0\leq k\leq n-a_n} (\xi(0, k+a_n) - \xi(0,k)) = \left[\frac{a_n}{2}\right] \quad a.s.$$
for all but finitely many n provided that $a_n = o(\log n)$.

Since the Invariance Principle 1 of Section 6.3 is valid with the rate $O(\log n)$, Theorem 11.10 implies

THEOREM 11.17 *Theorem 11.15 remains true replacing $\xi(x, n)$ by $M^+(n)$.*

The analogue of Theorem 11.11 for $\xi(x, n)$ is unknown except if $a_n \geq n^{1/2+\varepsilon}$.

The analogues of Theorems 11.12 and 11.13 can be obtained by the Invariance Principle for $\xi(x, n)$.

11.4 Strassen type theorems

Let
$$u_t(x) = b_t \eta(0, xt) \quad (0 \leq x \leq 1, t > 0)$$
and
$$U_n(x) = b_n \left(\xi(0, k) + n\left(x - \frac{k}{n}\right)(\xi(0, k+1) - \xi(0, k))\right) \text{ if } \frac{k}{n} \leq x < \frac{k+1}{n}$$

($k = 0, 1, 2, \ldots, n-1$; $n = 1, 2, \ldots$). We intend to characterize the limit points of the sequence $U_n(x)$ and those of $u_t(x)$. Since $U_n(x)$ ($0 \leq x \leq 1$) for any fixed n is a nondecreasing function, its limit points must also be nondecreasing.

Definition. Let $\mathcal{S}_M \subset \mathcal{S}$ be the set of nondecreasing elements of $\mathcal{S}$ (cf. Notations to the Strassen type theorems).

Then we formulate

THEOREM 11.18 (Csáki–Révész, 1983). *The sequence $\{U_n(x); 0 \leq x \leq 1\}$ and the net $\{u_t(x); 0 \leq x \leq 1\}$ are relatively compact in $C(0,1)$ with probability 1 and the sets of their limit points are $\mathcal{S}_M$.*

Proof. This result is a trivial consequence of Theorems 8.2 and 10.2.

Define the process $\rho(xn)$ ($0 \leq x \leq 1$; $n = 1, 2, \ldots$) by $\rho(xn) = \rho_k$ if $x = k/n$ ($k = 0, 1, 2, \ldots, n$) and linear between k/n and $(k+1)/n$. Then taking into account that ρ_n is the inverse of $\xi(0, n)$, i.e. $\xi(0, \rho_n) = n$, we obtain the following consequence of Theorem 11.18:

THEOREM 11.19 *The set of limit points of the functions*

$$\{2n^{-2}(\log\log n)\rho(xn); \ 0 \leq x \leq 1\} \quad (n \to \infty)$$

consists of those and only those functions $f(x)$ for which $f^{-1}(x) \in \mathcal{S}_M$.

It is also interesting to characterize the sets of limit points of the sequences $\xi(x, n)$ (resp. $\eta(x, t)$) when we consider them as functions of n (resp. t) and we choose a big but not too big x. In fact the Other LIL (cf. Section 5.3) tells us that

$$\xi(x_n, n) = 0 \quad \text{resp.} \quad \eta(x_t, t) = 0 \quad \text{i.o. a.s.}$$

if

$$x_n \geq (1+\varepsilon)\frac{\pi}{\sqrt{8}}\left(\frac{n}{\log\log n}\right)^{1/2} \quad \text{resp.} \quad x_t \geq (1+\varepsilon)\frac{\pi}{\sqrt{8}}\left(\frac{t}{\log\log t}\right)^{1/2}.$$

Hence we consider the case when x is smaller than the above limits, i.e. when $\xi(\cdot, \cdot)$ and $\eta(\cdot, \cdot)$ are strictly positive a.s. Now we formulate

THEOREM OF DONSKER AND VARADHAN (1977). *In the topology of $C(-\infty, +\infty)$ the set of limit points of the functions*

$$u_t^*(x) = (t\log\log t)^{-1/2}\eta\left(x\left(\frac{t}{\log\log t}\right)^{1/2}, t\right) \quad (t \to \infty)$$

resp.

$$U_n^*(x) = (n \log \log n)^{-1/2} \xi\left(\left[x\left(\frac{n}{\log \log n}\right)^{1/2}\right], n\right) \quad (n \to \infty)$$

consists of those and only those subprobability density functions $f(x)$ for which

$$\frac{1}{8}\int_{-\infty}^{\infty} \frac{(f'(x))^2}{f(x)} dx \leq 1.$$

Remark 1. Mueller (1983) gave a common generalization of the Theorem of Donsker–Varadhan and that of Wichura (cf. Section 8.4).

11.5 Stability

Intuitively it is clear that $\xi(x, n)$ is close to $\xi(y, n)$ if x is close to y. This Section is devoted to studying this problem.

THEOREM 11.20 (Csörgő–Révész, 1985/A).

$$\limsup_{N\to\infty} \frac{\xi(k, N) - \xi(0, N)}{(\xi(0, N) \log\log N)^{1/2}}$$
$$= \limsup_{N\to\infty} \frac{|\xi(k, N) - \xi(0, N)|}{(\xi(0, N) \log\log N)^{1/2}} = \limsup_{N\to\infty} \sup_{n\leq N} \frac{\xi(k, n) - \xi(0, n)}{(\xi(0, N) \log\log N)^{1/2}}$$
$$= \limsup_{N\to\infty} \sup_{n\leq N} \frac{|\xi(k, n) - \xi(0, n)|}{(\xi(0, N) \log\log N)^{1/2}} = 2(2k - 1)^{1/2} \quad a.s.$$

where $k = \pm 1, \pm 2, \ldots$.

THEOREM 11.21 (Csörgő–Révész, 1985/A).

$$\limsup_{N\to\infty} \frac{\xi(1, N) - \xi(0, N)}{N^{1/4}(\log\log N)^{3/4}}$$
$$= \limsup_{N\to\infty} \frac{|\xi(1, N) - \xi(0, N)|}{N^{1/4}(\log\log N)^{3/4}} = \limsup_{N\to\infty} \sup_{n\leq N} \frac{\xi(1, n) - \xi(0, n)}{N^{1/4}(\log\log N)^{3/4}}$$
$$= \limsup_{N\to\infty} \sup_{n\leq N} \frac{|\xi(1, n) - \xi(0, n)|}{N^{1/4}(\log\log N)^{3/4}} = \left(\frac{128}{27}\right)^{1/4} \quad a.s.$$

Remark 1. Since for any $x \in \mathbb{Z}^1$,

$$\xi(x, n) = \xi(0, n) \quad \text{i.o. a.s.}$$

STRONG THEOREMS OF THE LOCAL TIME

the study of the liminf of $|\xi(x,n) - \xi(0,n)|$ is not interesting. The limsup properties of $\xi(x,n) - \xi(0,n)$ follow trivially from Theorem 11.10.

Theorem 11.20 stated that $\xi(x,n)$ is close to $\xi(0,n)$ for any fixed x if n is big enough. The next two Theorems claim that in a weaker sense $\xi(x,n)$ is nearly equal to $\xi(0,n)$ in a long interval around 0.

THEOREM 11.22 (Csáki–Földes, 1987). *Put*

$$g(t) = \frac{t^{1/2}}{(\log t)(\log \log t)^\rho}.$$

Then

$$\lim_{n \to \infty} \sup_{|x| \leq g(n)} \left| \frac{\xi(x,n)}{\xi(0,n)} - 1 \right| = 0 \quad a.s. \quad if \quad \rho > 2 \qquad (11.5)$$

and

$$\limsup_{n \to \infty} \sup_{|x| \leq g(n)} \left| \frac{\xi(x,n)}{\xi(0,n)} - 1 \right| \geq 1 \quad a.s. \quad if \quad \rho \leq 1.$$

THEOREM 11.23 (Csáki–Földes, 1987). *Put*

$$h_1(n) = \frac{-M^-(n)}{\log n (\log \log n)^\rho}, \quad h_2(n) = \frac{M^+(n)}{\log n (\log \log n)^\rho}.$$

Then

$$\lim_{n \to \infty} \sup_{h_1(n) \leq x \leq h_2(n)} \left| \frac{\xi(x,n)}{\xi(0,n)} - 1 \right| = 0 \quad a.s. \quad if \quad \rho > \frac{5}{2}$$

and

$$\limsup_{n \to \infty} \sup_{ch_1(n) \leq x \leq ch_2(n)} \left| \frac{\xi(x,n)}{\xi(0,n)} - 1 \right| = \infty \quad a.s. \quad if \quad \rho = 0$$

where c is any positive constant.

Remark 2. The Theorem of Hirsch says that $\xi(x,n) = 0$ i.o. a.s. if $x \geq n^{1/2}(\log n)^{-1}$. Hence it is clear that (11.5) can be true only if $g(n) > n^{1/2}(\log n)^{-1}$. Theorem 11.22 tells us that $g(n)$ must be smaller than this trivial upper estimate. The behaviour of $\xi(M^+(n) - j, n)$ for small j will be described in Theorem 12.27. It implies that $\xi(M^+(n) - j, n)$ is much, much smaller than $\xi(0,n)$. Theorem 11.23 gives the longest interval, depending on $M^+(n)$ and $M^-(n)$, where $\xi(x,n)$ is stable.

In order to prove Theorem 11.20 we present a few lemmas.

LEMMA 11.1 *Let*

$$\alpha_i = \alpha_i(k) = \xi(k, \rho_i) - \xi(k, \rho_{i-1}) - 1 \quad (i = 1, 2, \ldots, \ k = 1, 2, \ldots).$$

Then

$$\mathbf{E}\alpha_1 = 0, \quad \mathbf{E}\alpha_1^2 = 4k - 2, \tag{11.6}$$

$$\lim_{n \to \infty} \mathbf{P}\left\{ n^{-1/2}(\alpha_1(k) + \alpha_2(k) + \cdots + \alpha_n(k)) \leq x(4k-2)^{1/2} \right\}$$
$$= (2\pi)^{-1/2} \int_{-\infty}^{x} e^{-u^2/2} du, \quad -\infty < x < \infty, \tag{11.7}$$

$$\lim_{n \to \infty} \mathbf{P}\left\{ n^{-1/2} \sup_{j \leq n}(\alpha_1(k) + \alpha_2(k) + \cdots + \alpha_j(k)) \leq x(4k-2)^{1/2} \right\}$$
$$= \left(\frac{2}{\pi}\right)^{1/2} \int_0^x e^{-u^2/2} du, \quad x > 0, \tag{11.8}$$

and

$$\limsup_{n \to \infty} \frac{\alpha_1(k) + \alpha_2(k) + \cdots + \alpha_n(k)}{(n \log \log n)^{1/2}} = 2(2k-1)^{1/2} \quad a.s. \tag{11.9}$$

Proof. (11.6) is a trivial consequence of Theorem 9.7. (11.7), (11.8) and (11.9) follow from Theorems 2.9, 2.12 and the LIL of Khinchine of Section 4.4 respectively.

The following two lemmas are simple consequences of (11.9).

LEMMA 11.2 *Let $\{\mu_n\}$ be any sequence of positive integer valued r.v.'s with $\lim_{n \to \infty} \mu_n = \infty$ a.s. Then*

$$\limsup_{n \to \infty} \frac{\alpha_1(k) + \alpha_2(k) + \ldots + \alpha_{\mu_n}(k)}{(\mu_n \log \log \mu_n)^{1/2}} \leq 2(2k-1)^{1/2} \quad a.s.$$

LEMMA 11.3 *Let $\{\nu_n\}$ be a sequence of positive integer valued r.v.'s with the following properties:*

(i) $\lim_{n \to \infty} \nu_n = \infty$ *a.s.*

(ii) *there exists a set $\Omega_0 \subset \Omega$ such that $\mathbf{P}\{\Omega_0\} = 0$ and for each $\omega \notin \Omega_0$ and $k = 1, 2, \ldots$ there exists an $n = n(\omega, k)$ for which $\nu_{n(\omega,k)} = k$.*

Then

$$\limsup_{n \to \infty} \frac{\alpha_1(k) + \alpha_2(k) + \cdots + \alpha_{\nu_n}(k)}{(\nu_n \log \log \nu_n)^{1/2}} = 2(2k-1)^{1/2} \quad a.s.$$

STRONG THEOREMS OF THE LOCAL TIME   129

Utilizing Lemma 11.3 with $\nu_n = \xi(0,n)$ and the trivial inequality $\alpha_1(k) + \alpha_2(k) + \ldots + \alpha_{\xi(0,n)}(k) \leq \xi(k,n) - \xi(0,n) \leq \alpha_1(k) + \alpha_2(k) + \ldots + \alpha_{\xi(0,n)+1}(k) + 1$, we obtain Theorem 11.20.

As far as the proof of Theorem 11.21 is concerned we only present a proof of the statement

$$\limsup_{N \to \infty} \frac{\xi(1,N) - \xi(0,N)}{N^{1/4}(\log \log N)^{3/4}} = \left(\frac{128}{27}\right)^{1/4} \quad \text{a.s.}$$

The other statements of Theorem 11.21 are proved along similar lines.

The proof of Theorem 11.21 is based on the following result of Dobrushin (1955).

THEOREM 11.24

$$\lim_{n \to \infty} \mathbf{P}\left\{\frac{\xi(1,n) - \xi(0,n)}{n^{1/4}} \leq \sqrt{2}x\right\} = \frac{2}{\pi}\int_{-\infty}^{x}\int_{0}^{\infty} \exp\left(-\frac{y^2}{2z^2} - \frac{z^4}{2}\right) dz\,dy.$$

Remark 3. One can also prove that (cf. Theorem 14.1 and (14.17))

$$\lim_{n \to \infty} \mathbf{P}\left\{\frac{\eta(1,n) - \eta(0,n)}{n^{1/4}} \leq 2x\right\} = \frac{2}{\pi}\int_{-\infty}^{x}\int_{0}^{\infty} \exp\left(-\frac{y^2}{2z^2} - \frac{z^4}{2}\right) dz\,dy.$$

This fact together with Theorem 11.24 implies that the rate in the uniform Invariance Principle (Theorem 10.1) cannot be true with rate $n^{1/4}$.

Dobrushin also notes that if N_1 and N_2 are independent normal $(0,1)$ r.v.'s then the density function g of $|N_1|^{1/2} N_2$ is

$$g(y) = \frac{2}{\pi}\int_0^\infty \exp\left(-\frac{y^2}{2z^2} - \frac{z^4}{2}\right) dz.$$

Hence Theorem 11.24 can be reformulated by saying that

$$2^{-1/2} n^{-1/4}(\xi(1,n) - \xi(0,n)) \xrightarrow{\mathcal{D}} |N_1|^{1/2} N_2 \quad (n \to \infty). \tag{11.10}$$

In fact this statement is not very surprising since on replacing n by $\xi(0,n)$ and k by 1 in (11.7), intuitively it is clear that

$$\frac{\alpha_1(1) + \alpha_2(1) + \cdots + \alpha_{\xi(0,n)}(1)}{\sqrt{2\xi(0,n)}} \sim \frac{\xi(1,n) - \xi(0,n)}{\sqrt{2\xi(0,n)}} \xrightarrow{\mathcal{D}} N_2 \quad (n \to \infty). \tag{11.11}$$

To find an exact proof of (11.11) is not simple at all. We will study this question in Chapter 14.

Also, by Theorem 9.12
$$n^{-1/4}(\xi(0,n))^{1/2} \xrightarrow{D} |N_1|^{1/2} \quad (n \to \infty). \tag{11.12}$$
Intuitively it is again clear (for an exact formulation see Chapter 14) that
$$\frac{\xi(1,n) - \xi(0,n)}{\sqrt{2\xi(0,n)}} \quad \text{and} \quad \frac{\xi(0,n)}{\sqrt{n}} \quad \text{are asymptotically independent.}$$
$$\tag{11.13}$$
Hence (11.11), (11.12) and (11.13) together imply (11.10). The proof of Dobrushin is not based on this idea. Following his method, however, a slightly stronger version of his Theorem 11.24 can be obtained.

THEOREM 11.25 *Let $\{x_n\}$ be any sequence of positive numbers such that $x_n = o(\log n)$. Then*
$$\mathbf{P}\left\{\frac{\xi(1,n) - \xi(0,n)}{n^{1/4}} < -\sqrt{2}x_n\right\} \approx \frac{2}{\pi} \int_{-\infty}^{-x_n} \int_0^\infty \exp\left(-\frac{y^2}{2z^2} - \frac{z^4}{2}\right) dz\, dy$$
and
$$\mathbf{P}\left\{\frac{\xi(1,n) - \xi(0,n)}{n^{1/4}} > \sqrt{2}x_n\right\} \approx \frac{2}{\pi} \int_{x_n}^\infty \int_0^\infty \exp\left(-\frac{y^2}{2z^2} - \frac{z^4}{2}\right) dz\, dy.$$

The following lemma describes some properties of the density function $g(y)$. Its proof requires only standard analytic methods, the details will be omitted.

LEMMA 11.4

(i) *There exists a positive constant C such that for any $y \in \mathbb{R}^1$*
$$g(y) \leq C y^{1/3} \exp\left(-\frac{3}{2^{5/3}} y^{4/3}\right). \tag{11.14}$$

(ii) *For any $\varepsilon > 0$ there exists a $C = C(\varepsilon) > 0$ such that*
$$g(y) \geq C \exp\left(-\frac{y^{4/3}}{2-\varepsilon} \frac{3}{2^{2/3}}\right).$$

(iii) *Let $\{a_n\}$ be a sequence of positive numbers with $a_n \uparrow \infty$. Then for any $\varepsilon > 0$ there exist a $C_1 = C_1(\varepsilon) > 0$ and a $C_2 = C_2(\varepsilon) > 0$ such that*
$$C_1 \exp\left(-\frac{a_n^{4/3}}{2-\varepsilon} \frac{3}{2^{2/3}}\right) \leq \int_{a_n}^\infty g(y)\, dy \leq C_2 \exp\left(-\frac{a_n^{4/3}}{2+\varepsilon} \frac{3}{2^{2/3}}\right).$$

STRONG THEOREMS OF THE LOCAL TIME   131

By Theorem 11.25 and (iii) of Lemma 11.4 we have

LEMMA 11.5 *For any $\varepsilon > 0$ there exist a $C_1 = C_1(\varepsilon) > 0$ and a $C_2 = C_2(\varepsilon) > 0$ such that*

$$\mathbf{P}\left\{\frac{(\xi(1,0) - \xi(0,n))}{n^{1/4}} \geq (1+2\varepsilon)\left(\frac{128}{27}\right)^{1/4}(\log\log n)^{3/4}\right\} \leq \frac{C_2}{(\log n)^{1+\varepsilon}}$$

and

$$\mathbf{P}\left\{\frac{(\xi(1,n) - \xi(0,n))}{n^{1/4}} \geq (1-2\varepsilon)\left(\frac{128}{27}\right)^{1/4}(\log\log n)^{3/4}\right\} \geq \frac{C_1}{(\log n)^{1-\varepsilon}}.$$

Now we prove

LEMMA 11.6

$$\limsup_{n\to\infty} \frac{\xi(1,n) - \xi(0,n)}{n^{1/4}(\log\log n)^{3/4}} \geq \left(\frac{128}{27}\right)^{1/4} \quad a.s.$$

Proof. Let

$$n_k = [\exp(k\log k)],$$
$$d_k = \left(\frac{128}{27}\right)^{1/4} n_k^{1/4}(\log\log n_k)^{3/4},$$
$$\zeta(n) = \xi(1,n) - \xi(0,n),$$
$$\xi(x,(m,n)) = \xi(x,n) - \xi(x,m) \quad (m < n),$$
$$\zeta(m,n) = \xi(1,(m,n)) - \xi(0,(m,n)),$$
$$A_k = \{\zeta(n_k) \geq (1-2\varepsilon)d_k\},$$
$$\beta_k = (1-2\varepsilon)d_k.$$

By Lemma 11.5
$$\mathbf{P}\{A_k\} \geq C(k\log k)^{-(1-\varepsilon)}. \tag{11.15}$$

Let $j < k$ and consider

$$\mathbf{P}\{A_k A_j\}$$
$$= \sum_{l=\beta_j}^{\infty} \mathbf{P}\{A_k, \zeta(n_j) = l\} = \sum_x \sum_{l=\beta_j}^{\infty} \mathbf{P}\{A_k, \zeta(n_j) = l, S_{n_j} = x\}$$
$$= \sum_x \sum_{l=\beta_j}^{\infty} \mathbf{P}\{A_k \mid \zeta(n_j) = l, S_{n_j} = x\}\mathbf{P}\{\zeta(n_j) = l, S_{n_j} = x\}$$

$$= \sum_x \sum_{l=\beta_j}^{\infty} \mathbf{P}\{\zeta(n_j,n_k) \geq \beta_k - l \mid S_{n_j} = x\}\mathbf{P}\{\zeta(n_j) = l, S_{n_j} = x\}$$

$$\leq \sum_{l=\beta_j}^{\infty} \sup_x \mathbf{P}\{\zeta(n_j,n_k) \geq \beta_k - l \mid S_{n_j} = x\} \sum_x \mathbf{P}\{\zeta(n_j) = l, S_{n_j} = x\}$$

$$= \sum_{l=\beta_j}^{\infty} \mathbf{P}\{\zeta(n_k - n_j) \geq \beta_k - l\}\mathbf{P}\{\zeta(n_j) = l\}$$

$$\leq \sum_{l=\beta_j}^{\infty} \mathbf{P}\{\zeta(n_k) \geq \beta_k - l\}\mathbf{P}\{\zeta(n_j) = l\}$$

$$\approx \int_A^{\infty} \mathbf{P}\{\zeta(n_k) \geq \beta_k - 2^{1/2}n_j^{1/4}y\}\mathbf{P}\{\zeta(n_j) = 2^{1/2}n_j^{1/4}y\}dy$$

$$\approx \int_A^{\infty} g(y) \int_{B(y)}^{\infty} g(z)dzdy,$$

where

$$A = \beta_j 2^{-1/2} n_j^{-1/4} = (1 - 2\varepsilon)2^{-1/2}\left(\frac{128}{27}\right)^{1/4}(\log\log n_j)^{3/4}$$

and

$$B(y) = \beta_k 2^{-1/2} n_k^{-1/4} - 2^{1/2}n_j^{1/4}y 2^{-1/2}n_k^{-1/4}$$
$$= (1 - 2\varepsilon)2^{-1/2}\left(\frac{128}{27}\right)^{1/4}(\log\log n_k)^{3/4} - y\left(\frac{n_j}{n_k}\right)^{1/4}.$$

Now a simple but tedious calculation shows that for any $\varepsilon > 0$ there exists a j_0 such that if $j_0 < j < k$, then

$$\mathbf{P}\{A_j A_k\} \leq (1 + \varepsilon)\mathbf{P}\{A_j\}\mathbf{P}\{A_k\}. \tag{11.16}$$

Here we omit the details of the proof of this fact, and sketch only the main idea behind it. Since $(n_j/n_k)^{1/4} \leq k^{-1/4}$ $(j = 1, 2, \ldots, k - 1)$, the lower limit of integration $B(y)$ above is nearly equal to

$$(1 - 2\varepsilon)2^{-1/2}\left(\frac{128}{27}\right)^{1/4}(\log\log n_k)^{3/4} \quad \text{if} \quad y \leq k^{1/4}, \quad \text{say.}$$

Hence for latter y values the integral $\int_{B(y)}^{\infty} g(z)dz$ is nearly equal to $\mathbf{P}\{A_k\}$. Similarly, the integral $\int_A^{\infty} g(y)dy$ gives $\mathbf{P}\{A_j\}$, and (11.16) follows, for in the case of $y > k^{1/4}$ the value of $g(y)$ is very small.

Now (11.15), (11.16) and the Borel–Cantelli lemma combined give Lemma 11.6.

We also have

LEMMA 11.7 Let
$$m_k = [\exp(k/\log^2 k)]$$
and
$$B_k = \{\xi(0, (m_k, m_{k+1})) \geq a_{k+1}\}$$
where
$$a_{k+1} = (1+\varepsilon)\left[(m_{k+1} - m_k)\left(\log \frac{m_{k+1}}{m_{k+1} - m_k} + 2\log\log m_{k+1}\right)\right]^{1/2}.$$
Then of the events B_k only finitely many occur with probability 1.

Proof. This lemma is an immediate consequence of Theorem 11.15.

LEMMA 11.8 Let
$$M_{k+1} = ((2+\varepsilon)m_{k+1}\log\log m_{k+1})^{1/2}$$
and
$$D_k = \left\{ \sup_{l \leq M_{k+1} - a_{k+1}} \sup_{j \leq a_{k+1}} |\alpha_l(1) + \alpha_{l+1}(1) + \cdots + \alpha_{l+j}(1)| \right. $$
$$\left. \geq \sqrt{2}\left[(2+\varepsilon)a_{k+1}\left(\log \frac{M_{k+1}}{a_{k+1}} + \log\log M_{k+1}\right)\right]^{1/2}\right\}.$$
Then of the events D_k only finitely many occur with probability 1.

Proof. Cf. Theorem 7.13.

A simple consequence of Lemmas 11.7, 11.8 and Theorem 11.25 is

LEMMA 11.9 Let $E_k =$
$$\left\{ \sup_{m_k \leq n \leq m_{k+1}} |\zeta(m_k, n)| \geq \left[2(2+\varepsilon)a_{k+1}\left(\log \frac{M_{k+1}}{a_{k+1}} + \log\log M_{k+1}\right)\right]^{1/2}\right\}.$$
Then of the events E_k only finitely many occur with probability 1.

LEMMA 11.10
$$\limsup_{n \to \infty} \frac{\xi(1,n) - \xi(0,n)}{n^{1/4}(\log\log n)^{3/4}} \leq \left(\frac{128}{27}\right)^{1/4} \quad a.s. \qquad (11.17)$$

Proof. Let

$$c_k = \left(\frac{128}{27}\right)^{1/4} m_k^{1/4}(\log\log m_k)^{3/4}, \quad F_k = \{\zeta(m_k) \geq (1+\varepsilon)c_k\}.$$

Then by Lemma 11.5 only finitely many of the events F_k occur with probability 1. Now observing that

$$\left[2(2+\varepsilon)a_{k+1}\left(\log\frac{M_{k+1}}{a_{k+1}} + \log\log M_{k+1}\right)\right]^{1/2} = o(c_k),$$

we have (11.17) by Lemma 11.9 and Lemma 11.10 is proved.

Also Lemmas 11.6 and 11.10 combined give Theorem 11.21.

The Theorems of the present Section (especially Theorem 11.20) suggest that $|\xi(i+1,n) - \xi(i,n)|$ is about $2(\xi(i,n))^{1/2}$ for any $i = 0, \pm 1, \ldots$ if n is large enough. Consequently somebody is interested in the quadratic variation

$$\Delta = \Delta(n; A, B) = \sum_{i=A}^{B} (\xi(i+1,n) - \xi(i,n))^2 \qquad (A < B)$$

of the local time might think that Δ is about $2\sum_{i=A}^{B}\xi(i,n)$. It is really so:

THEOREM 11.26 (Földes–Révész, 1993). *For any $\varepsilon > 0$ we have*

$$\lim_{N\to\infty} N^{-3/4}(\log N)^{-7/2-\varepsilon} \sup_{1\leq n \leq N} \sup_{B\geq 0} \left|\Delta(n;0,B) - 2\sum_{i=0}^{B}\xi(i,n)\right| = 0 \quad a.s.$$

Remark 4. In case $B < N^{1/4}$ this Theorem is only a triviality.

Remark 5. Since

$$\sum_{i=-\infty}^{+\infty} \xi(i,n) = n, \qquad (n = 1, 2, \ldots)$$

as a consequence of the above Theorem we obtain

$$\sum_{i=-\infty}^{+\infty} (\xi(i+1,n) - \xi(i,n))^2 = 2n + o(n^{3/4}(\log n)^{7/2+\varepsilon}) \quad a.s.$$

Chapter 12

Excursions

12.1 On the distribution of the zeros of a random walk

(9.11) and Theorem 11.1 are telling us in different forms that $\xi(0,n)$ converges to ∞ like $n^{1/2}$, i.e. the particle during its first n steps visits the origin practically $n^{1/2}$ times. Clearly these $n^{1/2}$ visits are distributed in $[0,n]$ in a very nonuniform way.

We have already met the Chung–Erdős theorem (Theorem 11.12) and the arcsine law (9.12) claiming that the zeros of $\{S_k\}$ are very nonuniformly distributed at least for some n. Now we give a few reformulations of the Chung–Erdős theorem in order to see how it describes the nonuniformness of the distribution of the zeros of $\{S_k\}$. First a few notations:

(i) let
$$R(n) = \max\{k: \ k > 1 \text{ for which there exists a } 0 < j < n - k \\ \text{such that } \xi(0, j+k) - \xi(0,j) = 0\}$$
be the length of the longest zero–free interval (longest excursion),

(ii) let
$$\hat{R}(n) = \max\{k: \ k > 1 \text{ for which there exists a } 0 < j < n - k \\ \text{such that } M^+(j+k) = M^+(j)\}$$
be the length of the longest flat interval of M_k^+ up to n,

(iii) let
$$\Psi(n) = \max\{k: \ 1 < k \leq n, \ S_k = 0\}$$
be the location of the last zero up to n,

(iv) let ζ_n be the number of those terms of $S_1, S_2, \ldots, S_n$ which are positive or which are equal to 0 but the preceding term of which is positive,

(v) let
$$\mu^+(n) = \inf\{k : 0 \leq k \leq n \text{ for which } S_k = M_n^+\}.$$

Now we can reformulate the Chung–Erdős theorem (Theorem 11.12) as follows:

THEOREM 12.1 *Let $f(x)$ be a nondecreasing function for which $\lim_{x \to \infty} f(x) = \infty$, $x/f(x)$ is nondecreasing and $\lim_{x \to \infty} x/f(x) = \infty$. Then*

$$n\left(1 - \frac{1}{f(n)}\right) \in \mathrm{UUC}(Y(n))$$

if and only if

$$\int_1^\infty \frac{dx}{x(f(x))^{1/2}} < \infty$$

where $Y(n)$ is any of the processes $R(n)$, $\hat{R}(n)$, $n - \Psi(n)$, $\zeta_n, n - \mu^+(n)$.

Proof. It is immediately clear that

$$\mathrm{UUC}(R(n)) = \mathrm{UUC}(n - \Psi(n))$$

and

$$\mathrm{UUC}(\hat{R}(n)) = \mathrm{UUC}(n - \mu^+(n)).$$

By Theorem 10.3 it is also clear that

$$\mathrm{UUC}(R(n)) = \mathrm{UUC}(\hat{R}(n)).$$

As far as the process ζ_n is concerned, the inequality

$$\mathrm{UUC}(\zeta_n) \subseteq \mathrm{UUC}(n - \Psi(n))$$

is trivial. However, the equality is not quite clear but following the original proof of Theorem 11.12 given by Chung and Erdős (1952) we get the required result.

The characterization of the lower classes of $n - \Psi(n)$ is trivial since we have

$$\Psi(n) = n \quad \text{i.o. a.s.}$$

The characterization of the lower classes of ζ_n is also trivial. In fact as a simple consequence of Theorem 12.1 we obtain

THEOREM 12.2 *Assume that $f(x)$ satisfies the conditions of Theorem 12.1. Then*
$$\frac{n}{f(n)} \in \text{LLC}(\zeta_n)$$
if and only if
$$\int_1^\infty \frac{dx}{x(f(x))^{1/2}} < \infty.$$

The characterization of the lower classes of $R(n)$ and $\hat{R}(n)$ is much harder. We have

THEOREM 12.3 (Csáki–Erdős–Révész, 1985). *Let $f(x)$ be a nondecreasing function for which*
$$f(x) \nearrow \infty, \quad \frac{x}{f(x)} \nearrow \infty \quad (x \to \infty).$$

Then
$$\beta \frac{n}{f(n)} \in \text{LLC}(Y^*(n))$$

if and only if
$$\sum_{n=1}^\infty \frac{f(n)}{n} e^{-f(n)} < \infty$$

where Y^* is any of the processes $R(n)$ and $\hat{R}(n)$ and $\beta = 0,85403\ldots$ is the root of the equation
$$\sum_{k=1}^\infty \frac{\beta^k}{k!(2k-1)} = 1.$$

Consequence 1.
$$\liminf_{n\to\infty} \frac{\log\log n}{n} R(n) = \liminf_{n\to\infty} \frac{\log\log n}{n} \hat{R}(n) = \beta \quad \text{a.s.}$$

Besides studying the length of longest excursion $R(n)$, it looks interesting to say something about the second, third, ... etc. longest excursions. Consider the sample $\rho_1, \rho_2 - \rho_1, \ldots, \rho_{\xi(0,n)} - \rho_{\xi(0,n)-1}, n - \rho_{\xi(0,n)}$ (the lengths of the excursions) and the corresponding ordered sample $R_1(n) = R(n) \geq R_2(n) \geq \ldots \geq R_{\xi(0,n)+1}(n)$. Now we present

THEOREM 12.4 *For any fixed $k = 1, 2, \ldots$ we have*
$$\liminf_{n\to\infty} \frac{\log\log n}{n} \sum_{j=1}^k R_j(n) = k\beta \quad \text{a.s.}$$

This theorem in some sense answers the question: How small can the r.v.'s $R_2(n), R_3(n), \ldots$ be? In order to obtain a more complete description of these r.v.'s we present the following:

Problem 1. Characterize the set of those nondecreasing functions $f(n)$ $(n = 1, 2, \ldots)$ for which

$$\mathbf{P}\left\{R_2(n) \geq \frac{n}{2}\left(1 - \frac{1}{f(n)}\right) \text{ i.o.}\right\} = 1.$$

Theorem 12.1 tells us that for some n nearly the whole random walk $\{S(k)\}_{k=0}^{n}$ is one excursion. Theorem 12.3 tells us that for some n the random walk consists of at least $\beta^{-1} \log \log n$ excursions. These results suggest the question: For what values of $k = k(n)$ will the sum $\sum_{j=1}^{k} R_j(n)$ be nearly equal to n? In fact we formulate two questions:

Question 1. For any $0 < \varepsilon < 1$ let $\mathcal{F}(\varepsilon)$ be the set of those functions $f(n)$ $(n = 1, 2, \ldots)$ for which

$$\sum_{j=1}^{f(n)} R_j(n) \geq n(1 - \varepsilon)$$

with probability 1 except finitely many n. How can we characterize $\mathcal{F}(\varepsilon)$?

Question 2. Let $\mathcal{F}(o)$ be the set of those functions $f(n)$ $(n = 1, 2, \ldots)$ for which

$$\lim_{n \to \infty} n^{-1} \sum_{j=1}^{f(n)} R_j(n) = 1 \quad \text{a.s.}$$

How can we characterize $\mathcal{F}(o)$?

Studying the first question we have

THEOREM 12.5 (Csáki–Erdős–Révész, 1985). *For any $0 < \varepsilon < 1$ there exists a $C = C(\varepsilon) > 0$ such that*

$$C \log \log n \in \mathcal{F}(\varepsilon).$$

Concerning Question 2, we have the following result:

THEOREM 12.6 *For any $C > 0$*

$$C \log \log n \notin \mathcal{F}(o)$$

and for any $h(n) \nearrow \infty$ $(n \to \infty)$

$$h(n) \log \log n \in \mathcal{F}(o).$$

Knight (1986) was interested in the distribution of the duration of the longest excursion of a Wiener process. In order to formulate his results introduce the following notations: for arbitrary $t > 0$ we set

$$t_0(t) = \sup\{s: \ s < t, \ W(s) = 0\},$$
$$t_1(t) = \inf\{s: \ s > t, \ W(s) = 0\},$$
$$d(t) = t_1(t) - t_0(t),$$
$$D(t) = \sup\{d(s): \ t_0(s) < t\},$$
$$E(t) = \sup\{d(s): \ s < t, \ t_1(s) < t\}.$$

Then we call $d(t)$ the duration of the excursion containing t. $D(t)$ (resp. $E(t)$) is the maximal duration of excursions starting by t (resp. ending by t).

Knight evaluated completely the Laplace transforms of the distributions of $D(t)$ and $E(t)$ and the distributions themselves over a finite interval. His results run as follows:

THEOREM 12.7 (Knight, 1986).

$$\mathbf{P}\left\{\frac{D(t)}{t} < x\right\} = 1 - F\left(\frac{1}{x}\right)$$

where

$$F(y) = \begin{cases} 2\pi^{-1}y^{1/2} & \text{if } y \leq 1, \\ \pi^{-1}\left(3 - y + \frac{4}{\sqrt{3}}\log y\right) & \text{if } 1 \leq y \leq 2 \end{cases}$$

and

$$\mathbf{P}\left\{\frac{E(t)}{t} < x\right\} = 1 - G\left(\frac{1}{x}\right)$$

where $G(1) = 0$,

$$G'(y) = \begin{cases} (y\pi)^{-1}(y-1)^{1/2} & \text{if } 1 \leq y < 2, \\ (y\pi)^{-1}g^*(y) & \text{if } 2 < y < 3, \end{cases}$$

$$g^*(y) = (y-1)^{-1/2} + \frac{4}{3}\left((y-1)^{1/2} - y + 1\right) + \frac{2}{3^{1/2}y}\log\left(\frac{y^{1/2}-1}{y^{1/2}+1}\right)$$

and

$$G(2) = \frac{2}{\pi} - \frac{1}{2}.$$

The multiple Laplace transform of $D(t)$ and some other characteristics of a Wiener process were investigated by Csáki–Földes (1988/A). A very different characterization of the distribution of the zeros of $\{S_n\}$ is due to Erdős and Taylor (1960/A), who proved

THEOREM 12.8

$$\lim_{n\to\infty} \frac{1}{\log n} \sum_{k=1}^{n} \rho_k^{-1/2} = \pi^{-1/2} \quad a.s.$$

Remark 1. (9.8) and Theorem 11.6 claim that ρ_k converges to infinity like k^2. However, these two results are also claiming that the fluctuation of $k^{-2}\rho_k$ can be and will be very large. Theorem 12.8, via investigating the logarithmic density of $\rho_k^{1/2}$, also tells us that ρ_k behaves like k^2.

Let us mention a result of Lévy (1948) that is very similar to the above theorem.

THEOREM 12.9

$$\lim_{n\to\infty} \frac{1}{\log n} \sum_{k=1}^{n} \frac{I(S_k)}{k} = \frac{1}{2} \quad a.s.$$

where

$$I(S_k) = \begin{cases} 1 & \text{if } S_k \geq 0, \\ 0 & \text{if } S_k < 0. \end{cases}$$

Remark 2. Theorems 12.1 and 12.2 imply that

$$\liminf_{n\to\infty} \frac{1}{n} \sum_{k=1}^{n} I(S_k) = 0 \quad a.s.$$

and

$$\limsup_{n\to\infty} \frac{1}{n} \sum_{k=1}^{n} I(S_k) = 1 \quad a.s.$$

Hence the sequence $I(S_k)$ does not have a density in the ordinary sense but by Theorem 12.9 its logarithmic density is $1/2$.

It is natural to ask what happens if in Theorem 12.9 the indicator function $I(\cdot)$ of $(-\infty, 0)$ is replaced by the indicator function of an arbitrary Borel set of $\mathbb{R}^1$. We obtain

THEOREM 12.10 (Brosamler, 1988; Fisher, 1987 and Schatte, 1989). *There is a **P**-null set $N \subset \Omega$ such that for all $\omega \notin N$ and for all Borel set $A \subset \mathbb{R}^1$ with $\lambda(\partial A) = 0$ we have*

$$\lim_{n\to\infty} \frac{1}{\log n} \sum_{k=1}^{n} k^{-1} I_A(k^{-1/2} S_k) = \frac{1}{(2\pi)^{1/2}} \int_A e^{-u^2/2} du$$

where ∂A is the boundary of A and

$$I_A(x) = \begin{cases} 1 & \text{if } x \in A, \\ 0 & \text{if } x \notin A. \end{cases}$$

For a Strassen type generalization of Theorem 12.10, cf. Brosamler (1988) and Lacey–Philipp (1990).

For the sake of completeness we also mention

THEOREM 12.11 (Weigl, 1989).

$$\lim_{n\to\infty} \mathbf{P}\left\{\frac{1}{c(\log n)^{1/2}}\left(\sum_{k=1}^n k^{-1}I(S_k) - \frac{1}{2}\log n\right) < x\right\} = \Phi(x)$$

where

$$c = \left(\frac{1}{4\pi}\int_0^\infty A(y)dy\right)^{1/2},$$
$$A(y) = y^{-3/2}(\log(1+2y))^2 \quad (0 < y < \infty)$$

and $I(\cdot)$ is defined in Theorem 12.9.

12.2 Local time and the number of long excursions (Mesure du voisinage)

The definition of the local time of a Wiener process (cf. Section 9.3) is extrinsic in the sense that one cannot recover the local time $\eta(0,T)$ from the random set $A_T = \{t : 0 \leq t \leq T, W(t) = 0\}$. Lévy called attention to the necessity of an intrinsic definition.

He proposed the following: Let $N(h,x,t)$ be the number of excursions of $W(\cdot)$ away from x that are greater than h in length and are completed by time t. Then the "mesure du voisinage" of W at time t is $\lim_{h\searrow 0} h^{1/2}N(h,x,t)$, and the connection between η and N is given by the following result of P. Lévy (cf. Itô and McKean 1965, p. 43).

THEOREM 12.12 *For all real x and for all positive t we have*

$$\lim_{h\searrow 0} h^{1/2}N(h,x,t) = \sqrt{\frac{2}{\pi}}\eta(x,t) \quad a.s.$$

Perkins (1981) proved that Theorem 12.12 holds uniformly in x and t. Csörgő and Révész (1986) proved a stronger version of Perkins' result. Their results can be summarized in the following four theorems.

THEOREM 12.13 *For any fixed $t' > 0$ we have*

$$\lim_{h \searrow 0} h^{-1/4}(\log h^{-1})^{-1} \sup_{(x,t) \in \mathbb{R}^1 \times [0,t']} \left| h^{1/2} N(h,x,t) - \sqrt{\frac{2}{\pi}} \eta(x,t) \right| = 0 \quad a.s.$$

The connection between N and η is also investigated in the case when a Wiener process through a long time t is observed and the number of long (but much shorter than t) excursions is considered. We have

THEOREM 12.14 *For some $0 < \alpha < 1$ let $0 < a_t < t^\alpha$ ($t > 0$) be a nondecreasing function of t so that a_t/t is nonincreasing. Then*

$$\lim_{t \to \infty} \frac{(a_t/t)^{1/4}}{\log(t/a_t)} \sup_{x \in \mathbb{R}^1} \left| N(a_t, x, t) - \sqrt{\frac{2}{\pi a_t}} \eta(x,t) \right| = 0 \quad a.s.$$

The proofs of Theorems 12.13 and 12.14 are based on two large deviation type inequalities which are of interest on their own.

THEOREM 12.15 *For any $K > 0$ and $t' > 0$ there exist a $C = C(K, t') > 0$ and a $D = D(K, t') > 0$ such that*

$$\mathbf{P}\left\{ \frac{(\log h^{-1})^{-3/4}}{h^{1/4}} \sup_{(x,t) \in \mathbb{R}^1 \times [h,t']} \left| h^{1/2} N(h,x,t) - \sqrt{\frac{2}{\pi}} \eta(x,t) \right| \geq C \right\} \leq Dh^K,$$

where $h < t'$.

THEOREM 12.16 *For any $K > 0$ there exist a $C = C(K) > 0$ and a $D = D(K) > 0$ such that*

$$\mathbf{P}\left\{ \frac{(a_t/t)^{1/4}}{(\log(t/a_t))^{3/4}} \sup_{x \in \mathbb{R}^1} \left| N(a_t, x, t) - \sqrt{\frac{2}{\pi a_t}} \eta(x,t) \right| \geq C \right\} \leq D \left(\frac{a_t}{t} \right)^K,$$

where $0 < a_t < t$.

It is natural to ask about the analogues of the above theorems for random walk.

Clearly for any $x = 0, \pm 1, \pm 2, \ldots$ the number of excursions away from x completed by n is equal to the local time $\xi(x, n)$, i.e.

$$M(x, n) = \{\text{the number of excursions away from } x \text{ completed by } n\}$$
$$= \max\{i : \rho_i(x) \leq n\} = \xi(x, n).$$

Hence we consider the following problem: knowing the number of long excursions (longer than $a = a_n$) away from x completed by n, what can be

EXCURSIONS

said about $\xi(x,n)$? Let $M(a,x,n)$ be the number of excursions away from x longer than a and completed by n. Our main result says that observing the sequence $\{M(a_n,x,n)\}_{n=1}^{\infty}$ with some $a_n = [n^{\alpha}]$ ($0 < \alpha < 1/3$) the local time sequence $\{\xi(x,n)\}_{n=1}^{\infty}$ can be relatively well estimated. In fact we have

THEOREM 12.17 Let $a_n = [n^{\alpha}]$ with $0 < \alpha < 1/3$. Then

$$\lim_{n\to\infty} \frac{(a_n/n)^{1/4}}{\log(n/a_n)} \sup_{x \in \mathbb{Z}^1} |M(a_n,x,n) - \xi(x,n)P(a_n)| = 0 \quad a.s.$$

where

$$P(a) = \mathbf{P}\{\rho_1 > a\}.$$

The proof of this theorem is based on

THEOREM 12.18 For any $K > 0$ there exist a $C = C(K) > 0$ and a $D = D(K) > 0$ such that

$$\mathbf{P}\left\{\frac{(a_n/n)^{1/4}}{(\log(n/a_n))^{3/4}} \sup_{x \in \mathbb{Z}^1} |M(a_n,x,n) - \xi(x,n)P(a_n)| \geq C\right\} \leq Dn^{-K},$$

where $a_n = [n^{\alpha}]$ ($0 < \alpha < 1/3$).

Remark 1. Very likely Theorems 12.17 and 12.18 remain true assuming only that $0 < \alpha < 1$.

In order to prove Theorem 12.18, first we prove the following

LEMMA 12.1 Let n and a be positive integers and $C > 0$. Then

$$\mathbf{P}\left\{\left|\frac{M(a,x,\rho_n(x)) - nP(a)}{(nP(a)(1-P(a))\log nP(a))^{1/2}}\right| \geq C^{1/2}\right\} \leq 2(P(a)n)^{-2C/9},$$

provided that

$$C \log(nP(a)) < nP(a)(1-P(a)). \tag{12.1}$$

Proof. Clearly $M(a,x,\rho_n(x))$ is binomially distributed with parameters n and $P(a)$. Hence the Bernstein inequality (Theorem 2.3) easily implies Lemma 12.1.

Proof of Theorem 12.18. Since by (9.10)

$$P(a) = \left(\frac{2}{\pi a}\right)^{1/2} (1 + O(a^{-1})),$$

condition (12.1) holds true if $a \leq n^\rho (\rho < 2)$ and n is large enough, Lemma 12.1 can be reformulated as follows: for any $K > 0$ and $0 < \psi < \rho < 2$ there exist a $C = C(\psi, \rho, K) > 0$ and $D = D(\psi, \rho, K) > 0$ such that

$$\mathbf{P}\left\{\left|\frac{M(a,x,\rho_n(x)) - \xi(x,\rho_n(x))P(a)}{n^{1/2}\left(\frac{2}{\pi a}\right)^{1/4}(\log n/\sqrt{a})^{1/2}}\right| \geq C\right\} \leq Dn^{-K}, \qquad (12.2)$$

provided that $n^\psi < a < n^\rho$.

(12.2) in turn implies

$$\mathbf{P}\left\{\sup_{n^\psi < a < n^\rho}\left|\frac{M(a,x,\rho_n(x)) - \xi(x,\rho_n(x))P(a)}{n^{1/2}\left(\frac{2}{\pi a}\right)^{1/4}(\log n/\sqrt{a})^{1/2}}\right| > C\right\} \leq Dn^{-K}, \qquad (12.3)$$

and for any $K > 0$, $0 < \psi < \rho < 2$ and $0 < \gamma < \delta < \infty$ there exist a $C = C(\gamma, \delta, \psi, \rho, K)$ and a $D = D(\gamma, \delta, \psi, \rho, K)$ such that

$$\mathbf{P}\left\{\sup_{N^\gamma < n < N^\delta}\sup_{n^\psi < a < n^\rho}\left|\frac{M(a,x,\rho_n(x)) - \xi(x,\rho_n(x))P(a)}{n^{1/2}\left(\frac{2}{\pi a}\right)^{1/4}(\log n/\sqrt{a})^{1/2}}\right| > C\right\} \leq DN^{-K}. \qquad (12.4)$$

Then by a slight generalization of (9.11) (or applying the exact distribution of $\xi(0,n)$, cf. Theorem 9.3) for any $K > 0$ there exist a $C = C(K) > 0$ and a $D = D(K) > 0$ such that

$$\mathbf{P}\left\{\rho_n(x) < \frac{n^2}{C \log n}\right\} \leq Dn^{-K} \qquad (12.5)$$

for any $x \in \mathbb{Z}^1$ or equivalently

$$\mathbf{P}\left\{\xi(x,n) \geq C(n \log n)^{1/2}\right\} \leq Dn^{-K}. \qquad (12.6)$$

Let m be a fixed positive integer and assume that the event

$$A_m = \left\{m^\beta \leq \xi(x,m) \leq C(m \log m)^{1/2}\right\} \quad (0 < \beta < 1/2)$$

holds true. Then replacing m by $\rho_n(x)$ (more exactly assuming that $\xi(x,m) = n$, i.e. $\rho_n(x) \leq m < \rho_{n+1}(x)$) we obtain

$$J = J(m,x)$$
$$= \left|\frac{M(a_m,x,m) - \xi(x,m)P(a_m)}{m^{1/4}a_m^{-1/4}(\log m a_m^{-1})^{3/4}}\right| = \left|\frac{M(a_m,x,\rho_n(x)) - nP(a_m)}{m^{1/4}a_m^{-1/4}(\log m a_m^{-1})^{3/4}}\right|$$

where by the assumption that A_m holds true we have

$$m^\beta \leq \xi(x,m) = n \leq C(m\log m)^{1/2},$$

i.e.
$$\frac{n^2}{2C^2\log n} \leq m \leq n^{1/\beta}.$$

Hence

$$J \leq \sup_{m^\beta \leq n \leq C(m\log m)^{1/2}} \left|\frac{M(a_m,x,\rho_n(x)) - nP(a_m)}{\left(\frac{n^2}{2C^2\log n}\right)^{1/4} a_m^{-1/4}\left(\log \frac{n^2}{a_m 2C^2 \log n}\right)^{3/4}}\right|$$

$$\leq 4 \sup_{m^\beta \leq n \leq C(m\log m)^{1/2}} \sup_{\frac{n^{2\alpha}}{(2C^2\log n)^\alpha} \leq a \leq n^{\alpha/\beta}} \left|\frac{M(a,x,\rho_n(x)) - nP(a)}{n^{1/2}a^{-1/4}\left(\log \frac{n}{a}\right)^{1/2}}\right|. \quad (12.7)$$

Observe that if $\xi(x,m) < m^\beta$ then

$$J \leq \frac{m^\beta}{m^{1/4}a_m^{-1/4}(\log m a_m^{-1})^{3/4}} \to 0 \quad \text{if} \quad \beta < \frac{1-\alpha}{4}.$$

Consequently
$$\mathbf{P}\left\{J > C, \xi(x,m) < m^\beta\right\} = 0 \quad (12.8)$$

if m is large enough and $\beta < (1-\alpha)/4$. Hence by (12.8), (12.7) and (12.4) we obtain

$$\mathbf{P}\{J > C\} = \mathbf{P}\{J > C, A_m\} + \mathbf{P}\{J > C, \xi(x,m) < m^\beta\}$$
$$+ \mathbf{P}\{J > C, \xi(x,m) > C(m\log m)^{1/2}\}$$
$$\leq \mathbf{P}\{J > C, A_m\} + \mathbf{P}\{\xi(x,m) > C(m\log m)^{1/2}\}$$
$$\leq \mathbf{P}\{J > C, A_m\} + Dm^{-K} \leq 2Dm^{-K}$$

if m is large enough, $\beta < (1-\alpha)/4$ and $\alpha/\beta < 2$. β can be chosen in such a way if $0 < \alpha < 1/3$. Consequently we also have that

$$\mathbf{P}\{\sup_x J(m,x) > C\} \leq Dm^{-K}$$

for any $K > 0$ if C, D are large enough and $0 < \alpha < 1/3$. Hence the proof of Theorem 12.18 is complete.

Theorem 12.17 is a trivial consequence of Theorem 12.18.

Note that if $\alpha > 1/5$ then $P(a)$ can be replaced by $(2/\pi a)^{1/2}$. Hence we also obtain

THEOREM 12.19 *Let* $a_n = [n^\alpha]$ *with* $1/5 < \alpha < 1/3$. *Then*

$$\lim_{n \to \infty} \frac{(a_n/n)^{1/4}}{\log(n/a_n)} \sup_{x \in \mathbb{Z}^1} \left| M(a_n, x, n) - \xi(x, n) \left(\frac{2}{\pi a_n} \right)^{1/2} \right| = 0 \quad a.s.$$

THEOREM 12.20 *For any* $K > 0$ *there exist a* $C = C(K) > 0$ *and a* $D = D(K) > 0$ *such that*

$$\mathbf{P} \left\{ \frac{(a_n/n)^{1/4}}{(\log(n/a_n))^{3/4}} \sup_{x \in \mathbb{Z}^1} \left| M(a_n, x, n) - \xi(x, n) \left(\frac{2}{\pi a_n} \right)^{1/2} \right| \geq C \right\} \leq \frac{D}{n^K}$$

where $a_n = [n^\alpha]$ $(1/5 < \alpha < 1/3)$.

12.3 Local time and the number of high excursions

The previous section gave a method to evaluate the local time of a Wiener process resp. random walk in $[0, t]$ having the number of long excursions in this interval. A natural question is: can the local time be evaluated having the number of high excursions. A positive answer of this question was given by Khoshnevisan (1994). He succeeded to give a very exact rate of convergence in this problem.

Let $u_\varepsilon(x, t)$ be the total number of times before time t that $W(\cdot)$ upcrossed the interval $[x, x + \varepsilon]$. Equivalently $u_\varepsilon(x, t)$ is the number of excursions away from x up to t which are higher than ε.

THEOREM 12.21 (Khoshnevisan, 1994) *For every* $t > 0$ *and* $x \in \mathbb{R}^1$ *we have*

$$\limsup_{\varepsilon \to 0} \frac{|2\varepsilon u_\varepsilon(x, t) - \eta(x, t)|}{(\varepsilon \log \log \varepsilon^{-1})^{1/2}} = 2(\eta(x, t))^{1/2}$$

and

$$\limsup_{\varepsilon \to 0} \sup_{x \in \mathbb{R}^1} \frac{|2\varepsilon u_\varepsilon(x, t) - \eta(x, t)|}{(\varepsilon \log \varepsilon^{-1})^{1/2}} = 2(\eta(t))^{1/2}.$$

The analogue question for random walk can be answered similarly applying the method of the previous section and Lemma 3.1 instead of (9.10). Here we present only a weak form of the answer. Let $\mathcal{M}(a, x, n)$ be the number of excursions away from x higher than a and completed by n.

EXCURSIONS

THEOREM 12.22 *For any $x = 0, \pm 1, \pm 2, \ldots$*

$$\lim_{n \to \infty} \frac{1}{\log n} \left(\frac{a_n}{n}\right)^{1/2} \left| \mathcal{M}(a_n, x, n) - \frac{1}{2a_n} \xi(x, n) \right| = 0 \quad a.s.$$

provided that $a_n = n^\alpha$ $(0 < \alpha < 1/2)$.

12.4 The local time of high excursions

Theorem 9.7 described the distribution of the local time $\xi(k, \rho_1)$ of the excursion $\{S_0, S_1, \ldots, S_{\rho_1}\}$. Now we are interested in the properties of $\xi(k, \rho_1)$ when k is big, i.e. when k is close to $M^+(\rho_1)$, the height of the excursion $\{S_0, S_1, \ldots, S_{\rho_1}\}$. We are especially interested in the limit distribution of $\xi(k, \rho_1)$ when k is close to $M^+(\rho_1) = n$ and $n \to \infty$. First we present the following

THEOREM 12.23 *For any $n = 1, 2, \ldots$ and $l = 1, 2, \ldots$ we have*

$$\mathbf{P}\{M^+(\rho_1) = n, \xi(n, \rho_1) = l\} = n^{-2} 2^{-l-1} \left(1 - \frac{1}{n}\right)^{l-1}$$

and if $n \to \infty$ then

$$\mathbf{P}\{\xi(n, \rho_1) = l \mid M^+(\rho_1) = n\} = \frac{n+1}{2n} \left(\frac{n-1}{2n}\right)^{l-1} \to 2^{-l},$$

$$\mathbf{E}(\xi(n, \rho_1) \mid M^+(\rho_1) = n) = \frac{2n}{n+1} \to 2,$$

$$\mathbf{E}\left(\left(\xi(n, \rho_1) - \frac{2n}{n+1}\right)^2 \mid M^+(\rho_1) = n\right) = \frac{2(n-1)n}{(n+1)^2} \to 2.$$

Proof. By Lemma 3.1 the probability that the excursion $\{S_0, S_1, \ldots, S_{\rho_1}\}$ hits n is $(2n)^{-1}$, i.e. $\mathbf{P}\{M^+(\rho_1) \geq n\} = (2n)^{-1}$. The probability that after the arrival time $\rho_1(n)$ the particle turns back but hits n once more before arriving at 0 is $1/2(1 - 1/n)$. Hence the probability of having $l - 1$ negative excursions away from n before ρ_1 is $(1/2(1 - 1/n))^{l-1}$. Finally $(2n)^{-1}$ is the probability that after $l - 1$ excursions the particle returns to 0.

In order to study the properties of $\xi(M^+(\rho_1) - j, \rho_1)$, first we investigate the distribution of $\xi(M^+(\rho_1) - j, \rho_1(M^+(\rho_1)))$. (Note that $\rho_1(M^+(\rho_1))$ is the first hitting of the level $M^+(\rho_1)$.)

LEMMA 12.2 *For any $l = 1, 2, \ldots$, $n = 1, 2, \ldots$, $j = 1, 2, \ldots, n-1$*

$$\mathbf{P}\{M^+(\rho_1) \geq n, \xi(n-j, \rho_1(n)) = l\}$$

$$= \frac{1}{2}\frac{1}{n-j}\sum_{u=0}^{l-1}\binom{l-1}{u}\left(\frac{1}{2}\left(1-\frac{1}{n-j}\right)\right)^u \left(\frac{1}{2}\left(1-\frac{1}{j}\right)\right)^{l-1-u}\frac{1}{2j}$$

$$= \frac{1}{4j(n-j)}\left(1 - \frac{n}{2j(n-j)}\right)^{l-1}$$

and if $n \to \infty$ then

$$\mathbf{P}\{\xi(n-j, \rho_1(n)) = l \mid M^+(\rho_1) \geq n\}$$

$$= \frac{n}{2j(n-j)}\left(1 - \frac{n}{2j(n-j)}\right)^{l-1} \to \frac{1}{2j}\left(1 - \frac{1}{2j}\right)^{l-1},$$

$$\mathbf{E}(\xi(n-j, \rho_1(n)) \mid M^+(\rho_1) \geq n) = \frac{2j(n-j)}{n} \to 2j,$$

$$\mathbf{E}\left(\left(\xi(n-j, \rho_1(n)) - \frac{2j(n-j)}{n}\right)^2 \mid M^+(\rho_1) \geq n\right)$$

$$= \frac{2j(n-j)}{n}\frac{2j(n-j)-n}{n} \to 4j^2 - 2j.$$

Proof. By Lemma 3.1

$$\frac{1}{2}\frac{1}{n-j} = \mathbf{P}\{M^+(\rho_1) \geq n-j\}.$$

Further,

$$\left(\frac{1}{2}\left(1-\frac{1}{n-j}\right)\right)^u \left(\frac{1}{2}\left(1-\frac{1}{j}\right)\right)^{l-1-u}$$

is the probability that after $\rho_1(n-j)$ the particle makes u negative excursions away from $n-j$ (none of them reaches 0) and $l-1-u$ positive excursions away from $n-j$ (none of them reaches n) in a given order. Finally $(2j)^{-1}$ is the probability that after the $l-1$ excursions the particle goes to n.

LEMMA 12.3 *For any $n = 2, 3, \ldots$, $l = 2, 3, \ldots$, $j = 1, 2, \ldots, n-1$*

$$\mathbf{P}\{\xi(n-j, \rho_1) = l \mid M^+(\rho_1) = n, \xi(n, \rho_1) = 1\}$$

$$= \mathbf{P}\{U_1 + U_2 = l\} = \left(\frac{n}{2j(n-j)}\right)^2 (l-1)\left(1 - \frac{n}{2j(n-j)}\right)^{l-2}$$

$$\to \frac{l-1}{4j^2}\left(1 - \frac{1}{2j}\right)^{l-2} \quad (n \to \infty)$$

EXCURSIONS   149

where U_1 and U_2 are i.i.d.r.v.'s with

$$\mathbf{P}\{U_i = m\} = \frac{n}{2j(n-j)}\left(1 - \frac{n}{2j(n-j)}\right)^{m-1} \quad (i = 1, 2; \; m = 1, 2, \ldots). \tag{12.9}$$

Further,

$$\mathbf{E}(\xi(n-j, \rho_1) \mid M^+(\rho_1) = n, \xi(n, \rho_1) = 1\} = \frac{4j(n-j)}{n} \to 4j \quad (n \to \infty)$$

and

$$\mathbf{E}\left(\left(\xi(n-j, \rho_1) - \frac{4j(n-j)}{n}\right)^2 \mid M^+(\rho_1) = n, \xi(n, \rho_1) = 1\right)$$

$$= \frac{4j(n-j)}{n} \frac{2j(n-j) - n}{n} \to 8j^2 - 4j \quad (n \to \infty).$$

Proof. Since

$$\xi(n-j, \rho_1) = \xi(n-j, \rho_1(n)) + (\xi(n-j, \rho_1) - \xi(n-j, \rho_1(n))) \tag{12.10}$$

and by Lemma 12.2 the conditional distribution of $\xi(n-j, \rho_1(n))$ and that of $\xi(n-j, \rho_1) - \xi(n-j, \rho_1(n))$ (given $\{M^+(\rho_1) = n, \xi(n, \rho_1) = 1\}$) are equal to the distribution of U_1, we obtain Lemma 12.3 realizing that the two terms of the right-hand side of (12.10) are conditionally independent.

LEMMA 12.4 For any $n = 2, 3, \ldots$, $j = 1, 2, \ldots, n-1$ and $l = 0, 1, 2, \ldots$

$$\mathbf{P}\{\xi(j, \rho_1) = l, M^+(\rho_1) \leq n \mid X_1 = 1\}$$

$$= \begin{cases} 1 - \dfrac{1}{j} & \text{if } l = 0, \\ \dfrac{1}{2j^2}\left(1 - \dfrac{n+1}{2j(n+1-j)}\right)^{l-1} & \text{if } l \geq 1, \end{cases}$$

$$\mathbf{P}\{\xi(j, \rho_1) = l \mid M^+(\rho_1) \leq n, X_1 = 1\}$$

$$= \begin{cases} \left(1 - \dfrac{1}{j}\right)\dfrac{n+1}{n} & \text{if } l = 0, \\ \dfrac{n+1}{2j^2 n}\left(1 - \dfrac{n+1}{2j(n+1-j)}\right)^{l-1} & \text{if } l \geq 1, \end{cases}$$

and if $n \to \infty$ then

$$\mathbf{E}(\xi(j,\rho_1) \mid M^+(\rho_1) \leq n, X_1 = 1) = \frac{2}{n(n+1)}(n+1-j)^2 \to 2,$$

$$\mathbf{E}\left(\left(\xi(j,\rho_1) - \frac{2}{n(n+1)}(n+1-j)^2\right)^2 \mid M^+(\rho_1) \leq n, X_1 = 1\right)$$
$$= \frac{4}{n(n+1)^2}(n+1-j)^3\left(2j - \frac{n+1}{2(n+1-j)} - \frac{n+1-j}{n}\right) \to 8j - 6.$$

Proof is essentially the same as that of Lemma 12.2.

LEMMA 12.5 *For any $n = 2, 3, \ldots$, $j = 1, 2, \ldots, n-1$, $k = 1, 2, \ldots$ and $l = 2, 3, \ldots$*

$$\mathbf{P}\{\xi(n-j,\rho_1) = l \mid M^+(\rho_1) = n, \xi(n,\rho_1) = k\}$$
$$= \mathbf{P}\{U_1 + V_1 + V_2 + \cdots + V_{k-1} + U_2 = l\}$$

where $U_1, V_1, V_2, \ldots, V_{k-1}, U_2$ are independent r.v.'s with

$$\mathbf{P}\{U_i = m\} = \frac{n}{2j(n-j)}\left(1 - \frac{n}{2j(n-j)}\right)^{m-1} \quad (i = 1, 2; \ m = 1, 2, \ldots)$$

$$\mathbf{P}\{V_i = m\} = \begin{cases} \left(1 - \dfrac{1}{j}\right)\dfrac{n}{n-1} & \text{if} \quad m = 0, \\ \dfrac{n}{2j^2(n-1)}\left(1 - \dfrac{n}{2j(n-j)}\right)^{m-1} & \text{if} \quad m \geq 1, \end{cases}$$

$$\mathbf{E}(\xi(n-j,\rho_1) \mid M^+(\rho_1) = n, \xi(n,\rho_1) = k)$$
$$= \frac{4j(n-1-j)}{n-1} + (k-1)\frac{2}{n(n-1)}(n-j)^2 \to 4j + 2(k-1),$$

$$\mathbf{E}((\xi(n-j,\rho_1) - \mathbf{E}\xi(n-j,\rho_1))^2 \mid M^+(\rho_1) = n, \xi(n,\rho_1) = k)$$
$$= \frac{4j(n-1-j)}{n-1}\frac{2j(n-1-j)-n+1}{n-1}$$
$$+ \frac{4(k-1)}{n(n-1)^2}(n-j)^3\left(2j - \frac{n}{2(n-j)} - \frac{n-j}{n-1}\right)$$
$$\to 8j^2 - 4j + (k-1)(8j-6) = 8j^2 + (2k-3)4j - 6(k-1).$$

Proof. Clearly

$$\xi(n-j,\rho_1) = \xi(n-j,\rho_1(n)) + \sum_{i=1}^{k-1}(\xi(n-j,\rho_{i+1}(n)) - \xi(n-j,\rho_i(n)))$$
$$+ (\xi(n-j,\rho_1) - \xi(n-j,\rho_k(n)))$$

where the terms

$$\xi(n-j,\rho_1(n)), \quad \xi(n-j,\rho_{i+1}(n)) - \xi(n-j,\rho_i(n)) \quad (i=1,2,\ldots,k-1)$$

and

$$\xi(n-j,\rho_1) - \xi(n-j,\rho_k(n))$$

are independent. Lemma 12.3 tells us that the conditional distribution of $\xi(n-j,\rho_1(n))$ and $\xi(n-j,\rho_1) - \xi(n-j,\rho_k(n))$ is equal to the distribution of U_1 and U_2. Lemma 12.4 is telling us that the conditional distribution of $\xi(n-j,\rho_{i+1}(n)) - \xi(n-j,\rho_i(n))$ ($i=1,2,\ldots,k-1$) is equal to the distribution of $V_1, V_2, \ldots, V_{k-1}$. Hence we have Lemma 12.5.

Theorem 12.21 and Lemmas 12.3, 12.4 and 12.5 combined imply

THEOREM 12.24 *For any $j = 1,2,\ldots,n-1$; $n = 2,3,\ldots$*

$$\mathbf{E}(\xi(n-j,\rho_1) \mid M^+(\rho_1) = n)$$
$$= \frac{4j(n-j)}{n} + \left(\frac{2n}{n+1} - 1\right)\frac{2}{n(n+1)}(n+1-j)^2 \to 4j+2,$$

$$\mathbf{E}\left((\xi(n-j,\rho_1) - \mathbf{E}\xi(n-j,\rho_1))^2 \mid M^+(\rho_1) = n\right) \to 8j^2 + 4j - 6.$$

Further, for any $j = 0,1,2,\ldots$ and $K > 0$ there exist a $C_1 = C_1(K,j) > 0$ and a $C_2 = C_2(K,j) > 0$ such that

$$\mathbf{P}\{\xi(n-j,\rho_1) > C_1 \log n \mid M^+(\rho_1) = n\} \leq C_2 n^{-K}$$

and for any $\alpha > 0$ and $K > 0$ there exist a $C_1 = C_1(\alpha,K) > 0$ and a $C_2 = C_2(\alpha,K) > 0$ such that

$$\mathbf{P}\{\xi(n - \alpha \log n, \rho_1) > C_1 \log^2 n \mid M^+(\rho_1) = n\} \leq C_2 n^{-K}.$$

We also obtain the following:

Consequence 1. For any $j = 0,1,2,\ldots,n$ and n large enough we have

$$\mathbf{P}\{\xi(n-j,\rho_1) \geq 6j^2 + 4j + 2 \mid M^+(\rho_1) = n\} \leq \frac{1}{4j^2}.$$

Proof. By Chebyshev inequality and Theorem 12.22 we have

$$\mathbf{P}\{\xi(n-j,\rho_1) \geq \lambda(8j^2+4j-6)^{1/2}+4j+2 \mid M^+(\rho_1)=n\} \leq \frac{1}{\lambda^2}.$$

Taking $\lambda = 2j$ and observing that

$$2j(8j^2+4j-6)^{1/2}+4j+2 \leq 6j^2+4j+2$$

we obtain the above inequality.

12.5 How many times can a random walk reach its maximum?

Let $\chi(n)$ be the number of those places where the maximum of the random walk $S_0, S_1, \ldots, S_n$ is reached, i.e. $\chi(n)$ is the largest positive integer for which there exists a sequence of integers $0 \leq k_1 < k_2 < \ldots < k_{\chi(n)} \leq n$ such that

$$S(k_1) = S(k_2) = \cdots = S(k_{\chi(n)}) = M^+(n). \qquad (12.11)$$

Csáki (personal communication) evaluated the exact distribution of $\chi(n)$. In fact he obtained

THEOREM 12.25 *For any $k = 0, 1, 2, \ldots, [n/2]$; $n = 1, 2, \ldots$ we have*

$$\mathbf{P}\{\chi(n) \geq k+1\} = 2^{-k}\mathbf{P}\{M^+_{n-k} \geq k\}.$$

Proof. Consider the sequence

$$\{X(k_1+2), X(k_1+3), \ldots, X(k_2)\},$$
$$\{X(k_2+2), X(k_2+3), \ldots, X(k_3)\}, \ldots,$$
$$\{X(k_{\chi(n)-1}+2), X(k_{\chi(n)-1}+3), \ldots, X(k_{\chi(n)})\},$$
$$\{X(1), X(2), \ldots, X(k_1)\},$$
$$\{X(k_{\chi(n)}+1), X(k_{\chi(n)}+2), \ldots, X_n\}$$

where $X(l) = X_l = S(l+1) - S(l)$ and $k_1, k_2, \ldots, k_{\chi(n)}$ are defined by (12.11). Let S^*_j ($j = 0, 1, 2, \ldots, n - \chi(n) + 1$) be the sum of the first j of the above given random variables in the given order. Then $\{S^*_j\}$ is a random walk and $\chi(n) \geq k+1$ if and only if $\max_{0 \leq j \leq n-k} S^*_j \geq k$ which implies the Theorem.

Now we prove a strong law.

EXCURSIONS   153

THEOREM 12.26

$$\lim_{n\to\infty} \frac{\max_{1\leq k\leq n} \chi(k)}{\lg n} = \frac{1}{2} \quad a.s.,$$

consequently

$$\limsup_{n\to\infty} \frac{\chi(n)}{\lg n} = \frac{1}{2} \quad a.s.$$

and trivially

$$\chi(n) = 1 \quad i.o. \ a.s.$$

Proof. Consider the sequence

$$\xi_0^* = \xi(0, \rho_1(1)) + 1, \quad \xi_1^* = \xi(1, \rho_1(2)), \quad \xi_2^* = \xi(2, \rho_1(3)), \ldots$$

Then

$$\max_{k\leq n} \chi(k) = \max_{i\leq M_n^+} \xi_i^*.$$

Clearly

$$\mathbf{P}\{\xi_i^* = k\} = 2^{-k} \quad (k = 1, 2, \ldots; \ i = 0, 1, \ldots)$$

and the random variables ξ_i^* ($i = 0, 1, \ldots$) are independent. Hence for any $L = 1, 2, \ldots$ and $K = 1, 2, \ldots$ we have

$$\mathbf{P}\{\max_{i\leq L} \xi_i^* \leq K\} = \left(1 - \frac{1}{2^K}\right)^{L+1}. \tag{12.12}$$

Choosing

$$K = K_n = \frac{1-\varepsilon}{2} \lg n \quad \text{and} \quad L = L_n = \frac{n^{1/2}}{(\lg n)^2}$$

we obtain

$$\mathbf{P}\left\{\max_{i\leq n^{1/2}(\lg n)^{-2}} \xi_i^* \leq \frac{1-\varepsilon}{2} \lg n\right\}$$

$$= \left(1 - \frac{1}{n^{(1-\varepsilon)/2}}\right)^{n^{1/2}(\lg n)^{-2}} \leq \exp(-n^{\varepsilon/4}),$$

if n is large enough. Hence

$$\liminf_{n\to\infty} \frac{\max_{i\leq n^{1/2}(\lg n)^{-2}} \xi_i^*}{\lg n} \geq \frac{1-\varepsilon}{2} \quad a.s.$$

Since $M_n^+ \geq n^{1/2}(\lg n)^{-2}$ a.s. (cf. Theorem of Hirsch, Section 5.3) for all but finitely many n we get

$$\liminf_{n \to \infty} \frac{\max_{i \leq n} \chi(i)}{\lg n} \geq \frac{1}{2} \quad \text{a.s.}$$

Similarly, choosing

$$K = K_n = \frac{1+\varepsilon}{2} \lg n \quad \text{and} \quad L = L_n = n^{1/2} \lg n$$

we get

$$\mathbf{P}\{\max_{i \leq L} \xi_i^* > K\} = 1 - \left(1 - \frac{1}{n^{(1+\varepsilon)/2}}\right)^{n^{1/2} \lg n} \approx \frac{\lg n}{n^{\varepsilon/2}}.$$

Let $n_j(T) = j^T$. Then we have

$$\max_{i \leq L} \xi_i^* \leq K \quad \text{a.s.}$$

for all but finitely many j where $K = K_{n_j(T)}$ and $L = L_{n_j(T)}$. Let $j^T \leq N \leq (j+1)^T$. Then

$$\max_{i \leq L_N} \xi_i^* \leq K_{n_{j+1}(T)} \leq \frac{1+2\varepsilon}{2} \lg N \quad \text{a.s.}$$

if N is large enough, which in turn implies the Theorem.

LEMMA 12.6 *Let*

$$\mathcal{M}_k = \max\{S(\rho_k), S(\rho_k + 1), \ldots, S(\rho_{k+1})\} \quad (k = 0, 1, \ldots, n-1)$$

and let $0 \leq \mathcal{M}_{1:n} \leq \mathcal{M}_{2:n} \leq \ldots \leq \mathcal{M}_{n:n} = M^+(\rho_n)$ *be the ordered sample obtained from the sample* $\mathcal{M}_0, \mathcal{M}_1, \ldots, \mathcal{M}_{n-1}$. *Then for any* $0 < \varepsilon < 1$ *we have*

$$\mathcal{M}_{n:n} - \mathcal{M}_{n:n-1} \geq n^\varepsilon \quad \text{a.s.}$$

for all but finitely many n.

Proof. Let

$$A_n = A_n(\alpha, \varepsilon)$$
$$= \bigcup_{i=0}^{n} \bigcup_{j=0}^{n} \left\{ \frac{n}{(\log n)^\alpha} \leq \mathcal{M}_i, \mathcal{M}_j \leq n(\log n)^\alpha, |\mathcal{M}_i - \mathcal{M}_j| \leq n^\varepsilon \right\}.$$

Observe that for any i fixed

$$\mathbf{P}\left\{\frac{n}{(\log n)^\alpha} \leq \mathcal{M}_i \leq n(\log n)^\alpha\right\} \leq \mathbf{P}\left\{\frac{n}{(\log n)^\alpha} \leq \mathcal{M}_i\right\} = \frac{(\log n)^\alpha}{2n}$$

and for any i, j, m with $n(\log n)^{-\alpha} \leq m \leq n(\log n)^\alpha$, $0 \leq i \neq j \leq n$ we have

$$\mathbf{P}\{|\mathcal{M}_i - \mathcal{M}_j| \leq n^\varepsilon \mid \mathcal{M}_j = m\} \leq O\left(n^\varepsilon \frac{(\log n)^{2\alpha}}{n^2}\right) = O\left(\frac{(\log n)^{2\alpha}}{n^{2-\varepsilon}}\right).$$

Hence

$$\mathbf{P}\{A_n\} \leq O\left(\binom{n}{2}\frac{(\log n)^{3\alpha}}{n^{3-\varepsilon}}\right) \leq O\left(\frac{(\log n)^{3\alpha}}{n^{1-\varepsilon}}\right).$$

Let T be a positive integer with $T(1-\varepsilon) > 1$. Then only finitely many of the events A_{n^T} will occur with probability 1. Let $n^T \leq N \leq (n+1)^T$. Then

$$A_N \subset A_{(n+1)^T}(2\alpha, \varepsilon).$$

Consequently only finitely many of the events A_n will occur with probability one.

Since

$$n(\log n)^{-3} \leq \mathcal{M}_{n:n} \leq n(\log n)^3 \quad \text{a.s.}$$

(cf. the LIL, the Other LIL and Theorem 11.6) we obtain the Lemma.

Lemma 12.6 and Theorem 12.22 combined imply

THEOREM 12.27 *For any $C > 0$ there exists a $D = D(C) > 0$ such that*

$$\sup_{j \leq C \log n} \xi(M^+(n) - j, n) \leq D \log^3 n \quad \text{a.s.}$$

for all but finitely many n.

In this section as well as in Section 12.3 we investigated the local time of big values. Many efforts were devoted to studying the local time of small values. Here we mention only the following:

THEOREM 12.28 (Földes–Puri, 1989). *Let*

$$\hat{\rho}_N = \min\{k : |S_k| = N\}$$

and

$$\Xi((-\alpha N, \alpha N), \hat{\rho}_N) = \#\{k : 0 \leq k \leq \hat{\rho}_N, |S_k| \leq \alpha N\}.$$

Then for any $0 < \alpha \leq 1$ we have

$$\limsup_{N \to \infty} \frac{\Xi((-\alpha N, \alpha N), \hat{\rho}_N) c_0^2(\alpha)}{2\alpha^2 N^2 \log \log N} = 1 \quad \text{a.s.} \tag{12.13}$$

and

$$\liminf_{N\to\infty} \frac{\Xi((-\alpha N, \alpha N), \hat{\rho}_N)}{\alpha^2 N^2} 2\log\log N = 1 \quad a.s. \tag{12.14}$$

where $c_0(\alpha)$ is the unique root of the equation

$$u \tan u = \frac{\alpha}{1-\alpha}$$

in the interval $(0, \pi/2]$.

Note that in case $\alpha = 1$, $c_0(\alpha) = \pi/2$. Hence (12.13) (resp. (12.14)) are equivalent with the Other LIL (resp. LIL of Khinchine).

Chapter 13

Frequently and Rarely Visited Sites

13.1 Favourite sites

The random set $\mathcal{F}_n = \{x : \xi(x,n) = \xi(n)\}$ will be called the set of favourite points of the random walk $\{S(n)\}$ at time n. The largest favourite points will be denoted by $f_n = \max\{x : x \in \mathcal{F}_n\}$.

Of the properties of $\{f_n\}$ it is trivial that $f_n \leq u(n)$ with probability 1 except for finitely many n if $u(n) \in \text{UUC}(S_n)$. Hence we have a trivial result saying that f_n cannot be very large. The next theorem claims that f_n will occasionally be large.

THEOREM 13.1 (Erdős–Révész, 1984). *For any $\varepsilon > 0$*

$$f_n \geq (1-\varepsilon)b_n^{-1}$$

with probability 1 infinitely often.

Having this result, one can conjecture that f_n will be larger than any function $l(n)$ i.o. with probability 1 if $l(n) \in \text{ULC}(S_n)$.

However, it is not the case. Conversely, we have

THEOREM 13.2 (Erdős–Révész, 1984).

$$f_n \leq (n(2\log_2 n + 3\log_3 n + 2\log_4 n + 2\log_5 n + 2\log_6 n))^{1/2}$$

with probability 1 except for finitely many n.

It also looks interesting to investigate the small favourite points. Let $g_n = \min\{|x| : x \in \mathcal{F}_n\}$. Bass and Griffin (1985) proved that g_n cannot be very small. In fact

THEOREM 13.3

$$\liminf_{n\to\infty} \frac{g_n}{n^{1/2}(\log n)^{-\gamma}} = \begin{cases} \infty & \text{if } \gamma > 11, \\ 0 & \text{if } \gamma < 2. \end{cases}$$

Shi and Tóth (2000) proposed the

Question. Find the value of the constant γ_0 such that with probability 1

$$\liminf_{n\to\infty} \frac{|f_n|}{n^{1/2}(\log n)^{-\gamma}} = \begin{cases} 0 & \text{if } \gamma < \gamma_0, \\ \infty & \text{if } \gamma > \gamma_0. \end{cases}$$

They also remark that "there is a good reason to expect that γ_0 would be in $[1, 2]$."

They also investigated the limit distribution of f_n and proved

THEOREM 13.4

$$\lim_{n\to\infty} \mathbf{P}\{n^{-1/2} f_n < x\} = U(x)$$

where $U(x)$ is the distribution function of the random variable U defined by

$$\eta(U, 1) = \sup_{x \in \mathbb{R}^1} \eta(x, 1).$$

The exact form of $U(x)$ is unknown. However Shi and Tóth formulated the following:

Conjecture. There exists a constant $\nu > 1$ such that

$$0 < \liminf_{x\to\infty} x^\nu e^{x^2/2} \mathbf{P}\{U > x\} \leq \limsup_{x\to\infty} x^\nu e^{x^2/2} \mathbf{P}\{U > x\} < \infty.$$

Erdős–Révész (1984) proposed to study the cardinality $\#|\mathcal{F}_n|$ of $\mathcal{F}_n$. Everyone can see immediately that $\#|\mathcal{F}_n| = 1$ and $\#|\mathcal{F}_n| = 2$ i.o. with probability 1. This problem is still open. The strongest available result is

THEOREM 13.5 (Tóth, 2001) *Let*

$$f(r) = \#\{n : \#|\mathcal{F}_n| = r\}.$$

Then

$$\mathbf{E}f(4) < \infty, \tag{13.1}$$
$$\mathbf{E}f(3) = \infty.$$

(13.1) *clearly implies*

$$\mathbf{P}\{\#|\mathcal{F}_n| \geq 4 \quad i.o.\} = 0.$$

Tóth and Werner (1997) proposed to investigate the problem of favourite edges instead of favourite points. Let

$$\bar{S}_i = \frac{S_{i-1} + S_i + 1}{2}$$

and let

$$\ell(x,n) = \#\{j \in [1,n], \bar{S}_j = x\}$$

be the local time of the edge $[x-1, x]$. Now let

$$K_n = \{x : \ell(x,n) = \max_{y \in \mathbb{Z}^1} \ell(y,n)\}$$

be the set of favourite edges at time n and put

$$k(r) = \#\{i : i \geq 1, \#|K_i| \geq r\}.$$

THEOREM 13.6 (Tóth–Werner, 1997) *Almost surely $k(4) < \infty$. Moreover $\mathbf{E}k(4) < \infty$.*

We ask about the joint behaviour of f_n and $\xi(n)$. If f_n and $\xi(n)$ were asymptotically independent, then one would expect that the limit set of $\{f_n/(2n \log \log n)^{1/2}, \xi(n)/(2n \log \log n)^{1/2}\}$ should be the half-disc $\{(x,y) : y \geq 0, x^2 + y^2 \leq 1\}$. However, the correct answer shows that things do not go exactly like this.

THEOREM 13.7 (Csáki–Révész–Shi, 2000) *With probability 1, the random sequence*

$$\left(\frac{f_n}{(2n \log \log n)^{1/2}}, \frac{\xi(n)}{(2n \log \log n)^{1/2}}\right)$$

is relatively compact, whose limit set is identical to the triangle

$$\{(x,y) : y \geq 0, |x| + y \leq 1\}.$$

We also study the jump sizes of favourite sites. Let $\ell(0) = 0$ and

$$\ell(n+1) = \begin{cases} \ell(n) & \text{if } S(n+1) \notin \mathcal{F}_{n+1}, \\ S(n+1) & \text{if } S(n+1) \in \mathcal{F}_{n+1}. \end{cases}$$

THEOREM 13.8 (Csáki–Révész–Shi, 2000) *With probability 1,*

$$\limsup_{n \to \infty} \frac{|\ell(n+1) - \ell(n)|}{(2n \log \log n)^{1/2}} = 1.$$

This theorem tells us that the extraordinarily large jumps of the favourite site are asymptotically comparable to the size of the range of the random walk.

Here we present a few unsolved problems (Erdős–Révész, 1984 and 1987).

1. Consider the random sequence $\{\nu_n\}$ for which $|\mathcal{F}_{\nu_n}| \geq 2$. What can we say about the sequence $\{\nu_n\}$? Can we say, for example, that $\lim_{n \to \infty} \nu_n/n = \infty$ with probability 1?

2. How can the properties of the sequence $|f_{n+1} - f_n|$ be characterized? Is it true that $\limsup_{n \to \infty} |f_{n+1} - f_n| = \infty$? If yes, what is the rate of convergence?

3. Let $\alpha(n)$ be the number of different favourite values up to n, i.e. $\alpha(n) = \#|\sum_{k=1}^n \mathcal{F}_k|$. We guess that $\alpha(n)$ is very small, i.e. $\alpha(n) < (\log n)^c$ for some $c > 0$, but we cannot prove it. Hence we ask: how can one describe the limit behaviour of $\alpha(n)$?

4. We also ask how long a point can stay as a favourite value, i.e. let $1 \leq i = i(n) < j = j(n) \leq n$ be two integers for which

$$\# \left| \prod_{k=i}^{j} \mathcal{F}_k \right| \geq 1$$

and $j - i = \beta(n)$ is as large as possible. The question is to describe the limit behaviour of $\beta(n)$.

5. Further if x was a favourite value once, can it happen that the favourite value moves away from x but later returns to x again, i.e. do sequences $a_n < b_n < c_n$ of positive random integers exist such that

$$\mathcal{F}_{a_n} \mathcal{F}_{b_n} = \emptyset \quad \text{and} \quad \mathcal{F}_{a_n} \mathcal{F}_{c_n} \neq \emptyset \quad (n = 1, 2, \ldots)?$$

6. By the arcsine law we learned that the particle spends a long time on one half of the line and only a short time on the other half with a large probability. We ask whether the favourite value is located on the same side where the particle has spent the long time. For example let $0 < n_1 = n_1(\omega) < n_2 \ldots$ be a random sequence of integers for which

$$n_k^{-1} \sum_{j=1}^{n_k} I(S_j) \to 1 \quad (k \to \infty)$$

where
$$I(S_j) = \begin{cases} 1 & \text{if } S_j \geq 0, \\ 0 & \text{if } S_j < 0. \end{cases}$$

Then we conjecture that $f_{n_k} \to \infty$ as $k \to \infty$ a.s.

13.2 Rarely visited sites

It is easy to see that for infinitely many n almost all paths assume every value at least twice which they assume at all, i.e. let $\delta_n^{(r)} = 0$ if $\xi(0,n) \neq r-1$ and $\delta_n^{(r)} = 1$ if $\xi(0,n) = r-1$ and let

$$f_r(n) = \#\{k: \ k \neq 0, \ \xi(k,n) = r\} + \delta_n^{(r)}$$

be the number of points visited exactly r-times up to n. Then

$$\mathbf{P}\{f_1(n) = 0 \text{ i.o.}\} = 1.$$

However, for some n the number of points visited exactly once might be very large.

THEOREM 13.9 (Major, 1988)

$$\limsup_{n\to\infty} \frac{f_1(n)}{\log^2 n} = C \quad a.s.$$

where $0 < C < \infty$ but its exact value is unknown.

Another interesting result on $f_1(n)$ is

THEOREM 13.10 (Newman, 1984).

$$\mathbf{E} f_1(n) = 2 \quad (n = 1, 2, \ldots).$$

Proof. Since $f_1(1) = 2$, we only prove that $\mathbf{E} f_1(n+1) = \mathbf{E} f_1(n)$ for $n > 0$. Consider the walk $S_k^* = S_{k+1} - X_1$ ($k = 0, 1, 2, \ldots$) and let $f_1^*(n)$ be the number of points visited exactly once by S_k^* up to n. Then

$$f_1(n+1) = \begin{cases} f_1^*(n) + 1 & \text{if } \xi(0, n+1) = 0, \\ f_1^*(n) - 1 & \text{if } \xi(0, n+1) = 1, \\ f_1^*(n) & \text{if } \xi(0, n+1) > 1. \end{cases}$$

Theorem 9.3 implies that $\mathbf{P}\{\xi(0, n+1) = 0\} = \mathbf{P}\{\xi(0, n+1) = 1\}$. Hence we have the Theorem.

Let
$$g_r(n) = \sum_{j=1}^{r} f_j(n)$$
be the number of those visited points which are visited at most r times up to n. Our question is: might be $g_r(n) = 0$ i.o. $(r \geq 2)$.

This question was answered by Tóth (1996/A).

THEOREM 13.11 *For any positive integer n*

$$\mathbf{P}\{g_r(n) = 0 \quad i.o.\} = 1. \qquad (13.2)$$

Clearly (13.2) tells us that for infinitely many n each visited point is visited at least r times. We suggest the following problem: characterize those sequences $\{r(n),\ n=1,2,\ldots\}$ for which (13.2) remains valid replacing r by $r(n)$. It seems to be clear that (13.2) holds true if $r(n)$ goes to infinity slowly enough. It is not hard to give an upper bound for the speed of the sequences $\{r(n)\}$ which might satisfy (13.2).

THEOREM 13.12 *Let $r(n) = a \log n$ $(a > 1/2)$. Then*

$$g_{r(n)} > 0 \qquad a.s.$$

for all but finitely many n.

Proof. It is a trivial consequence of Theorem 12.26.

Chapter 14

An Embedding Theorem

14.1 On the Wiener sheet

Let $\{X_{i,j},\ i = 1, 2, \ldots,\ j = 1, 2, \ldots\}$ be a double array of i.i.d.r.v.'s with

$$\mathbf{P}\{X_{i,j} = 1\} = \mathbf{P}\{X_{i,j} = -1\} = \frac{1}{2}$$

and define $S_{0,n} = S_{m,0} = 0\ (n = 0, 1, 2, \ldots;\ m = 0, 1, 2, \ldots)$,

$$S_{n,m} = \sum_{j=1}^{m} \sum_{i=1}^{n} X_{i,j}, \quad (n = 1, 2, \ldots;\ m = 1, 2, \ldots).$$

The arrays $\{S_{n,m}\}$ and $\{X_{i,j}\}$ are called *random fields*. Some properties of $\{S_{n,m}\}$ can be obtained as simple consequences of the corresponding properties of the random walk, some properties of $\{S_{n,m}\}$ are essentially different. Here we mention one example of both types. Just like in the one-dimensional case we have

$$\lim_{\substack{n \to \infty \\ m \to \infty}} \mathbf{P}\left\{\frac{S_{n,m}}{\sqrt{nm}} < x\right\} = \Phi(x).$$

However,

$$\limsup_{\substack{n \to \infty \\ m \to \infty}} b(nm) S_{n,m} = 2^{1/2} \quad \text{a.s.}$$

This latter result is due to Zimmermann (1972) (see also Csörgő–Révész, 1981).

On the same way as the Wiener process was defined (Section 6.2) a continuous analogue of $\{S_{n,m},\ n = 0, 1, 2, \ldots;\ m = 0, 1, 2, \ldots\}$ can be defined. This continuous random field will be called *Wiener sheet (two-parameter Wiener process)* and will be denoted by

$$\{W(x,y),\ x \geq 0,\ y \geq 0\}.$$

Among the properties of the Wiener sheet we mention

(i) $W(\cdot,\cdot)$ is a Gaussian process,

(ii) $W(0,y) = W(x,0) = 0$,

(iii) $\mathbf{E}W(x_1,y_1)W(x_2,y_2) = \min(x_1,x_2)\min(y_1,y_2)$,

(iv) $W(x,y)$ is continuous a.s.,

(v) for any $x_0 > 0$, the one-dimensional process $\{x_0^{-1/2}W(x_0,y),\ y \geq 0\}$ is a Wiener process,

(vi) for any $y_0 > 0$, the one-dimensional process $\{y_0^{-1/2}W(x,y_0),\ x \geq 0\}$ is a Wiener process.

For some further study and a detailed definition of the Wiener sheet we refer to Csörgő–Révész (1981). A very useful and detailed new study of the multiparameter processes is due to Khoshnevisan (2002).

14.2 The theorem

We have already seen that the study of the processes $\xi(x,n)$ (resp. $\eta(x,t)$) is relatively easy when x is fixed and we let only n (resp. t) vary. The main reason of this fact is the following trivial:

LEMMA 14.1 *For any integer x*

$$\xi(x,\rho_1) - \xi(x,\rho_0) = \xi(x,\rho_1), \xi(x,\rho_2) - \xi(x,\rho_1), \xi(x,\rho_3) - \xi(x,\rho_2),\ldots$$

are i.i.d.r.v.'s with

$$\mathbf{E}(\xi(x,\rho_k) - \xi(x,\rho_{k-1})) = 1, \quad \mathbf{E}(\xi(x,\rho_k) - \xi(x,\rho_{k-1}) - 1)^2 = 4x - 2$$

(cf. Theorem 9.7).

In order to formulate the analogue of Lemma 14.1 for $\eta(\cdot,\cdot)$ let

$$\rho_0^* = 0, \quad \rho_u^* = \inf\{t:\ t \geq 0,\ \eta(0,t) \geq u\} \quad (u > 0). \tag{14.1}$$

Then we have

LEMMA 14.2 *For any $x \in \mathbb{R}^1$, $\eta(x,\rho_u^*)$ is a process of independent increments in $u(u \geq 0)$, i.e. for any $0 \leq u_1 < u_2 < \ldots < u_k$ ($k = 1,2,\ldots$), the r.v.'s*

$$\eta(x,\rho_{u_1}^*), \eta(x,\rho_{u_2}^*) - \eta(x,\rho_{u_1}^*),\ldots,\eta(x,\rho_{u_k}^*) - \eta(x,\rho_{u_{k-1}}^*)$$

AN EMBEDDING THEOREM

are independent with

$$\mathbf{E}(\eta(x,\rho^*_{u_j}) - \eta(x,\rho^*_{u_{j-1}})) = u_j - u_{j-1},$$

$$\mathbf{E}(\eta(x,\rho^*_{u_j}) - \eta(x,\rho^*_{u_{j-1}}) - u_j + u_{j-1})^2 = 4x(u_j - u_{j-1})$$

where $j = 1, 2, \ldots, k$.

Consider the process

$$\mathcal{L}(x,u) = \eta(x,\rho^*_u) - \eta(0,\rho^*_u) = \eta(x,\rho^*_u) - u. \qquad (14.2)$$

Then we have

(i)
$$\mathbf{E}\mathcal{L}(x,u) = 0, \quad \mathbf{E}\mathcal{L}^2(x,u) = 4xu,$$

(ii) $\{\mathcal{L}(x,u); u \geq 0\}$ is a strictly stationary process of independent increments in u for any $x \in \mathbb{R}^1$.

One can also prove that

(iii) $\mathcal{L}(x,u)$ has a finite moment generating function in a neighbourhood of the origin.

By the Invariance Principle 2 (cf. Section 6.3) this fact easily implies that for any $x \in \mathbb{R}^1$ the process $\mathcal{L}(x,u)$ can be approximated by a Wiener process $W^*_x(\cdot)$ with rate $O(\log u)$, i.e.

$$|W^*_x(u) - \mathcal{L}(x,u)| = O(\log u).$$

Having a fixed x this result gives an important tool to describe the properties of $\mathcal{L}(x,u)$.

What can we say about $\mathcal{L}(x,u)$ when u is fixed and x is varying? It is easy to prove that for any fixed u $\{\mathcal{L}(x,u), x \geq 0\}$ has orthogonal increments and it is a martingale in x. This observation suggests the question:

Can the process $\mathcal{L}(x,u)$ be approximated by a two-parameter Wiener process?

Since by the LIL $\eta(x,u) = 0$ a.s. if $x \geq ((2+\varepsilon)u\log\log u)^{1/2}$ and u is big enough, we have $\mathcal{L}(x,u) = -u$ for any x big enough. This clearly shows that the structure of $\mathcal{L}(x,u)$ is quite different from that of $W(x,u)$ whenever x is big. Hence we modify the above question as follows:

Can the process $\mathcal{L}(x,u)$ be approximated by a Wiener sheet provided that u is big but x is not very big?

The answer to this question is positive. In fact we have

THEOREM 14.1 (Csáki–Csörgő–Földes–Révész, 1989). *There exists a probability space with*

(i) *a standard Wiener process* $\{W(t), t \geq 0\}$, *its two-parameter local time process* $\{\eta(x,t), x \in \mathbb{R}^1, t > 0\}$ *and the inverse process* ρ_u^* *of* $\eta(0,t)$ *defined by* (14.1),

(ii) *a two-time parameter Wiener process* $\{W(x,u); x \geq 0, u \geq 0\}$

such that

$$\sup_{0 \leq x \leq Au^\delta} |\mathcal{L}(x,u) - 2W(x,u)| = O\left(u^{(1+\delta)/2-\varepsilon}\right) \quad a.s. \quad (u \to \infty) \quad (14.3)$$

where $\mathcal{L}(x,u)$ *is defined by* (14.2), A *is an arbitrary positive constant and*

$$0 \leq \delta \leq 7/100, \qquad 0 < \varepsilon < 1/72 - \delta/7.$$

This theorem is certainly a useful tool for studying the properties of $\mathcal{L}(x,u)$ or $\eta(x, \rho_u^*)$. Unfortunately it does not say much about $\eta(x,t)$. However, we can continue Theorem 14.1 as follows:

THEOREM 14.2 *On the probability space of Theorem* 14.1 *we can also define a process* $\hat{\rho}_u^*$ *such that*

$$\{\hat{\rho}_u^*, \ u \geq 0\} \stackrel{\mathcal{D}}{=} \{\rho_u^*, \ u \geq 0\}, \quad (14.4)$$

$$|\rho_u^* - \hat{\rho}_u^*| = O\left(u^{15/8}\right) \quad a.s. \quad (u \to \infty), \quad (14.5)$$

$\{\hat{\rho}_u^*, \ u \geq 0\}$ *and* $\{W(x,u); \ x \geq 0, \ u \geq 0\}$ *are independent.* (14.6)

Having the process $\{\hat{\rho}_u^*, \ u \geq 0\}$ we can proceed as follows:
Define the local time process $\hat{\eta}(0,t)$ by

$$\hat{\eta}(0, \hat{\rho}_u^*) = u \quad (u \geq 0).$$

By the continuity properties of $\eta(0,t)$ (cf. Theorem 11.7) we have

$$\lim_{u \to \infty} \frac{|\eta(0, \rho_u^*) - \eta(0, \hat{\rho}_u^*)|}{|\rho_u^* - \hat{\rho}_u^*|^{1/2} \log u} = 0 \quad a.s.$$

Thus by Theorem 14.2 we conclude that the local time process $\{\hat{\eta}(0,t) \ t \geq 0\}$ has the following properties:

$$\{\hat{\eta}(0,t); \ t \geq 0\} \stackrel{\mathcal{D}}{=} \{\eta(0,t); \ t \geq 0\}, \quad (14.7)$$

$$|\hat{\eta}(0,t) - \eta(0,t)| \text{ is small a.s. } (t \to \infty), \quad (14.8)$$

AN EMBEDDING THEOREM

$\{\hat{\eta}(0,t); t \geq 0\}$ and $\{W(x,u); x \geq 0, u \geq 0\}$ are independent. (14.9)

(14.7) (resp. (14.9)) follows immediately from (14.4) (resp. (14.6)). In order to see (14.8) it is enough to show that

$$| \hat{\eta}(0, \hat{\rho}_u^*) - \eta(0, \hat{\rho}_u^*) | = | u - \eta(0, \hat{\rho}_u^*) |$$

is small, which in turn follows from the fact that

$$|\eta(0, \hat{\rho}_u^*) - \eta(0, \rho_u^*)| = |\eta(0, \hat{\rho}_u^*) - u|$$

is small. Now (14.3), (14.8), and the continuity of $W(\cdot, \cdot)$ imply

$$|\eta(x,t) - \eta(0,t) - 2W(x, \hat{\eta}(0,t))| \text{ is small a.s.}$$

where $\hat{\eta}(0,t)$ satisfies (14.8) and (14.9).

A precise version of the above sketched idea implies

THEOREM 14.3 (Csáki–Csörgő–Földes–Révész, 1989). *There exists a probability space with*

(i) *a standard Wiener process $\{W(t); t \geq 0\}$ and its two-parameter local time process $\{\eta(x,t); x \in \mathbb{R}^1, t \geq 0\}$,*

(ii) *a two-time parameter Wiener process $\{W(x,u); x \geq 0, u \geq 0\}$,*

(iii) *a process $\{\hat{\eta}(0,t); t \geq 0\} \stackrel{\mathcal{D}}{=} \{\eta(0,t); t \geq 0\}$*

such that

$$\sup_{0 \leq x \leq At^{\delta/2}} |\eta(x,t) - \eta(0,t) - 2W(x, \hat{\eta}(0,t))|$$
$$= O\left(t^{(1+\delta)/4 - \varepsilon/2}\right) \quad a.s. \quad (t \to \infty),$$

$$|\hat{\eta}(0,t) - \eta(0,t)| = O\left(t^{15/32} \log^2 t\right) \quad a.s. \quad (t \to \infty),$$

$\{\hat{\eta}(0,t); t \geq 0\}$ *and* $\{W(x,u); x \geq 0, u \geq 0\}$ *are independent*

where

$$A > 0, \quad 0 \leq \delta < 7/100, \quad 0 < \varepsilon < 1/72 - \delta/7.$$

14.3 Applications

In order to show how the above theorem can be used in the study of the properties of $\eta(\cdot,\cdot)$, first we list a few simple properties of the vector valued process
$$\{(2W(x,\hat{\eta}(0,t)), t^{-1/4}\hat{\eta}(0,t)); \ t \geq 0\}$$
which can be obtained by standard methods of proof.

Namely for any $x > 0$ and $t > 0$ we have

$$\frac{W(x,\hat{\eta}(0,t))}{(x\hat{\eta}(0,t))^{1/2}} \overset{\mathcal{D}}{=} N_1, \tag{14.10}$$

$$t^{-1/2}\hat{\eta}(0,t) \overset{\mathcal{D}}{=} |N_2|, \tag{14.11}$$

$$\frac{W(x,\hat{\eta}(0,t))}{x^{1/2}t^{1/4}} \overset{\mathcal{D}}{=} N_1|N_2|^{1/2} \tag{14.12}$$

where N_1, N_2 are independent normal $(0,1)$ r.v.'s. Note that since $\hat{\eta}(0,t)$ and $W(x,u)$ are independent (cf. Theorem 14.3), (14.12) follows from (14.10) and (14.11).

Also, for any $x > 0$, the set of limit points of

$$U_t = \frac{W(x,\hat{\eta}(0,t))}{\sqrt{2x\hat{\eta}(0,t)\log\log t}} \quad (t \to \infty)$$

is the interval $[-1,1]$ a.s. The set of limit points of

$$V_t = \frac{\hat{\eta}(0,t)}{\sqrt{2t\log\log t}}$$

is the interval $[0,1]$ a.s. Applying the fact that the processes $\{\hat{\eta}(0,t); \ t \geq 0\}$ and $\{W(x,u); \ x \geq 0, \ u \geq 0\}$ are independent, we get that the set of limit points of (U_t, V_t) is the semidisc $\{(u,v): \ v \geq 0, \ u^2 + v^2 \leq 1\}$. The set of limit points of

$$U_t V_t^{1/2} = \frac{W(x,\hat{\eta}(0,t))}{2^{3/4}x^{1/2}t^{1/4}(\log\log t)^{3/4}}$$

is the interval $[0, 2^{1/2}3^{-3/4}]$ a.s. for any $x > 0$, that is,

$$\limsup_{t\to\infty} \frac{W(x,\hat{\eta}(0,t))}{x^{1/2}t^{1/4}(\log\log t)^{3/4}} = 2^{5/4}3^{-3/4} \quad \text{a.s.} \tag{14.13}$$

For any $K > 0$ the usual LIL implies

$$\limsup_{t\to\infty} \sup_{0<x\leq Kt^\delta} \frac{W(x,\hat{\eta}(0,t))}{\sqrt{2K\hat{\eta}(0,t)t^\delta \log\log t}} = 1 \quad \text{a.s.}$$

AN EMBEDDING THEOREM

Applying again the independence of U_t and V_t we obtain

$$\limsup_{t\to\infty} \sup_{0<x\leq Kt^\delta} \left(\frac{27}{32}\right)^{1/4} \frac{W(x,\hat\eta(0,t))}{(Kt^\delta)^{1/2}t^{1/4}(\log\log t)^{3/4}} = 1 \quad \text{a.s.} \quad (14.14)$$

for any $K > 0$ and $\delta \geq 0$.

Consequently, by Theorem 14.3 and by (14.10), (14.11), (14.12), (14.13), (14.14) respectively we obtain

$$\frac{\eta(x,t) - \eta(0,t)}{2\sqrt{x\eta(0,t)}} \xrightarrow{\mathcal{D}} N_1 \quad \text{for any} \quad x > 0 \quad (t\to\infty), \quad (14.15)$$

$$t^{-1/2}\eta(0,t) \xrightarrow{\mathcal{D}} |N_2| \quad (t>0), \quad (14.16)$$

$$\frac{\eta(x,t) - \eta(0,t)}{2x^{1/2}t^{1/4}} \xrightarrow{\mathcal{D}} N_1|N_2|^{1/2} \quad t\to\infty, \quad \text{for any} \quad x > 0, \quad (14.17)$$

$$\limsup_{t\to\infty} \frac{\eta(x,t) - \eta(0,t)}{2\sqrt{2x\eta(0,t)\log\log t}} = 1 \quad \text{a.s. for any} \quad x > 0, \quad (14.18)$$

$$\limsup_{t\to\infty} \frac{\eta(x,t) - \eta(0,t)}{x^{1/2}t^{1/4}(\log\log t)^{3/4}} = \frac{4}{3}6^{1/4} \quad \text{a.s. for any} \quad x > 0, \quad (14.19)$$

$$\limsup_{t\to\infty} \sup_{0<x\leq Kt^\delta} \frac{\eta(x,t) - \eta(0,t)}{2\sqrt{2Kt^\delta\eta(0,t)\log\log t}}$$
$$= \limsup_{t\to\infty} \sup_{0<x\leq Kt^\delta} \frac{3}{4}6^{-1/4}\frac{\eta(x,t) - \eta(0,t)}{(Kt^\delta)^{1/2}t^{1/4}(\log\log t)^{3/4}} = 1 \quad \text{a.s.} \quad (14.20)$$

for any $K > 0$ and $0 \leq \delta < 7/200$.

For the direct proofs of (14.18) and (14.19) see Csáki–Földes (1988/B).

Chapter 15

A Few Further Results

15.1 On the location of the maximum of a random walk

Let $\mu(n)$ $(n = 1, 2, \ldots)$ be the location of the maximum of the absolute value of a random walk $\{S_n\}$, i.e. $\mu(n)$ is defined by

$$M(n) = \max_{0 \leq k \leq n} |S_k| = S(\mu(n)) \quad \text{and} \quad \mu(n) \leq n. \tag{15.1}$$

If there are more than one integer satisfying (15.1), then the smallest one will be considered as $\mu(n)$. Since $\mu(n) = n$ i.o. a.s., the characterization of the upper classes of $\mu(n)$ is trivial. In order to get some idea about the lower classes we can argue as follows.

Since $\lim_{n \to \infty} \mu(n) = \infty$ a.s. by the LIL for any $\varepsilon > 0$

$$|S(\mu(n))| \leq (1 + \varepsilon)(2\mu(n) \log \log \mu(n))^{1/2}$$

with probability 1 if n is big enough. By the Other LIL for any $\varepsilon > 0$

$$|S(\mu(n))| = M(n) \geq (1 - \varepsilon) \left(\frac{\pi^2}{8} \frac{n}{\log \log n} \right)^{1/2},$$

consequently

$$\frac{\pi^2}{8} \frac{n}{\log \log n} \leq (1 + \varepsilon) 2\mu(n) \log \log \mu(n)$$

and

$$\mu(n) \geq (1 - \varepsilon) \frac{\pi^2}{16} \frac{n}{(\log \log n)^2} \quad \text{a.s.} \tag{15.2}$$

if n is big enough.

We ask:

Question 1. Can $\mu(n)$ attain the lower bound of (15.2)? The answer is negative. In fact we have

THEOREM 15.1 (Csáki–Földes–Révész, 1987).
$$\liminf_{n\to\infty} \frac{(\log\log n)^2}{n}\mu(n) = \frac{\pi^2}{4} \quad a.s.$$

Now we formulate our

Question 2. If

$$\mu(n) \leq (1+\varepsilon)\frac{\pi^2}{4}\frac{n}{(\log\log n)^2} \quad \text{for some } \varepsilon > 0 \tag{15.3}$$

then by the LIL

$$|S(\mu(n))| \leq (1+\varepsilon)b^{-1}(\mu(n))$$
$$\leq (1+2\varepsilon)\left(\frac{\pi^2}{4}\frac{n}{(\log\log n)^2}\right)^{1/2}(2\log\log n)^{1/2}$$
$$= (1+2\varepsilon)\frac{\pi}{\sqrt{2}}\left(\frac{n}{\log\log n}\right)^{1/2}. \tag{15.4}$$

We ask: Can $|S(\mu(n))|$ attain the upper bound of (15.4) if $\mu(n)$ is as small as possible, i.e. if (15.3) holds? The answer is negative again. In fact we have

THEOREM 15.2 (Csáki–Földes–Révész, 1987). *Let*

$$\mathcal{F} = \mathcal{F}_\varepsilon = \left\{n:\ n>1,\ \mu(n) \leq (1+\varepsilon)\frac{\pi^2}{4}\frac{n}{(\log\log n)^2}\right\}.$$

Then for any $\delta > 0$ there exists an $\varepsilon = \varepsilon(\delta) > 0$ such that

$$(1-\delta)\frac{\pi}{2} \leq \liminf_{\substack{n\in\mathcal{F}\\n\to\infty}} \left(\frac{\log\log n}{n}\right)^{1/2} M(n)$$
$$\leq \limsup_{\substack{n\in\mathcal{F}\\n\to\infty}} \left(\frac{\log\log n}{n}\right)^{1/2} M(n) \leq (1+\delta)\frac{\pi}{2} \quad a.s.$$

This theorem roughly says that if $\mu(n) \sim (\pi^2/4)(n/(\log\log n)^2)$ (i.e. $\mu(n)$ is as small as possible) then

$$|S(\mu(n))| = M(n) \sim \frac{\pi}{2}\left(\frac{n}{\log\log n}\right)^{1/2}.$$

Question 3. Intuitively it is clear that $M(n)$ can be (and will be) small if $\mu(n)$ is small. Theorem 15.2 somewhat contradicts this feeling. It says that if $\mu(n)$ is as small as possible, then $M(n)$ will be small but not as small as possible without having any condition about $\mu(n)$. It will be $(\pi/2)(n/\log\log n)^{1/2}$ instead of $(\pi/\sqrt{8})(n/\log\log n)^{1/2}$ which is the smallest possible value of $M(n)$ by the Other LIL. We ask: How small can $\mu(n)$ be, if $M(n)$ is as small as possible? The answer is:

THEOREM 15.3 (Csáki–Földes–Révész, 1987). *For any $L > 0$ there exists an $\varepsilon = \varepsilon(L) > 0$ such that with probability 1 the inequalities*

$$M(n) < (1+\varepsilon)\left(\frac{\pi^2}{8}\frac{n}{\log\log n}\right)^{1/2}$$

and

$$\mu(n) < L\frac{n}{(\log\log n)^2}$$

cannot occur simultaneously if n is big enough. However, if $g(n)$ is a positive function with $g(n) \nearrow \infty$ then for almost all $\omega \in \Omega$ and $\varepsilon > 0$ there exists a sequence $0 < n_1 = n_1(\omega,\varepsilon) < n_2 = n_2(\omega,\varepsilon) < \ldots$ such that

$$\mu(n_k) \leq g(n_k)\frac{n_k}{(\log\log n_k)^2} \quad \text{and} \quad \lim_{n\to\infty}\left(\frac{\log\log n_k}{n_k}\right)^{1/2} M(n_k) = \frac{\pi}{\sqrt{8}}.$$

Question 4. Instead of Question 3 one can ask: How big can $\mu(n)$ be, if $M(n)$ is as small as possible? The answer to this question is unknown.

The following theorem gives a joint generalization of the above three theorems. It also contains the LIL and the Other LIL (cf. Sections 4.4 and 5.3).

THEOREM 15.4 (Csáki–Földes–Révész, 1987). *Let*

$$a(n) = \frac{(\log\log n)^2}{n}\mu(n),$$

$$b(n) = \left(\frac{\log\log n}{n}\right)^{1/2} M(n),$$

$$K = \left\{(x,y): x \geq \frac{\pi^2}{4},\ x^{1/2}\left(1-\left(1-\frac{\pi^2}{4x}\right)^{1/2}\right)^{1/2}\right.$$

$$\left.\leq y \leq x^{1/2}\left(1+\left(1-\frac{\pi^2}{4x}\right)^{1/2}\right)^{1/2}\right\}$$

$$= \left\{(x,y): x > 0,\ y > 0,\ \frac{y^2}{2x}+\frac{\pi^2}{8y^2} \leq 1\right\}.$$

Then the set of limit points of the sequence $(a(n), b(n))$ as $(n \to \infty)$ is K with probability 1.

Remark 1. This theorem clearly does not imply that $(a(n), b(n)) \in K$ or even $(a(n), b(n))$ belongs to a neighbourhood of K if n is big enough. However, $(a(n), b(n))$ belongs to a somewhat larger set $K_\varepsilon \supset K$ if n is big enough. In fact we have

THEOREM 15.5 (Csáki–Földes–Révész, 1987). *Let*

$$K_\varepsilon = \left\{ (x, y) : x > 0, \ y > 0, \ \frac{y^2}{2x} + \frac{\pi^2}{8y^2} \leq 1 + \varepsilon \right\} \quad (\varepsilon > 0).$$

Then for any $\varepsilon > 0$

$$(a(n), b(n)) \in K_\varepsilon \quad a.s.$$

for all but finitely many n.

In order to formulate a simple consequence of Theorem 15.4 let $R^*(n)$ be the length of the longest flat interval of $\{M(k), 0 \leq k \leq n\}$, i.e. $R^*(n)$ is the largest positive number for which there exists a positive integer α such that

$$0 < \alpha < \alpha + R^*(n) < n$$

and

$$M(\alpha) = M(\alpha + R^*(n)).$$

Then by Theorem 15.1 (or 15.4) we have

THEOREM 15.6 (Csáki–Földes–Révész, 1987).

$$\liminf_{n \to \infty} \frac{(\log \log n)^2}{n}(n - R^*(n)) = \frac{\pi^2}{4} \quad a.s.$$

Equivalently for any $\varepsilon > 0$

$$n - (1 - \varepsilon)\frac{\pi^2}{4} \frac{n}{(\log \log n)^2} \in \mathrm{UUC}\{R^*(n)\}$$

and

$$n - (1 + \varepsilon)\frac{\pi^2}{4} \frac{n}{(\log \log n)^2} \in \mathrm{ULC}\{R^*(n)\}.$$

As far as the lower classes of $R^*(n)$ are concerned we have

A FEW FURTHER RESULTS

THEOREM 15.7 (Csáki–Földes–Révész, 1987).

$$\liminf_{n\to\infty} \frac{\log\log n}{n} R^*(n) = \beta$$

where β is the root of the equation

$$\sum_{k=1}^{\infty} \frac{\beta^k}{k!(2k-1)} = 1$$

(cf. Theorem 12.3). Equivalently for any $\varepsilon > 0$

$$(1+\varepsilon)\beta \frac{n}{\log\log n} \in \mathrm{LUC}\{R^*(n)\}$$

and

$$(1-\varepsilon)\beta \frac{n}{\log\log n} \in \mathrm{LLC}\{R^*(n)\}.$$

Remark 2. In Theorems 12.1 and 12.3 we investigated the length of the longest flat interval of $M^+(n)$. Comparing our results regarding the upper classes we obtain the intuitively clear fact that the longest flat interval of $M^+(n)$ can be (and will be) longer than that of $M(n)$. Comparing the known results regarding the lower classes no difference can be obtained.

About the proofs of the above theorems we mention that they are based on the following:

THEOREM 15.8 (Imhof, 1984). *Let $\mathcal{M}(t)$ be the location of the maximum of a Wiener process and let $u_t(x,y)$ be the joint density of the process $(t^{-1}\mathcal{M}(t), t^{-1/2}m(t))$. Then*

$$u_t(x,y) = \frac{2y}{\pi x^{3/2}(1-x)^{1/2}} \sum_{i=0}^{\infty} (-1)^i (2i+1) \exp\left(-\frac{(2i+1)^2 y^2}{2x}\right)$$

$$\times \sum_{j=-\infty}^{\infty} (-1)^j \exp\left(-\frac{2j^2 y^2}{1-x}\right).$$

15.2 On the location of the last zero

Let $\Psi(n)$ be the location of the last zero of a random walk $\{S_k, k \leq n\}$, i.e.

$$\Psi(n) = \max\{k: \ 0 \leq k \leq n, \ S_k = 0\}.$$

Theorem 12.1 claims that: if $g(n)$ is a nondecreasing sequence of positive numbers then

$$\frac{n}{g(n)} \in \mathrm{LLC}(\Psi(n))$$

if and only if
$$\sum_{n=1}^{\infty} n^{-1}(g(n))^{-1/2} < \infty.$$
Consequently for any $\varepsilon > 0$
$$\frac{n}{(\log n)^{2+\varepsilon}} \in \text{LLC}(\Psi(n)) \quad \text{and} \quad \frac{n}{(\log n)^2} \in \text{LUC}(\Psi(n)).$$

Since $\Psi(n) = n$ i.o. a.s. and $\Psi(n) \leq n$ the description of the upper classes of $\Psi(n)$ is trivial.

Here we wish to investigate the properties of the sequence $\{\Psi(n)\}$ for those n's only for which S_n is very big or $M(n)$ is very small. It looks very likely that if S_n is very big (e.g. $S_n \geq (2n \log \log n)^{1/2}$) then $\Psi(n)$ is very small. In the next theorems it turns out that this conjecture is not exactly true. In order to formulate our results, introduce the following notations.

Let $f(n) = n^{1/2} g(n) \in \text{ULC}(S_n)$ with $g(n) \uparrow \infty$. Define the infinite random set of integers
$$\mathcal{Z} = \mathcal{Z}(f) = \{n : S_n \geq f(n)\}.$$

Furthermore, let $\alpha(n), \beta(n)$ be sequences of positive numbers satisfying the following conditions:

$$\alpha(n) \text{ is nonincreasing,}$$
$$0 < \alpha(n) < 1,$$
$$\beta(n) \downarrow 0,$$
$$n\alpha(n) \uparrow \infty, \quad n\beta(n) \uparrow \infty.$$

Then we have

THEOREM 15.9 (Csáki–Grill, 1988).
$$n\alpha(n) \in \text{UUC}(\Psi(n), n \in \mathcal{Z})$$
if and only if
$$I_1(\alpha, g) = \sum_{n=1}^{\infty} \frac{1}{(\alpha(n)(1-\alpha(n)))^{1/2}} \frac{1}{n} \exp\left(-\frac{g^2(n)}{2(1-\alpha(n))}\right) < \infty.$$

Further,
$$n\beta(n) \in \text{LLC}(\Psi(n), n \in \mathcal{Z})$$
if and only if
$$I_2(\alpha, g) = \sum_{n=1}^{\infty} \frac{1}{n} g^2(n) \beta^{1/2}(n) \exp\left(-\frac{g^2(n)}{2}\right) < \infty.$$

A FEW FURTHER RESULTS

Remark 1. $n\alpha(n) \in \text{UUC}(\Psi(n), n \in \mathcal{Z})$ means that $n\alpha(n) \geq \Psi(n)$ a.s. for all but finitely many such n for which $n \in \mathcal{Z}$. In other words the inequalities $S_n \geq f(n)$ and $\Psi(n) \geq n\alpha(n)$ simultaneously hold with probability 1 only for finitely many n.

In order to illuminate the meaning of the above theorem we present two examples.

Example 1. Let $f(n) = ((2-\varepsilon)n \log \log n)^{1/2}$ ($0 < \varepsilon < 2$). Then we obtain that the inequalities

$$S_n \geq ((2-\varepsilon)n \log \log n)^{1/2} \quad \text{and} \quad \Psi(n) \geq \frac{\varepsilon}{2}(1+\varepsilon)n$$

hold with probability 1 only for finitely many n. However,

$$S_n \geq ((2-\varepsilon)n \log \log n)^{1/2} \quad \text{and} \quad \Psi(n) \geq \frac{\varepsilon}{2}(1-\varepsilon)n \quad \text{i.o. a.s.}$$

The above two statements also follow from Strassen's theorem (Section 8.1). Further,

$$S_n \geq ((2-\varepsilon)n \log \log n)^{1/2} \quad \text{and} \quad \Psi(n) \leq n(\log n)^{-\eta} \quad \text{i.o. a.s.}$$

if and only if $\eta \leq \varepsilon$. Note the surprising fact that $\Psi(n) \leq n(\log n)^{-1}$ i.o. a.s. but there are only finitely many n for which $\Psi(n) \leq n(\log n)^{-1}$ and $S_n \geq ((3/2)n \log \log n)^{1/2}$ (say) simultaneously hold.

Example 2. Let $f(n) = (2n \log \log n)^{1/2}$. Then we obtain that for any $\varepsilon > 0$ the inequalities

$$S_n \geq (2n \log \log n)^{1/2} \quad \text{and} \quad \Psi(n) \geq \left(\frac{3}{2}+\varepsilon\right) \frac{\log_3 n}{\log_2 n} n$$

hold with probability 1 only for finitely many n. However,

$$S_n \geq (2n \log \log n)^{1/2} \quad \text{and} \quad \Psi(n) \geq \frac{3}{2} \frac{\log_3 n}{\log_2 n} \quad \text{i.o. a.s.}$$

Further,

$$S_n \geq (2n \log \log n)^{1/2} \quad \text{and} \quad \Psi(n) \leq n(\log \log n)^{-\eta} \quad \text{i.o. a.s.}$$

if and only if $\eta \leq 4$.

Now we turn to our second question, i.e. we intend to study the behaviour of Ψ for those n's for which $M(n)$ is very small (nearly equal to $\pi n^{1/2}(8 \log \log n)^{-1/2}$, cf. the Other LIL (5.9)). In this case we can expect that $\Psi(n)$ is not very small. The next theorem shows that this feeling is true in some sense. In order to formulate it introduce the following notations. Let $\gamma(n)$ and $\delta(n)$ be sequences of positive numbers satisfying the following conditions:

$$0 < \gamma(n), \delta(n) \leq 1,$$
$$\delta(n) \text{ is nonincreasing}, \quad \delta(n)n^{1/2} \uparrow \infty,$$
$$\gamma(n) \text{ is monotone}, \quad \gamma(n)n \uparrow \infty,$$
$$\gamma(n)(\delta(n))^{-2} \text{ is monotone},$$
$$\lim_{n \to \infty} \frac{\gamma(2n)}{\gamma(n)} = \lim_{n \to \infty} \frac{\delta(2n)}{\delta(n)} = 1.$$

Then we have

THEOREM 15.10 (Grill, 1987/B).

$$\mathbf{P}\{\Psi(n) \leq n\gamma(n), M(n) \leq \delta(n)n^{1/2} \ i.o.\} = 0 \quad or \quad 1$$

depending on whether $I_3(\gamma, \delta) < \infty$ *or* $I_3(\gamma, \delta) = \infty$ *where*

$$I_3(\gamma, \delta) = \sum_{n=1}^{\infty} \frac{1}{n(\delta(n))^2} \left(1 + \frac{\delta^2(n)}{\gamma(n)}\right)^{-1/2} \exp\left(-\frac{(4 - 3\gamma(n))\pi^2}{8(\delta(n))^2}\right).$$

Consequence. The limit points of the sequence

$$\{n^{-1}\Psi(n), \pi^{-1}n^{-1/2}(8\log\log n)^{1/2}M(n)\}$$

are the set

$$\{(x, y) : y^2 \geq 4 - 3x, \ 0 \leq x \leq 1\}.$$

Example 3. The inequalities

$$\Psi(n) \leq \gamma n$$

and

$$M(n) \leq (4 - 3\gamma - \varepsilon)^{1/2}\pi \left(\frac{n}{8\log\log n}\right)^{1/2} \quad (0 < \gamma < 1)$$

hold with probability 1 only for finitely many n, i.e. if

$$\Psi(n) \leq \gamma n \quad \text{then} \quad M(n) \geq (4 - 3\gamma - \varepsilon)^{1/2}\pi \left(\frac{n}{8\log\log n}\right)^{1/2} \quad \text{a.s.}$$

A FEW FURTHER RESULTS

for all but finitely many n. Similarly if

$$M(n) \leq (4 - 3\gamma - \varepsilon)^{1/2} \pi \left(\frac{n}{8 \log \log n}\right)^{1/2} \quad \text{then} \quad \Psi(n) \geq \gamma n \quad \text{a.s.}$$

for all but finitely many n. This means that if $M(n)$ is very small then $\Psi(n)$ cannot be too small. For example, choosing $(4 - 3\gamma - \varepsilon)^{1/2} = 1 + \delta$ we have: if

$$M(n) \leq (1 + \delta) \pi \left(\frac{n}{8 \log \log n}\right)^{1/2} \quad \text{then} \quad \Psi(n) \geq (1 - \delta)n \quad \text{a.s.}$$

for all but finitely many n.

Having the above result we formulate the following

Conjecture. For any $\delta > 0$ and for almost all $\omega \in \Omega$ there exists a sequence of integers $0 < n_1 = n_1(\omega, \delta) < n_2 = n_2(\omega, \delta) < \ldots$ such that

$$M(n_i) \leq (1 + \delta) \pi \left(\frac{n_i}{8 \log \log n_i}\right)^{1/2} \quad \text{and} \quad \Psi(n_i) = n_i \quad (i = 1, 2, \ldots).$$

The proofs of the above two theorems of this paragraph are based on the evaluation of the joint distribution of $\Psi(n)$ and S_n. Here we present such a result formulated to Wiener process.

THEOREM 15.11 (Csáki–Grill, 1988). *Let $x > 0$, $0 < y < 1$. Then*

$$\mathbf{P}\{\psi(t) \geq ty, W(t) \geq xt^{1/2}\} = \frac{(1-y)^{3/2}}{x^2 y^{1/2}} \exp\left(-\frac{x^2}{2(1-y)}\right)$$

where $\psi(t)$ is the last zero of $W(s)$, $0 \leq s \leq t$, i.e.

$$\psi(t) = \sup\{s : 0 \leq s \leq t, W(s) = 0\}.$$

15.3 The Ornstein–Uhlenbeck process and a theorem of Darling and Erdős

Consider the Gaussian process $\{V(t) = t^{-1/2} W(t); \ 0 < t < \infty\}$. Then $\mathbf{E}V(t) = 0$, $\mathbf{E}V^2(t) = 1$ and $\mathbf{E}V(t)V(s) = \sqrt{s/t}$, $s < t$. The form of this covariance function immediately suggests that, in order to get a stationary Gaussian process out of $V(t)$, we should consider

$$U_\alpha(t) = V(e^{\alpha t}), \quad -\infty < t < +\infty \quad (\alpha \text{ fixed} > 0).$$

This latter process is a stationary Gaussian process, for $\mathbf{E}U_\alpha(t)U_\alpha(s) = e^{-\alpha|t-s|/2}$, and it is called *Ornstein–Uhlenbeck process*. We will use the notation $U(t) = U_2(t)$, and mention, without proof, the following:

THEOREM 15.12 (Darling–Erdös, 1956).

$$\lim_{T\to\infty} \mathbf{P}\{\sup_{0\le t\le T} U(t) \le a(y,T)\} = \exp(-e^{-y}), \tag{15.5}$$

$$\lim_{T\to\infty} \mathbf{P}\{\sup_{0\le t\le T} |U(t)| \le a(y,T)\} = \exp(-2e^{-y}), \tag{15.6}$$

where for any $-\infty < y < \infty$

$$a(y,T) = \left(y + 2\log T + \frac{1}{2}\log\log T - \frac{1}{2}\log\pi\right)(2\log T)^{-1/2}.$$

We also mention a large deviation type theorem of Qualls and Watanabe (1972) (cf. also Bickel–Rosenblatt, 1973).

THEOREM 15.13 *For any $T > 0$ we have*

$$\lim_{x\to\infty} \frac{2(2\pi)^{1/2} x^3 e^{x^2/2}}{T} \mathbf{P}\{\sup_{0\le t\le T} U(t) > x\} = 1.$$

Applying their own invariance-principle method Darling and Erdős (1956) also proved

THEOREM 15.14

$$\lim_{n\to\infty} \mathbf{P}\{\max_{1\le k\le n} k^{-1/2} S_k \le a(y,\log n)\} = \exp(-e^{-y})$$

and

$$\lim_{n\to\infty} \mathbf{P}\{\max_{1\le k\le n} k^{-1/2} |S_k| \le a(y,\log n)\} = \exp(-2e^{-y})$$

for any $-\infty < y < \infty$.

A strong characterization of the behaviour of $\max_{1\le k\le n} k^{-1/2} S_k$ is given in the following:

THEOREM 15.15

$$\lim_{n\to\infty} \frac{\max_{1\le k\le n} k^{-1/2} S_k}{(2\log\log n)^{1/2}} = 1 \quad a.s.$$

Proof. The LIL of Khinchine implies that

$$\limsup_{n\to\infty} \frac{\max_{1\le k\le n} k^{-1/2} S_k}{(2\log\log n)^{1/2}} \le 1 \quad a.s.$$

A FEW FURTHER RESULTS

Applying Theorem 5.3 we obtain that for any n big enough with probability 1 there exists a
$$n^{1-\delta_n} \leq \kappa = \kappa_n(\omega) \leq n$$
with
$$\delta_n = C \frac{\log_3 n}{(\log_2 n)^{1/2}}, \quad 2^{-2} \leq C \leq 2^{14},$$
such that
$$S_\kappa \geq b_\kappa^{-1}.$$
Hence
$$\max_{1\leq k \leq n} \frac{S_k}{\sqrt{k}} \geq \frac{S_\kappa}{\sqrt{\kappa}} \geq \sqrt{2\log\log \kappa} \geq \sqrt{2\log\log n^{1-\delta_n}} \geq (1-\varepsilon)\sqrt{2\log\log n}$$
and we have Theorem 15.15.

It looks also interesting to study the limit behaviour of the sequence
$$L_n = \max_{0 \leq j < n} \max_{1 \leq k \leq n-j} k^{-1/2}(S_{j+k} - S_j) \quad (n=1,2,\ldots).$$

We prove

THEOREM 15.16
$$1 \leq \liminf_{n\to\infty} \frac{L_n}{(2\log n)^{1/2}} \leq \limsup_{n\to\infty} \frac{L_n}{(2\log n)^{1/2}} = K < \infty \quad a.s.$$
where the exact value of K is unknown.

Proof. Let $a_n = [(\log n)^\alpha]$ ($\alpha > 1$). Then by Theorem 7.13 (see also Section 7.3) for any $\varepsilon > 0$ we have
$$\frac{L_n}{(2\log n)^{1/2}} \geq (2\log n)^{-1/2}(\max_{0\leq j<n} a_n^{-1/2}(S_{j+a_n} - S_j)) \geq 1-\varepsilon \quad a.s.$$
which proves the lower part of the Theorem.

In order to prove its upper part the following result of Hanson and Russo (1983/A (5.11 a)) will be utilized:
If $a_n = f(n) \log n$ with $f(n) \nearrow \infty$ then
$$\limsup_{n\to\infty} \sup_{0\leq j<n} \sup_{a_n \leq k \leq n-j} \frac{k^{-1/2}(S_{j+k} - S_j)}{(2\log n)^{1/2}} \leq 1 \quad a.s.$$

Applying again Theorem 7.13 we obtain
$$\limsup_{n\to\infty} \sup_{0\leq j<n} \sup_{2\log n \leq k \leq a_n} \frac{k^{-1/2}(S_{j+k}-S_j)}{\sqrt{2f(n)\log n}}$$
$$\leq \limsup_{n\to\infty} \sup_{0\leq j<n} \sup_{2\log n \leq k \leq a_n} \frac{S_{j+k}-S_j}{\sqrt{2a_n \log n}} \leq 1 \quad a.s.$$

and clearly
$$\limsup_{n\to\infty} \sup_{0\leq j<n} \sup_{1\leq k\leq 2\log n} \frac{S_{j+k}-S_j}{\sqrt{2k\log n}} \leq 1 \quad \text{a.s.}$$

Since $f(n)$ may converge to infinity arbitrarily slowly we obtain the Theorem by the Zero-One Law.

In the first Edition of this book on the value of K of Theorem 15.16 we presented the Conjecture: $K=1$.

Shao (1995) gave an affirmative answer of this Conjecture.

Let $\nu(n) = \nu(n,S)$ resp. $\nu(T) = \nu(T,W)$ be the smallest integer resp. the smallest positive real number for which
$$(\nu(n))^{-1/2} S_{\nu(n)} = \max_{1\leq k\leq n} k^{-1/2} S_k$$

resp.
$$(\nu(T))^{-1/2} W(\nu(T)) = \max_{1\leq t\leq T} t^{-1/2} W(t).$$

It looks an interesting question to characterize the properties of $\nu(n,S)$ resp. $\nu(T,W)$. Clearly
$$\nu(n,S) = n \quad \text{resp.} \quad \nu(T,W) = T \quad \text{i.o. a.s.}$$

On $\text{LLC}(\nu(\cdot,\cdot))$ we have

THEOREM 15.17 *For any $\varepsilon > 0$*
$$\exp((\log n)^{1-\varepsilon}) \in \text{LLC}(\nu(n,\cdot)).$$

Proof. By Theorem 15.15 and the LIL of Khinchine we have
$$(1-\varepsilon)(2\log\log n)^{1/2} \leq (\nu(n))^{-1/2} S_{\nu(n)} \leq (1+\varepsilon)(2\log\log \nu(n))^{1/2},$$
which implies Theorem 15.17.

Remark 1. Since the Invariance Principle (cf. Section 6.2) only implies that
$$|\max_{1\leq k\leq n} k^{-1/2} S_k - \max_{1\leq t\leq n} t^{-1/2} W(t)| \leq O(1) \quad \text{a.s.},$$
we cannot get Theorem 15.14 from Theorem 15.12 by the Invariance Principle. However, applying Theorem 15.17 we obtain
$$|\max_{1\leq k\leq n} k^{-1/2} S_k - \max_{1\leq t\leq n} t^{-1/2} W(t)| \leq O\left(\exp(-(\log n)^{1-\varepsilon})\right) \quad \text{a.s.}$$
for any $\varepsilon > 0$. Hence we obtain Theorem 15.14 via Theorem 15.12 and the Invariance Principle.

Studying the strong behaviour of $U(t)$ Qualls and Watanabe (1972) proved

THEOREM 15.18 *For any $\varepsilon > 0$*

$$(2\log T)^{1/2} + \left(\frac{3}{2} + \varepsilon\right)\frac{\log\log T}{(2\log T)^{1/2}} \in \text{UUC}(\sup_{0\leq t \leq T} U(t))$$

and

$$(2\log T)^{1/2} + \left(\frac{3}{2} - \varepsilon\right)\frac{\log\log T}{(2\log T)^{1/2}} \in \text{ULC}(\sup_{0\leq t \leq T} U(t)).$$

15.4 A discrete version of the Itô formula

Itô (1942) defined and studied the so-called Itô-integral

$$\int_0^t f(W(s))dW(s)$$

where $f(\cdot)$ is a continuously differentiable function. Here we do not give the definition but we mention an important property of this integral, the celebrated

ITÔ-FORMULA (Itô, 1942).

$$\int_0^t f(W(s))dW(s) = \int_0^{W(t)} f(\lambda)d\lambda - \int_0^t \frac{f'(W(s))}{2}ds. \qquad (15.7)$$

In fact (15.7) is a special case of the so-called Itô-formula. Here we are interested to find the analogue of (15.7) for random walk. In fact we prove

THEOREM 15.19 (Szabados, 1989). *Let $f(k)$ ($k \in \mathbb{Z}^1$) be an arbitrary function and define*

$$g(k) = \begin{cases} \dfrac{f(0)}{2} + \sum_{j=1}^{k-1} f(j) + \dfrac{f(k)}{2} & \text{if } k \geq 1, \\ 0 & \text{if } k = 0, \\ -\dfrac{f(k)}{2} - \sum_{j=k+1}^{-1} f(j) - \dfrac{f(0)}{2} & \text{if } k \leq -1. \end{cases}$$

Then for any $n = 0, 1, 2, \ldots$ we have

$$\sum_{i=0}^n f(S_i)X_{i+1} = g(S_{n+1}) - \frac{1}{2}\sum_{i=0}^n \frac{f(S_{i+1}) - f(S_i)}{X_{i+1}}. \qquad (15.8)$$

Remark 1. The function $g(\cdot)$ can be considered as the discrete analogue of the integral $\int_0^n f(\lambda)d\lambda$, $(f(S_{i+1}) - f(S_i))/X_{i+1}$ is the natural discrete version of $f'(S_i)$ and $\sum_{i=1}^n f(S_i)X_{i+1}$ can be considered as the discrete Itô-integral.

Proof of Theorem 15.19. We get

$$g(S_{i+1}) - g(S_i) = f(S_i)X_{i+1} + \frac{f(S_{i+1}) - f(S_i)}{2X_{i+1}}. \tag{15.9}$$

(In order to check (15.9) consider the cases corresponding to $S_i = 0$, $S_i > 0$, $S_i < 0$; $X_{i+1} = 1$, $X_{i-1} = -1$ separately.) Summing up (15.9) from 0 to n we obtain (15.8).

Example 1. Let $f(t) = t$. Then by (15.7) we have

$$\int_0^t W(s)dW(s) = \frac{W^2(t)}{2} - \frac{t}{2} \tag{15.10}$$

and by (15.8)

$$\sum_{i=0}^n S_i X_{i+1} = g(S_{n+1}) - \frac{n+1}{2} = \frac{S_{n+1}^2}{2} - \frac{n}{2}. \tag{15.11}$$

(15.10) and (15.11) completely agree.

Example 2. Let $f(t) = t^2$. Then by (15.7) we have

$$\int_0^t W^2(s)dW(s) = \frac{W^3(t)}{3} - \int_0^t W(s)ds \tag{15.12}$$

and by (15.8)

$$\sum_{i=0}^n S_i^2 X_{i+1} = g(S_{n+1}) - \sum_{i=0}^n S_i - \frac{S_{n+1}}{2} = \frac{S_{n+1}^3}{3} - \sum_{i=0}^n S_i - \frac{S_{n+1}}{3}. \tag{15.13}$$

The term $-S_{n+1}/3$ of (15.13) is not expected by (15.12). However, we know that the orders of magnitude of the terms $\sum_{i=0}^n S_i^2 X_{i+1}$, $S_{n+1}^3/3$ and $\sum_{i=0}^n S_i$ are $n^{3/2}$, while that of S_{n+1} is $n^{1/2}$ only.

Remark 2. Applying the Invariance Principle 1 (15.7) can be obtained as a consequence of (15.8).

The celebrated Tanaka formula gives a representation of the local time of a Wiener process via Itô-integral.

TANAKA FORMULA (cf. McKean, 1969). *For any $x \in \mathbb{R}^1$ and $t > 0$ we have*

$$\eta(x,t) = |W(t) - x| - |x| - \int_0^t \text{sign}(W(s) - x)dW(s).$$

Here we are interested in giving the discrete analogue of this formula. In order to do so instead of $\xi(\cdot,\cdot)$ we consider a slightly modified version of the local time. Let

$$\xi^*(x,n) = \#\{k: \ 0 \leq k < n, \ S_k = x\}.$$

Then we have

THEOREM 15.20 (Csörgő–Révész, 1985). *For any $x \in \mathbb{Z}^1$ and $n = 1, 2, \ldots$*

$$\xi^*(x,n) = |S_n - x| - |x| - \sum_{k=0}^{n-1} \text{sign}(S_k - x)X_{k+1}. \tag{15.14}$$

Proof. Observe that

$$\sum_{k=0}^{\rho_1(x)-1} \text{sign}(S_k - x)X_{k+1} = \begin{cases} -|x| & \text{if } x \neq 0, \\ -1 & \text{if } x = 0, \end{cases}$$

$$\sum_{k=\rho_j(x)}^{\rho_{j+1}(x)-1} \text{sign}(S_k - x)X_{k+1} = -1 \quad (j = 1, 2, \ldots),$$

$$\sum_{k=\rho_\mu(x)}^{n-1} \text{sign}(S_k - x)X_{k+1} = |S_n - x| - 1$$

where

$$\mu = \begin{cases} \xi^*(x,n) & \text{if } x \neq 0, \\ \xi^*(x,n) - 1 & \text{if } x = 0. \end{cases}$$

The above three equations easily imply (15.14).

Remark 3. The Tanaka formula can be proved from (15.14) using Invariance Principle 1.

Chapter 16

Summary of Part I

	Exact distr.	Limit distr.	Upper classes	Lower classes	Strassen type theorems
S_n	Th. 2.1	Th.'s 2.9, 2.10	EFKP LIL Sect. 5.2	Th. 5.1	Strassen's Th. 1. Sect. 8.1
M_n	Th. 2.6	Th. 2.13	EFKP LIL Sect. 5.2	Th. of Chung Sect. 5.3	Th. 8.2; Wichura's Theorem Sect. 8.4
M_n^+	Th. 2.4	Th. 2.12	EFKP LIL Sect. 5.2	Th. of Hirsch Sect. 5.3	Th. 8.2
$\xi(0,n)$	Th. 9.3	(9.11)	Th. 11.1	Th. 11.1	Th. 11.18
		The limit behaviour of $\xi(0,n)$ is the same as that of M_n^+ by Th. 10.3			
ρ_n	Th. 9.3	(9.9)	Th. 11.6	Th. 11.6	Th. 11.19
		Since $\xi(0,\rho_n) = n$ a description of $\xi(0,n)$ gives a description of ρ_n			
$\xi(x,n)$	Th. 9.4	The limit behaviour is the same as that of $\xi(0,n)$ for fixed $x, n \to \infty$.			Th. of Donsker-Varadhan Sect. 11.4
$\Theta(n)$	Th. 9.5	The limit behaviour is the same as that of $\xi(0,n)/2$ (cf. (10.11)) or $M_n^+/2$ (cf. Th. 10.4).			
$\xi(n)$		Th. 9.14 (10.6)	Th. 11.5	Th. 11.4	
ζ_n	Th. 9.8	(9.12)	Th. 12.1	Th. 12.2	
$\Psi(n)$	Th. 9.9	(9.12)	Trivial	Th. 12.1	

CHAPTER 16

	Exact distr.	Limit distr.	Upper classes	Lower classes	Strassen type theorems
$R(n)$			Th. 12.1	Th. 12.3	
$\hat{R}(n)$			Th. 12.1	Th. 12.3	
		\multicolumn{4}{l}{The limit behaviour of $\hat{R}(n)$ is the same as that of $R(n)$ by Th. 10.3}
$R^*(n)$			Th. 15.6	Th. 15.7	
$\mu(n)$			Trivial	Th. 15.1	
$\mu^+(n)$	Th. 9.10	(9.12)	Trivial	Th. 12.1	
$\chi(n)$	Th. 12.25		Th. 12.26	Th. 12.26	
Z_n	Th. 2.7	Th. 2.14	Th. 7.2	Th. 7.3	

Replacing $S_n, M_n, M_n^+, \xi(0,n) \ldots$ by $W(t), m(t), m^+(t), \eta(0,t), \ldots$ respectively the above-mentioned results remain true by the Invariance Principle 1 (cf. Section 6.4) and Theorem 10.1, with the exception that there is no immediate analogue of $\Theta(n)$ and the natural analogue of $\chi(n)$ does not have any interest.

In some cases we also investigated the joint behaviour of the r.v.'s of the above table. A table for these results is

	M_n^+	M_n^-	$\mu(n)$	$\Psi(n)$
S_n	Th.'s 2.5, 2.6, 5.8	Th.'s 2.5, 2.6, 5.8		Th. 15.9
M_n			Th.'s 15.4, 15.5 15.8	Th.'s 15.10, 15.11
M_n^+		Th.'s 5.5, 5.6		

Clearly many of the results of Part I are not included in the above two tables. For example, the results about increments, the rate of convergence of Strassen-type theorems, the results on the stability of the local time, etc. are missing from the above tables. A summary on the increments of the Wiener process is given at the end of Section 7.2.

"While 10 or 11 dimensions doesn't sound much like the spacetime we experience, the idea was that the other 6 or 7 dimensions are curled up so small that we don't notice them; we are only aware of the remaining 4 large and nearly flat dimensions."

(S. Hawking: The Universe in a Nutshell)

II. SIMPLE SYMMETRIC RANDOM WALK IN $\mathbb{Z}^d$

Notations

1. Consider a random walk on the lattice $\mathbb{Z}^d$. This means that if the moving particle is in $x \in \mathbb{Z}^d$ in the moment n, then at the moment $n+1$ the particle can move with equal probabilities to any of the $2d$ neighbours of x independently of how the particle achieved x. (The neighbours of an $x \in \mathbb{Z}^d$ are those elements of $\mathbb{Z}^d$ whose $d-1$ coordinates coincide with the corresponding coordinates of x and one coordinate differs by $+1$ or -1.)

   Let $S_n = S(n)$ be the location of the particle after n steps (i.e. in the momemt n) and assume that $S_0 = 0$. Equivalently: $S_n = X_1 + X_2 + \cdots + X_n$ $(n = 1, 2, \ldots)$ where $X_1, X_2, \ldots$ is a sequence of independent, identically distributed random vectors with

   $$\mathbf{P}\{X_1 = e_i\} = \mathbf{P}\{X_1 = -e_i\} = \frac{1}{2d} \quad (i = 1, 2, \ldots d)$$

   where $e_1, e_2, \ldots, e_d$ are the orthogonal unit-vectors of $\mathbb{Z}^d$.

2. For any $x = (x_1, x_2, \ldots, x_d) \in \mathbb{R}^d$ let $\|x\|^2 = \sum_{i=1}^{d} x_i^2$.

3. $M_n = M(n) = \max_{0 \leq k \leq n} \|S_k\|$.

4. $\xi(x, n) = \#\{k : 0 < k \leq n, S_k = x\}$ $(x \in \mathbb{Z}^d, n = 1, 2, \ldots)$.

5. $\xi(n) = \max_{x \in \mathbb{Z}^d} \xi(x, n)$.

6. $\rho_1 = \min\{k : k > 0, S_k = 0\}$,
   $\rho_2 = \min\{k : k > \rho_1, S_k = 0\}$,
   $\ldots\ldots\ldots\ldots\ldots$,
   $\rho_n = \min\{k : k > \rho_{n-1}, S_k = 0\}$.

7. $p_{2k} = \mathbf{P}\{S_{2k} = 0\}$, $p_{2k+1} = 0$ $(k = 0, 1, 2, \ldots)$.

8. $q_{2k} = \mathbf{P}\{S_{2k} = 0, S_{2k-2} \neq 0, \ldots, S_4 \neq 0, S_2 \neq 0\} = \mathbf{P}\{\rho_1 = 2k\}$
   $= \mathbf{P}\{\xi(0, 2k) = 1, \xi(0, 2k-1) = 0\}$ $(k = 1, 2, \ldots)$.

9. $\gamma_{2k+1} = \gamma_{2k} = \mathbf{P}\{S_2 \neq 0, S_4 \neq 0, \ldots, S_{2k-2} \neq 0\}$
   $= \mathbf{P}\{\xi(0, 2k-2) = 0\} = q_{2k} + \gamma_{2k+2}$
   $= \sum_{j=k}^{\infty} q_{2j} + \lim_{n \to \infty} \gamma_{2n}$ $(k = 2, 3, \ldots)$ and $\gamma_2 = 1$.

10. $\gamma = \lim_{n\to\infty} \gamma_{2n} = \mathbf{P}\{\lim_{n\to\infty} \xi(0,n) = 0\} = \mathbf{P}\{\rho_1 = \infty\} = 1 - \sum_{j=1}^{\infty} q_{2j}.$

11. Let T_x be the first hitting time of x, i.e. $T_x = \min\{i:\ i \geq 1,\ S_i = x\}$ with convention that $T_x = \infty$ if there is no i with $S_i = x$. Let $T = T_0$. In general, for a subset A of $\mathbb{Z}^d$, let T_A denote the first time when the random walk visits A i.e. $T_A = \min\{i:\ i > 0,\ S_i \in A\}$.

12. Let $\gamma(x)$ be the probability that the random walk never visits x i.e. $\gamma(x) = \mathbf{P}\{T_x = \infty\}$.

13. $z_x = \mathbf{P}\{T < T_x\}$.

14. $s_x = \mathbf{P}\{T_x < T\}$.

15. $\mathcal{S}(1) = \{x:\ x \in \mathbb{Z}^d,\ \|x\| = 1\}$.

16. Let $W(t) = (W_1(t), W_2(t), \ldots, W_d(t))$, where $W_1(t), W_2(t), \ldots, W_d(t)$ are independent Wiener processes. Then the $\mathbb{R}^d$ valued process $W(t)$ is called a d-dimensional Wiener process.

17. $m(t) = \max_{0 \leq s \leq t} \|W(s)\|$.

Chapter 17

The Recurrence Theorem

This chapter is devoted to proving the

RECURRENCE THEOREM (Pólya, 1921).

$$\mathbf{P}\{S_n = 0 \ \ i.o.\} = \begin{cases} 1 & if \quad d \leq 2, \\ 0 & if \quad d > 2. \end{cases}$$

In the early sixties György Pólya gave a talk in Budapest where he told the story of this Theorem. He studied in the teens at the ETH (Federal Polytechnical School) in Zürich, where he had a roommate. It happened once that the roommate was visited by his fiancée. From politeness Pólya left the room and went for a walk on a nearby mountain. After some time he met the couple. Both the couple and Pólya continued their walks in different directions. However, they met again. When it happened the third time, Pólya had a bad feeling. The couple might think that he is spying on them. Hence he asked himself what is the probability of such meetings if both parties are walking randomly and independently. If this probability is big, then Pólya might claim that he is innocent.

By a simple generalization of the Recurrence Theorem it is easy to see that if the parties wander independently according to the law of the random walk then they meet infinitely often with probability 1.

This Theorem was proved for $d = 1$ in Section 3.1. Hence we concentrate on the case $d \geq 2$.

LEMMA 17.1 For any $n = 1, 2, \ldots$; $d = 1, 2, \ldots$

$$\mathbf{P}\{S_{2n} = 0\} = (2d)^{-2n} \sum_{n_1+n_2+\cdots+n_d=n} \frac{(2n)!}{(n_1!n_2!\ldots n_d!)^2}$$

$$= (2d)^{-2n} \binom{2n}{n} \sum_{n_1+n_2+\cdots+n_d=n} \left(\frac{n!}{n_1!n_2!\ldots n_d!}\right)^2.$$

Proof is trivial by a combinatorial argument.

LEMMA 17.2 For any $d = 1, 2, \ldots$ as $n \to \infty$ we have

$$\mathbf{P}\{S_{2n} = 0\} \approx 2\left(\frac{d}{4n\pi}\right)^{d/2}.$$

Further, in case $d = 2$

$$\mathbf{P}\{S_{2n} = 0\} = \frac{1}{n\pi} + O\left(\frac{1}{n^2}\right).$$

Proof can be obtained by the Stirling formula.

For later reference we give the following analogues of Lemmas 17.1 and 17.2.

LEMMA 17.3 *Let* $d = 2$. *Then*

$$\mathbf{P}\{S_{2n} = (x,y)\} = 4^{-2n} \sum_{k \in A_n(x,y)} \binom{2n}{k}\binom{k}{(k+x)/2}\binom{2n-k}{(2n-k+y)/2}$$

$$= 4^{-2n} \binom{2n}{n+(x+y)/2}\binom{2n}{n+(x-y)/2}$$

provided that

$$x + y \equiv 0 \pmod{2} \quad \text{and} \quad |x| + |y| \leq 2n$$

where

$k \in A_n(x,y)$ *if and only if* $k \equiv x \pmod{2}$ *and* $|x| \leq k \leq 2n - |y|$.

Proof is trivial by a combinatorial argument.

LEMMA 17.4 (Erdős–Taylor, 1960/A, (2.9) and (2.10)). *Let* $d = 2$, $x = (x_1, x_2)$ *and* $x_1 + x_2 \equiv 0 \pmod{2}$. *Then*

$$\mathbf{P}\{S_{2n} = x\} \begin{cases} = \dfrac{1}{\pi n} + (\|x\|^2 + 1)O(n^{-2}) & \text{if} \quad n > \|x\|^2, \\ \leq \left(\dfrac{1}{\pi n} + O(n^{-2})\right) \exp\left(-\dfrac{\|x\|^2}{2n}\right) & \text{if} \quad n \leq \|x\|^2. \end{cases}$$

Proof can be obtained by the Stirling formula.

Similarly one can obtain

LEMMA 17.5 *Let* $d \geq 2$, $x = (x_1, x_2, \ldots, x_d)$ *and* $x_1 + \cdots + x_d \equiv 0 \pmod{2}$. *Then*

$$\mathbf{P}\{S_{2n} = x\} \begin{cases} = O(n^{-d/2}) & \text{if} \quad n > \|x\|^2, \\ \leq O\left(n^{-d/2} \exp\left(-\dfrac{\|x\|^2}{2n}\right)\right) & \text{if} \quad n \leq \|x\|^2. \end{cases}$$

THE RECURRENCE THEOREM

A more precise version of Lemma 17.5 is the following:

LEMMA 17.6 (Lawler, 1991, Theorem 1.2.1). *Let $x = (x_1, x_2, \ldots, x_d)$, $x_1 + x_2 + \cdots + x_d \equiv n \pmod{2}$ and $d \geq 2$. Then*

$$\mathbf{P}\{S_n = x\} = 2\left(\frac{d}{2\pi n}\right)^{d/2} \exp\left(-\frac{d\|x\|^2}{2n}\right) + R_n(x)$$

where

$$R_n(x) \leq \min(O(n^{-(d+2)/2}), O(\|x\|^{-2} n^{-d/2})).$$

Lemma 17.6 by a nontrivial calculation implies

LEMMA 17.7 (Lawler, 1991, Theorem 1.5.4). *Let $d \geq 3$. Then*

$$\lim_{\|x\| \to \infty} \frac{\sum_{n=0}^{\infty} \mathbf{P}\{S_n = x\}}{a_d \|x\|^{2-d}} = 1$$

where

$$a_d = \frac{d}{2} \Gamma\left(\frac{d}{2} - 1\right) \pi^{-d/2} = \frac{2}{(d-2) w_d},$$

where w_d is the volume of the unit ball in $\mathbb{R}^d$.

LEMMA 17.8 *For any $d \geq 3$ there exists a positive constant C_d such that*

$$\mathbf{P}\{S_n = x \text{ for some } n\} = \mathbf{P}\{J(x) = 1\} = \frac{C_d + o(1)}{R^{d-2}} \quad (R \to \infty)$$

where $R = \|x\|$ and

$$J(x) = \begin{cases} 0 & \text{if } \xi(x, n) = 0 \text{ for every } n = 0, 1, 2, \ldots, \\ 1 & \text{otherwise.} \end{cases}$$

Proof. Clearly

$$\mathbf{P}\{S_n = x\} = \sum_{k=0}^{n} \mathbf{P}\{S_k = x, S_j \neq x, \ j = 0, 1, 2, \ldots, k-1\} \mathbf{P}\{S_{n-k} = 0\}$$

and

$$\sum_{n=0}^{\infty} \mathbf{P}\{S_n = x\}$$

$$= \sum_{n=0}^{\infty} \sum_{k=0}^{n} \mathbf{P}\{S_k = x, S_j \neq x, \ j = 0, 1, 2, \ldots, k-1\} \mathbf{P}\{S_{n-k} = 0\}$$

$$= \sum_{n=0}^{\infty} \mathbf{P}\{S_n = 0\} \sum_{k=0}^{\infty} \mathbf{P}\{S_k = x, S_j \neq x, \ j = 0, 1, 2, \ldots, k-1\}.$$

Since by Lemma 17.5
$$\sum_{n=0}^{\infty} \mathbf{P}\{S_n = x\} = (K_d + o(1))R^{2-d} \quad (R \to \infty)$$
and by Lemma 17.2
$$\sum_{n=0}^{\infty} \mathbf{P}\{S_n = 0\} < \infty,$$
we obtain
$$\mathbf{P}\{J(x) = 1\} = \sum_{k=0}^{\infty} \mathbf{P}\{S_k = x, S_j \neq x, \ j = 0, 1, 2, \ldots, k-1\}$$
$$= (C_d + o(1))R^{2-d}.$$
Hence we have the Lemma.

Remark 1. The proof of Lemma 17.8 shows that
$$C_d = \frac{a_d}{\sum_{n=0}^{\infty} \mathbf{P}\{S_n = 0\}}.$$

LEMMA 17.9 *In case $d = 2$*
$$\lim_{n \to \infty} \frac{\sum_{j=1}^{n} \sum_{k=1}^{n} \mathbf{P}\{S_{2j} = 0, S_{2k} = 0\}}{\left(\sum_{k=1}^{n} \mathbf{P}\{S_{2k} = 0\}\right)^2} = 2.$$

Proof. Clearly
$$\sum_{j=1}^{n} \sum_{k=1}^{n} \mathbf{P}\{S_{2j} = 0, S_{2k} = 0\}$$
$$= 2 \sum_{j=1}^{n} \sum_{k=1}^{n-j} \mathbf{P}\{S_{2j} = 0, S_{2j+2k} = 0\} + \sum_{j=1}^{n} \mathbf{P}\{S_{2j} = 0\}$$
$$= 2 \sum_{j=1}^{n} \sum_{k=1}^{n-j} \mathbf{P}\{S_{2j} = 0\} \mathbf{P}\{S_{2k} = 0\} + \sum_{j=1}^{n} \mathbf{P}\{S_{2j} = 0\}$$
$$\approx \frac{2}{\pi^2} \sum_{j=1}^{n} \sum_{k=1}^{n-j} j^{-1} k^{-1} \approx 2 \left(\frac{\log n}{\pi}\right)^2 \approx 2 \left(\sum_{k=1}^{n} \mathbf{P}\{S_{2k} = 0\}\right)^2.$$

THE RECURRENCE THEOREM

Hence we have Lemma 17.9.

LEMMA 17.10 *Let*

$$p = p(d) = \mathbf{P}\{\exists n : n \geq 1, S_n = 0\} \quad (d = 1, 2, \ldots)$$

Then

$$\mathbf{P}\{S_n = 0 \ \ i.o.\} = \begin{cases} 0 & \text{if } p < 1, \\ 1 & \text{if } p = 1. \end{cases}$$

Proof is trivial.

Proof 1 of the Recurrence Theorem. In case $d = 2$ Lemma 17.9 and Borel–Cantelli lemma 2* of Section 4.1 imply that

$$p \geq \mathbf{P}\{S_n = 0 \ \ i.o.\} \geq \frac{1}{2}.$$

Hence by the zero-one law (cf. Section 3.2) we have the Recurrence Theorem in case $d = 2$.

In case $d \geq 3$ it follows from Lemma 17.2 and Borel–Cantelli lemma 1 (cf. Section 4.1).

Remark 2. Lemma 17.9 is also true in the case $d = 1$. (The proof is essentially the same.) Hence we obtain a new proof of the recurrence theorem in the case $d = 1$. The third proof (cf. Section 3.2) applied in case $d = 1$ does not work in the case $d \geq 2$. The idea of the first proof can be applied in the case $d = 2$ but it requires hard work. The second proof can be used without any difficulty.

Proof 2 of the Recurrence Theorem. Introduce the following notations:

$$p_0 = 1,$$

$$p_{2k} = \mathbf{P}\{S_{2k} = 0\} \approx 2\left(\frac{d}{4k\pi}\right)^{d/2},$$

$$q_{2k} = \mathbf{P}\{S_{2k} = 0, S_{2k-2} \neq 0, S_{2k-4} \neq 0, \ldots, S_2 \neq 0\},$$

$$P(z) = \sum_{k=0}^{\infty} p_{2k} z^{2k},$$

$$Q(z) = \sum_{k=1}^{\infty} q_{2k} z^{2k}.$$

Between the sequences $\{p_{2k}\}$ and $\{q_{2k}\}$ one can easily see the following relations:

(0) $p_0 = 1$,
(i) $p_2 = q_2$,
(ii) $p_4 = q_4 + q_2 p_2$,
(iii) $p_6 = q_6 + q_4 p_2 + q_2 p_4$,
$\ldots\ldots\quad\ldots\ldots$,
(k) $p_{2k} = q_{2k} + q_{2k-2} p_2 + \cdots + q_2 p_{2k-2}$,
$\ldots\ldots\quad\ldots\ldots$

Multiplying the k-th equation by z^{2k} and adding them up to infinity we obtain

$$P(z) = P(z)Q(z) + 1 \quad \text{i.e.} \quad Q(z) = 1 - \frac{1}{P(z)}. \tag{17.1}$$

By Lemma 17.2 we obtain

$$\lim_{z \nearrow 1} P(z) \begin{cases} = \infty & \text{if } d \leq 2, \\ < \infty & \text{if } d \geq 3, \end{cases}$$

i.e.

$$Q(1) = \lim_{z \nearrow 1} Q(z) \begin{cases} = 1 & \text{if } d \leq 2, \\ < 1 & \text{if } d \geq 3. \end{cases}$$

Since $Q(1) = \sum_{k=1}^{\infty} q_{2k}$ is the probability that the particle eventually visits the origin we obtain

$$Q(1) = \sum_{k=1}^{\infty} q_{2k} = \begin{cases} 1 - \left(\sum_{k=0}^{\infty} p_{2k}\right)^{-1} < 1 & \text{if } d \geq 3, \\ 1 & \text{if } d = 2. \end{cases}$$

Hence we have the theorem.

Remark 3. In the case $d \geq 3$ the formula

$$\mathbf{P}\{S_n = 0 \text{ for some } n = 1, 2 \ldots\}$$
$$= 1 - \mathbf{P}\{\lim_{n \to \infty} \xi(0, n) = 0\} = \sum_{k=1}^{\infty} q_{2k} = 1 - \frac{1}{\sum_{k=0}^{\infty} p_{2k}} \tag{17.2}$$

is applicable to the evaluation of the probability that the particle returns to the origin. In fact by Lemma 17.2 we have

$$\sum_{k=1}^{\infty} q_{2k} = 1 - \frac{1}{\sum_{k=0}^{n-1} p_{2k} + \sum_{k=n}^{\infty} p_{2k}} = 1 - \frac{1}{\sum_{k=0}^{n-1} p_{2k}} + O(n^{1-d/2})$$

where $\sum_{k=0}^{n-1} p_{2k}$ can be numerically evaluated by Lemma 17.1. For example, in the case $d = 3$ one can obtain $\sum_{k=1}^{\infty} q_{2k} \sim 0.35$. In fact this method is not easily applicable for concrete calculations. Griffin (1990) gave a better version of it and evaluated the probability of recurrence for many values of d. For example, for $d = 3, 4, 20$ he calculated that the probabilities of recurrence are 0.340537, 0.193202, 0.026376. Note that in the case $d = 20$, $\mathbf{P}\{S_2 = 0\} = 0.025$. Hence the probability that the particle returns to the origin but not in the second step is 0.001376.

It also looks interesting to estimate the probability

$$\mathbf{P}\{2n \leq \rho_1 < \infty\} = 1 - \gamma - \sum_{j=1}^{n-1} q_{2j} = \mathbf{P}\{\xi(0, 2n-2) = 0\} - \gamma = \sum_{k=n}^{\infty} q_{2k}$$

that the particle returns to the origin but not in the first $2n - 2$ steps. We prove

LEMMA 17.11 *For any $d \geq 3$ and $\varepsilon > 0$ we have*

$$C(d)n^{(2-d)/2} \leq \mathbf{P}\{2n \leq \rho_1 < \infty\} \leq \frac{4+\varepsilon}{d-2}\left(\frac{d}{4\pi}\right)^{d/2} n^{(2-d)/2} \quad (17.3)$$

where $C(d)$ is a small enough positive constant.

Proof. The upper part of (17.3) is trivial. In fact by Lemma 17.2 we have

$$\sum_{k=n}^{\infty} q_{2k} \leq \sum_{k=n}^{\infty} p_{2k} \leq 2(1+\varepsilon)\left(\frac{d}{4\pi}\right)^{d/2} \sum_{k=n}^{\infty} k^{-d/2}$$

$$\leq 2(1+2\varepsilon)\left(\frac{d}{4\pi}\right)^{d/2} \frac{n^{1-d/2}}{d/2 - 1}$$

$$= (1+2\varepsilon)\frac{4}{d-2}\left(\frac{d}{4\pi}\right)^{d/2} n^{(2-d)/2}.$$

Now we turn to the proof of the lower part of (17.3). Clearly by Lemma 17.8 for any $r > 0$ we have

$$\mathbf{P}\{2n \leq \rho_1 < \infty\} \geq \mathbf{P}\{2n \leq \rho_1 < \infty, |S_{2n}| \leq rn^{1/2}\}$$

$$= \sum_{|k| \leq rn^{1/2}} \mathbf{P}\{2n \leq \rho_1 < \infty, S_{2n} = k\}$$

$$= \sum_{|k| \leq rn^{1/2}} \mathbf{P}\{\rho_1 \geq 2n, S_{2n} = k\}\mathbf{P}\{\exists j : j > 0, S_j = k\}$$

$$\geq \min_{|k|\leq rn^{1/2}} \mathbf{P}\{\exists j: j > 0, S_j = k\} \sum_{|k|\leq rn^{1/2}} \mathbf{P}\{\rho_1 \geq 2n, S_{2n} = k\}$$

$$= \frac{O(1)}{r^{d-2}n^{(d-2)/2}} \mathbf{P}\{\rho_1 \geq 2n, S_{2n} \leq rn^{1/2}\}$$

$$= \frac{O(1)}{r^{d-2}n^{(d-2)/2}} (\mathbf{P}\{\rho_1 \geq 2n\} + \mathbf{P}\{S_{2n} \leq rn^{1/2}\}$$

$$- \mathbf{P}\{\{\rho_1 \geq 2n\} \cup \{S_{2n} \leq rn^{1/2}\}\})$$

$$\geq \frac{O(1)}{r^{d-2}n^{(d-2)/2}} (\gamma + \mathbf{P}\{S_{2n} \leq rn^{1/2}\} - 1).$$

Choose r so big that for any $n > 0$

$$\mathbf{P}\{S_{2n} \leq rn^{1/2}\} \geq 1 - \frac{\gamma}{2}.$$

Hence we have the lower part of (17.3).

Having Lemma 17.11 it is natural to study the properties of the tail distribution of ρ_t. We prove

LEMMA 17.12 *Let $d \geq 3$.*

$$\frac{O(1)(1-\gamma)^t}{n^{d/2-1}} \leq \mathbf{P}\{n \leq \rho_t < \infty\} \leq O(1)(1-\gamma)^t \frac{t^{d/2}}{n^{d/2-1}}.$$

Proof. Let $e_1, e_2, \ldots$ be the lengths of excursions of $\{S_k\}$ away from 0 and let

$$E_t = \max(e_1, e_2, \ldots, e_t).$$

Note that the elements of the sequence $\{e_i\}$ are infinity except finitely many. Then

$$A := \mathbf{P}\{n \leq \rho_t < \infty\} = \mathbf{P}\{n \leq e_1 + e_2 + \cdots e_t < \infty\}$$

and by Lemma 17.11

$$A \leq \mathbf{P}\left\{\bigcap_{i=1}^{t}\{e_i < \infty\} \cup \left\{\exists: 1 \leq i \leq t, e_i \geq \frac{n}{t}\right\}\right\}$$

$$\leq O(1)t(1-\gamma)^t \left(\frac{n}{t}\right)^{(2-d)/2}.$$

Similarly by Lemma 17.11 we have

$$A \geq \mathbf{P}\left\{\bigcap_{i=1}^{t}\{e_i < \infty\}, \{e_1 \geq n\}\right\} \geq O(1)(1-\gamma)^t n^{(2-d)/2}.$$

THE RECURRENCE THEOREM

Hence we have Lemma 17.12.

It is hard to say anything (except some triviality) about the probability $\gamma(x)$ (the probability that the random walk never visits x).

LEMMA 17.13 *Let $d \geq 3$. Then*

$$\gamma \leq \gamma(x) < 1.$$

Proof. The inequality $\gamma(x) < 1$ follows from the Recurrence Theorem. The other part of the inequality is implied by

$$1 - \gamma(x) = \mathbf{P}\{\exists n : n \geq 0, \ S_n \in \mathcal{S}(1) + x\}(1 - \gamma) \leq 1 - \gamma.$$

The next Lemma tells us how the probabilities z_x and s_x depend on γ_x.

LEMMA 17.14

$$z_x = 1 - \frac{\gamma}{1 - (1 - \gamma(x))^2}, \qquad (17.4)$$

$$s_x = (1 - \gamma(x))(1 - z_x), \qquad (17.5)$$

$$z_x + s_x = 1 - \mathbf{P}\{T = T_x = \infty\}$$

$$= 1 - \frac{\gamma}{2 - \gamma(x)}, \qquad (17.6)$$

$$\mathbf{P}\{\xi(0, \infty) + \xi(x, \infty) = j\} = (1 - z_x - s_x)(z_x + s_x)^j \qquad (17.7)$$
$$(j = 0, 1, 2, \ldots).$$

Proof. Let $Z(A)$ denote the number of visits in the set A up to the first return to 0, i.e.

$$Z(A) = \sum_{n=1}^{T} I\{S_n \in A\}$$

where $I\{\cdot\}$ denotes the usual indicator function.

Let $A^{(x)} = \{0, x\}$ (i.e. a two-points set) and observe that

$$\mathbf{P}\{Z(A^{(x)}) = j + 1, \ T < \infty\} = \begin{cases} z_x & \text{if } j = 0, \\ s_x^2 z_x^{j-1} & \text{if } j > 0. \end{cases} \qquad (17.8)$$

Summing up (17.8) we get

$$\sum_{j=0}^{\infty} \mathbf{P}\{Z(A^{(x)}) = j + 1, \ T < \infty\} = z_x + \frac{s_x^2}{1 - z_x}$$

$$= \mathbf{P}\{T < \infty\} = 1 - \gamma. \qquad (17.9)$$

On the other hand, one can easily see that

$$\begin{aligned}1-\gamma = \mathbf{P}\{T<\infty\} &= \mathbf{P}\{T<T_x\}+\mathbf{P}\{T>T_x,\ T<\infty\}\\ &= \mathbf{P}\{T<T_x\}+\mathbf{P}\{T>T_x\}\mathbf{P}\{T_x<\infty\}\\ &= z_x + s_x(1-\gamma(x)).\end{aligned} \qquad (17.10)$$

Now (17.9) and (17.10) imply Lemma 17.14.

LEMMA 17.15

$$1-\gamma_x(n) := \mathbf{P}\{T_x<n\} = 1-\gamma(x)+\frac{O(1)}{n^{d/2-1}}, \qquad (17.11)$$

$$z_x(n) := \mathbf{P}\{T<\min(n,T_x)\} = z_x + \frac{O(1)}{n^{d/2-1}}, \qquad (17.12)$$

$$s_x(n) := \mathbf{P}\{T_x<\min(n,T)\} = s_x + \frac{O(1)}{n^{d/2-1}}, \qquad (17.13)$$

where $O(1)$ is uniform in x.

Proof. It is a trivial consequence of Lemma 17.11.

Chapter 18

Wiener Process and Invariance Principle

Let $W_1(t), W_2(t), \ldots, W_d(t)$ be a sequence of independent Wiener processes. The $\mathbb{R}^d$ valued process $W(t) = \{W_1(t), W_2(t), \ldots, W_d(t)\}$ is called a d-dimensional Wiener process. We ask how the random walk S_n can be approximated by $W(t)$. The situation is very simple if $d = 2$.

Consider a new coordinate system with axes $y = x$, $y = -x$. In this coordinate system

$$X_n = \begin{cases} (2^{-1/2}, 2^{-1/2}) & \text{if } X_n = (1,0) \text{ in the original system}, \\ (-2^{-1/2}, 2^{-1/2}) & \text{if } X_n = (0,1) \text{ in the original system}, \\ (-2^{-1/2}, -2^{-1/2}) & \text{if } X_n = (-1,0) \text{ in the original system}, \\ (2^{-1/2}, -2^{-1/2}) & \text{if } X_n = (0,-1) \text{ in the original system}. \end{cases}$$

Observe that the coordinates of X_n are independent r.v.'s in the new coordinate system (it is not so in the old one); hence by Invariance Principle 1 of Section 6.3 there exist two independent Wiener processes $W_1(t)$ and $W_2(t)$ such that

$$|2^{-1/2} W_1(n) - \hat{S}_n^{(1)}| = O(\log n) \quad \text{a.s.}$$

and

$$|2^{-1/2} W_2(n) - \hat{S}_n^{(2)}| = O(\log n) \quad \text{a.s.}$$

where $\hat{S}_n^{(1)}, \hat{S}_n^{(2)}$ are the independent coordinates of S_n in the new coordinate system. Consequently we have

THEOREM 18.1 *Let $d = 2$. Then on a rich enough probability space $\{\Omega, \mathcal{F}, \mathbf{P}\}$ one can define a Wiener process $W(\cdot) \in \mathbb{R}^2$ and a random walk $S_n \in \mathbb{Z}^2$ such that*

$$\|S_n - 2^{-1/2} W(n)\| = O(\log n) \quad \text{a.s.}$$

In the case $d > 2$ the above idea does not work. Instead we define

$$\mathcal{K}_n^{(i)} = \#\{k: \ 1 \leq k \leq n, \ X_k = e_i \text{ or } -e_i\} \quad (i = 1, 2, \ldots, d)$$

where e_i is the i-th unit vector in $\mathbb{Z}^d$. Then by the LIL we have

$$\left|\mathcal{K}_n^{(i)} - \frac{n}{d}\right| \leq (1+\varepsilon)\left(\frac{1}{d}\left(1 - \frac{1}{d}\right)\right)^{1/2} b_n^{-1} \quad \text{a.s.} \qquad (18.1)$$

for any $\varepsilon > 0$ and for all but finitely many n.

Let $S_n = (S_n^{(1)}, S_n^{(2)}, \ldots, S_n^{(d)})$ (where $S_n^{(i)}$ is the i-th coordinate of S_n in the original coordinate system). Then by Invariance Principle 1 there exist independent Wiener processes $W_1(\cdot), W_2(\cdot), \ldots, W_d(\cdot)$ such that

$$|S_n^{(i)} - W_i(\mathcal{K}_n^{(i)})| = O(\log \mathcal{K}_n^{(i)}) = O(\log n) \quad \text{a.s.}$$

for any $i = 1, 2, \ldots, d$. By (18.1) and Theorem 7.13 we have

$$\left| W_i(\mathcal{K}_n^{(i)}) - W_i\left(\frac{n}{d}\right) \right| \leq O\left(n^{1/4}(\log \log n)^{3/4}\right) \quad \text{a.s.}$$

Consequently we have

THEOREM 18.2 *On a rich enough probability space $\{\Omega, \mathcal{F}, \mathbf{P}\}$ one can define a Wiener process $W(\cdot) \in \mathbb{R}^d$ and a random walk $S_n \in \mathbb{Z}^d$ such that*

$$\left\| S_n - W\left(\frac{n}{d}\right) \right\| \leq O\left(n^{1/4}(\log \log n)^{3/4}\right) \quad \text{a.s.}$$

for any $d = 1, 2, \ldots$.

It is not hard to prove that

$$\mathbf{P}\{W(t) = 0 \text{ i.o.}\} = 1 \quad \text{if} \quad d = 1$$

and

$$\mathbf{P}\{W(t) = 0 \text{ for any } t > 0\} = 0 \quad \text{if} \quad d \geq 2.$$

Hence we can say that the Wiener process is not recurrent if $d \geq 2$. However, it turns out that it is recurrent in a weaker sense if $d = 2$.

THEOREM 18.3 (see e.g. Knight, 1981, Th. 4.3.8). *For any $\varepsilon > 0$ we have*

$$\mathbf{P}\{\|W(t)\| \leq \varepsilon \text{ i.o.}\} = 1 \quad \text{if} \quad d = 2, \tag{18.2}$$

$$\mathbf{P}\{\|W(t)\| \leq \varepsilon \text{ i.o.}\} = 0 \quad \text{if} \quad d \geq 3 \tag{18.3}$$

where i.o. means that there exists a random sequence $0 < t_1 = t_1(\omega, \varepsilon) < t_2 = t_2(\omega, \varepsilon) < \ldots$ such that $\lim_{n \to \infty} t_n = \infty$ a.s. and $\|W(t_n)\| \leq \varepsilon$ ($n = 1, 2, \ldots$).

The proof of Theorem 18.3 is very simple having the following deep lemma which is the analogue of Lemma 3.1.

LEMMA 18.1 (Knight, 1981, Theorem 4.3.8). *Let $0 < a < b < c < \infty$. Then*

$$\mathbf{P}\{\inf\{s : s > 0, \|W(t+s)\| = a\} < \inf\{s : s > 0, \|W(t+s)\| = c\} \mid B\}$$

$$= \begin{cases} \dfrac{c-b}{c-a} & \text{if } d = 1, \\ \dfrac{\log c - \log b}{\log c - \log a} & \text{if } d = 2, \\ \dfrac{c^{2-d} - b^{2-d}}{c^{2-d} - a^{2-d}} & \text{if } d \geq 3, \end{cases}$$

where $B = \{\|W(t)\| = b\}$.

Remark 1. Choosing $a = r$, $b = R$, $c = \infty$ in Lemma 18.1 we obtain: for any $d \geq 3$ and $\|u\| = R \geq r$ we have

$$\mathbf{P}\{W(t) \in Q(u,r) \text{ for some } t\} = \left(\frac{r}{R}\right)^{d-2} \tag{18.4}$$

where

$$Q(u,r) = \{x : x \in \mathbb{R}^d, \|x - u\| \leq r\}.$$

(18.4) is an analogue of Lemma 17.8 for Wiener process.

Remark 2. (18.3) is equivalent to

$$\lim_{t \to \infty} \|W(t)\| = \infty \quad \text{a.s.} \quad \text{if } d \geq 3. \tag{18.5}$$

The rate of convergence in (18.5) will be studied in Chapter 19.

In connection with (18.2) it is natural to ask how the set of those functions ε_t can be characterized for which

$$\mathbf{P}\{\|W(t)\| \leq \varepsilon_t \text{ i.o.}\} = 1. \tag{18.6}$$

This question was studied by Spitzer (1958), who proved

THEOREM 18.4 *Let $g(t)$ be a positive nonincreasing function. Then*

$$g(t)t^{1/2} \in \text{LLC}(\|W(t)\|) \quad (d = 2)$$

if and only if $\sum_{k=1}^{\infty} (k|\log g(k)|)^{-1} < \infty$.

Remark 3. Theorem 18.4 implies

$$t^{-(\log t)^\varepsilon} \in \begin{cases} \text{LLC}(\|W(t)\|) & \text{if } \varepsilon > 0, \\ \text{LUC}(\|W(t)\|) & \text{if } \varepsilon \leq 0. \end{cases}$$

The proof of Theorem 18.4 is based on the following:

LEMMA 18.2 (Spitzer, 1958). *For any $0 < t_1 < t_2 < \infty$ we have*

$$\mathbf{P}\{\min_{t_1 \leq T \leq t_2} \|W(t)\| < r\} \approx \frac{\log t_2 - \log t_1}{\log r^{-2}} \quad (r \to 0, \ d = 2).$$

Here we also mention a simple consequence of Theorems 2.12 and 2.13 (cf. also Theorem 6.3).

THEOREM 18.5 *For any $d = 1, 2, \ldots$ and $T > 0$ we have*

$$\mathbf{P}\{m(T) > uT^{1/2}\} = O(u^{-1} e^{-u^2/2}) \quad as \quad u \to \infty$$

and

$$\mathbf{P}\{m(T) < uT^{1/2}\} = \exp(-O(u^{-2})) \quad as \quad u \to 0.$$

Similarly for any $d = 1, 2, \ldots$ as $N \to \infty$ we have

$$\mathbf{P}\{M(N) > uN^{1/2}\} = \exp(-O(u^2)) \quad if \quad u \to \infty \quad but \quad u \leq N^{1/3}$$

and

$$\mathbf{P}\{M(N) < uN^{1/2}\} = \exp(-O(u^{-2})) \quad if \quad u \to 0 \quad but \quad u \geq N^{-1/3}.$$

Chapter 19

The Law of Iterated Logarithm

At first we present the analogue of the LIL of Khinchine of Section 4.4.

THEOREM 19.1

$$\limsup_{t\to\infty} b_t\|W(t)\| = 1 \quad a.s. \tag{19.1}$$

and

$$\limsup_{n\to\infty} b_n d^{1/2}\|S_n\| = 1 \quad a.s. \tag{19.2}$$

where $b_t = (2t \log \log t)^{-1/2}$.

Proof. By the LIL of Khinchine we obtain

$$\limsup_{t\to\infty} b_t\|W(t)\| \geq 1.$$

In order to obtain the upper estimate assume that there exists an $\varepsilon > 0$ such that

$$\limsup_{t\to\infty} b_t\|W(t)\| \geq 1+\varepsilon \quad a.s.$$

(Zero-One Law (cf. Section 3.2)). For the sake of simplicity let $d = 2$ and define $\phi_k = k \arccos \Theta$, $(k = 0, 1, 2, \ldots [2\pi(\arccos \Theta)^{-1}])$ where $\Theta = (1 + \varepsilon/2)(1 + \varepsilon)^{-1}$. If $b_t\|W(t)\| \geq 1 + \varepsilon$ then there exists a k such that

$$b_t|\cos\phi_k W_1(t) + \sin\phi_k W_2(t)| > 1 + \frac{\varepsilon}{2}. \tag{19.3}$$

Since $\cos\phi_k W_1(t) + \sin\phi_k W_2(t)$ $(t \geq 0)$ is a Wiener process, (19.3) cannot occur if t is large enough. Hence we have (19.1) in the case $d = 2$. The proof of (19.1) for $d \geq 3$ is essentially the same. (19.2) follows from (19.1) by Theorem 18.2.

Applying the method of proof of Strassen's theorem 2 of Section 8.1 we obtain the following stronger theorem:

THEOREM 19.2 *The process* $\{b_t W(t), t \geq 0\}$ *is relatively compact in* $\mathbb{R}^d$ *with probability 1 and the set of its limit points is*

$$C_d = \{x \in \mathbb{R}^d, \|x\| \leq 1\}.$$

The real analogue of Strassen's theorem can also be easily proved. It goes like this:

THEOREM 19.3 *The net $\{b_t W(xt),\ 0 \leq x \leq 1\}$ is relatively compact in $C(0,1) \times \cdots \times C(0,1) = (C(0,1))^d$ and the set of its limit points is $\mathcal{S}_d$ where $\mathcal{S}_d$ consists of those and only those $\mathbb{R}^d$ valued functions $f(x) = (f_1(x), \ldots, f_d(x))$ for which $f_i(0) = 0$, $f_i(\cdot)$ $(i = 1, 2 \ldots, d)$ is absolutely continuous in $[0, 1]$ and $\sum_{i=1}^{d} \int_0^1 (f_i'(x))^2 dx \leq 1$.*

We ask about the analogue of the EFKP LIL of Section 5.2. It is trivial to see that if $a(t) \in \mathrm{ULC}\{W(t)\}$ in the case $d = 1$ then the same is true for any d. However, the analogue statement for $\mathrm{UUC}\{W(t)\}$ is not true. As an example, we mention that Consequence 1 of Section 5.2 tells us that in the case $d = 1$ for any $\varepsilon > 0$

$$S_n \leq \left(2n \left(\log\log n + \left(\frac{3}{2} + \varepsilon\right) \log\log\log n\right)\right)^{1/2} \quad \text{a.s.}$$

for all but finitely many n. However, it turns out that in case $d > 1$ it is not true. In fact for any $d > 1$

$$d^{1/2} S_n \geq \left(2n \left(\log\log n + \frac{2+d}{2} \log\log\log n\right)\right)^{1/2} \quad \text{i.o. a.s.}$$

Now we formulate the general

THEOREM 19.4 (Orey–Pruitt, 1973). *Let $a(t)$ be a nonnegative nondecreasing continuous function. Then for any $d = 1, 2, \ldots$*

$$t^{1/2} a(t) \in \mathrm{UUC}(\|W(t)\|)$$

and

$$n^{1/2} a(n) \in \mathrm{UUC}(d^{1/2} \|S_n\|)$$

if and only if

$$\int_1^\infty \frac{(a(x))^d}{x} e^{-a^2(x)/2} dx < \infty. \tag{19.4}$$

Remark 1. The function

$$a(x) = \left(\sum_{k=2}^n a_k \log_k x\right)^{1/2} \quad (n = 2, 3, \ldots)$$

does not satisfy (19.4) if $a_2 = 2$, $a_3 = d + 2$, $a_k = 2$ for $4 \leq k \leq n$ but it does if a_n is increased by $\varepsilon > 0$ for any $n \geq 2$.

It was already mentioned (Chapter 18, Remark 2) that

$$\lim_{t\to\infty} \|W(t)\| = \infty \quad \text{a.s. if } d \geq 3. \tag{19.5}$$

Now we are interested in the rate of convergence in (19.5). This rate is called *rate of escape*. We present

THEOREM 19.5 (Dvoretzky–Erdős, 1950). *Let $a(t)$ be a nonincreasing, nonnegative function. Then*

$$t^{1/2}a(t) \in \text{LLC}(\|W(t)\|) \quad (d \geq 3)$$

and

$$n^{1/2}a(n) \in \text{LLC}(d^{1/2}\|S_n\|) \quad (d \geq 3)$$

if and only if

$$\sum_{n=1}^{\infty}(a(2^n))^{d-2} < \infty. \tag{19.6}$$

Remark 2. The function

$$a(x) = \left(\log x \log_2 x \ldots (\log_k x)^{1+\varepsilon}\right)^{-1/(d-2)}$$

does not satisfy (19.6) if $\varepsilon \leq 0$, but it does if $\varepsilon > 0$.

In case $d = 2$ we might ask for the analogue of Theorem of Chung of Section 5.3, i.e. we are interested in the liminf properties of

$$m(t) = \sup_{s \leq t} \|W(s)\| \quad \text{and} \quad M(n) = \max_{k \leq n} \|S_k\|.$$

This question seems to be unsolved.

Theorems 19.4 and 19.5 together imply: *there are infinitely many n for which*

$$\|S_n\| \geq d^{-1/2}b_n^{-1} \tag{19.7}$$

and for every n big enough

$$\|S_n\| \geq n^{1/2}(\log n)^{-\varepsilon-1/(d-2)} \quad (d \geq 3,\ \varepsilon > 0). \tag{19.8}$$

Erdős and Taylor (1960/A) proved that if a particle is very far away from the origin, i.e. (19.7) holds, then it may remain far away forever ($d \geq 3$). In fact we have the following:

THEOREM 19.6

$$\mathbf{P}\{\inf_{k \geq n} \|S_k\| \geq d^{-1/2}b_n^{-1} \text{ i.o.}\} = 1 \quad (d \geq 3).$$

Chapter 20

Local Time

20.1 $\xi(0,n)$ in $\mathbb{Z}^2$

The Recurrence Theorem of Chapter 17 clearly implies

$$\lim_{n\to\infty} \xi(0,n) \begin{cases} = \infty & \text{if } d \leq 2, \\ < \infty & \text{if } d \geq 3. \end{cases}$$

Hence we study the limit properties of $\xi(0,n)$ in the case $d=2$.

THEOREM 20.1 (Erdős–Taylor, 1960/A). *Let $d=2$. Then*

$$\lim_{n\to\infty} \mathbf{P}\{\xi(0,n) < x\log n\} = 1 - e^{-\pi x}$$

uniformly for $0 \leq x < (\log n)^{3/4}$ and

$$\mathbf{E}\frac{\xi(0,n)}{\log n} \to \frac{1}{\pi} \quad (n\to\infty).$$

The proof of this theorem is based on the following:

LEMMA 20.1 (Dvoretzky–Erdős 1950, Erdős–Taylor 1960/A).

$$\mathbf{P}\{\rho_1 > n\} = \mathbf{P}\{\xi(0,n) = 0\} = \frac{\pi}{\log n} + O((\log n)^{-2}) \quad (d=2).$$

Proof. Counting the last return to the origin (cf. also 9. of Notations) we have

$$\sum_{k=0}^{n-1} \mathbf{P}\{S_{2k} = 0\}\mathbf{P}\{\xi(0,2n-2k-2) = 0\} = 1 \quad (n=1,2,\ldots) \quad (20.1)$$

where $\xi(0,0) = 0$. Since by Lemma 17.2

$$\mathbf{P}\{S_{2k} = 0\} \approx (k\pi)^{-1}$$

we have
$$\sum_{k=0}^{n} \mathbf{P}\{S_{2k} = 0\} \approx \frac{\log n}{\pi}. \tag{20.2}$$

Since the sequence $\mathbf{P}\{\xi(0, 2n) = 0\}$ is nonincreasing by (20.1) and (20.2) we obtain
$$1 \geq \mathbf{P}\{\xi(0, 2n-2) = 0\} \sum_{k=0}^{n-1} \mathbf{P}\{S_{2k} = 0\} \approx \mathbf{P}\{\xi(0, 2n-2) = 0\} \frac{\log n}{\pi}$$

and
$$\mathbf{P}\{\xi(0, 2n) = 0\} \leq \frac{\pi + o(1)}{\log n}. \tag{20.3}$$

Similarly for any $0 \leq k \leq n$ by (20.1)
$$1 \leq \left(\sum_{j=0}^{k} \mathbf{P}\{S_{2j} = 0\}\right) \mathbf{P}\{\xi(0, 2n-2k-2) = 0\} + \sum_{j=k+1}^{n-1} \mathbf{P}\{S_{2j} = 0\}. \tag{20.4}$$

Thus, if k tends to infinity together with n, (20.4) yields
$$\mathbf{P}\{\xi(0, 2n-2k) = 0\} \frac{1 + o(1)}{\pi} \log k + \frac{1 + o(1)}{\pi} \log \frac{n}{k} \geq 1.$$

Taking $k = n - [n/\log n]$ we have
$$\mathbf{P}\{\xi(0, 2n) = 0\} \geq \frac{\pi + o(1)}{\log n}.$$

Hence we have the main term of Lemma 20.1. The remainder can be obtained similarly but with a more tedious calculation.

Proof of Theorem 20.1. Let $q = [x \log n] + 1$ and $p = [n/q]$. Then
$$\mathbf{P}\{\xi(0, n) \geq x \log n\} \geq \mathbf{P}\{\rho_q < n\} \geq \prod_{k=1}^{q} \mathbf{P}\{\rho_k - \rho_{k-1} < [n/q]\}$$
$$= (\mathbf{P}\{\rho_1 < p\})^q = (1 - \mathbf{P}\{\xi(0, p) = 0\})^q.$$

Assuming that $x < \log n (\log_2 n)^{-1-\varepsilon}$ by Lemma 20.1 we obtain
$$\mathbf{P}\{\xi(0, n) \geq x \log n\} \geq e^{-\pi x}(1 + o(1)) \quad (n \to \infty). \tag{20.5}$$

In fact we obtain that
$$\mathbf{P}\{\xi(0, n) \geq x \log n\} \geq e^{-\pi x}(1 + o((\log n)^{-1/4}))$$

LOCAL TIME

uniformly in x for $x < (\log n)^{3/4}$.

In order to get an upper estimate, let E_k ($k = 1, 2, \ldots, q$) be the event that precisely k of the variables $\rho_i - \rho_{i-1}$ ($i = 1, 2, \ldots, q$) are greater than or equal to n, while $q - k$ of them are less than n. Then

$$\{\xi(0,n) \geq x \log n\} \subset \prod_{k=1}^{q} \bar{E}_k.$$

Hence

$$\mathbf{P}\{\xi(0,n) < x \log n\}$$
$$\geq \sum_{k=1}^{q} \mathbf{P}\{E_k\} = \sum_{k=1}^{q} \binom{q}{k} (\mathbf{P}\{\rho_1 \geq n\})^k (1 - \mathbf{P}\{\rho_1 \geq n\})^{q-k}$$
$$= 1 - (1 - \mathbf{P}\{\rho_1 \geq n\})^q \geq 1 - e^{-\pi x}(1 + O(\log n)^{-1/4}) \quad (20.6)$$

by Lemma 20.1 uniformly in x for $x < (\log n)^{3/4}$.

(20.5) and (20.6) combined imply the first statement of Theorem 20.1. The second statement can be obtained similarly observing that by (20.6) and Lemma 20.1 for $x = x_n$ ($n = 2, 3, \ldots$) we have

$$\mathbf{P}\{\xi(0,n) \geq x \log n\} \leq \exp\left(-\pi x + O\left(\frac{x}{\log n}\right)\right).$$

Note that Theorem 20.1 easily implies

THEOREM 20.2 *Let $d = 2$. Then*

$$\lim_{n \to \infty} \mathbf{P}\left\{\rho_n < \exp\left(\frac{n}{z}\right)\right\} = \exp(-\pi z)$$

uniformly for $0 < z < n^{3/7}$.

Clearly for any fixed $x \in \mathbb{Z}^2$ the limit properties of $\xi(x, n)$ are the same as those of $\xi(0, n)$. For example, Theorem 20.1 and Lemma 20.1 remain true replacing $\xi(0, n)$ by $\xi(x, n)$. However, if instead of a fixed x a sequence x_n (with $\|x_n\| \to \infty$) is considered, the situation will be completely different. The following result gives some information about this case.

THEOREM 20.3 (Erdős–Taylor, 1960/A). *Let $d = 2$. Then*

$$\mathbf{P}\{\xi(x, n) = 0\} =$$

$$\begin{cases} \dfrac{2\log\|x\|}{\log n}\left(1+O\left(\dfrac{\log_3\|x\|}{\log\|x\|}\right)\right) & \text{if } 20 < \|x\| < n^{1/3}, \\ 1 - 2\dfrac{\log(n^{1/2}/\|x\|)}{\log n}\left(1+O\left(\dfrac{\log_2\left(\frac{\|x\|}{\sqrt{n}}\right)}{\log\left(\frac{\|x\|}{\sqrt{n}}\right)}\right)\right) & \text{if } n^{1/6} < \|x\| < \dfrac{n^{1/2}}{20}. \end{cases}$$

Proof. By (20.1) we have

$$\sum_{i=0}^{n} p_{2i}\gamma_{2n-2i+2} = 1 \tag{20.7}$$

and similarly

$$\mathbf{P}\{\xi(x, 2n) = 0\} + \sum_{k=1}^{n} \mathbf{P}\{S_{2k} = x\}\gamma_{2n-2k+2} = 1 \tag{20.8}$$

provided that $x = (x_1, x_2)$ with $x_1 + x_2 \equiv 0 \pmod{2}$. (20.7) and (20.8) combined imply

$$\mathbf{P}\{\xi(x, 2n) = 0\} - \gamma_{2n+2} = \sum_{k=1}^{n}(p_{2k} - \mathbf{P}\{S_{2k} = x\})\gamma_{2n-2k+2}.$$

Consequently for $1 < k_1 < k_2 < n$ we get

$$\mathbf{P}\{\xi(x, 2n) = 0\} - \gamma_{2n+2}$$
$$\leq \gamma_{2n-2k_1+2}\sum_{k=1}^{k_1} p_{2k} + \gamma_{2n-2k_2+2}\sum_{k=k_1+1}^{k_2}(p_{2k} - \mathbf{P}\{S_{2k} = x\})$$
$$+ \sum_{k=k_2+1}^{n}(p_{2k} - \mathbf{P}\{S_{2k} = x\}).$$

Now in the case $400 < \|x\|^2 < n^{2/3}$ put $k_1 = \|x\|^2$ and $k_2 = [n^{4/5}]$ then by Lemmas 20.1, 17.2 and 17.4 we obtain

$$\mathbf{P}\{\xi(x, 2n) = 0\} - \gamma_{2n+2}$$
$$\leq \left(\dfrac{\pi}{\log(2n - 2k_1)} + O\left(\dfrac{1}{(\log n)^2}\right)\right)\sum_{k=1}^{k_1}\left(\dfrac{1}{k\pi} + O\left(\dfrac{1}{k^2}\right)\right)$$
$$+ \left(\dfrac{\pi}{\log(2n - 2k_2)} + O\left(\dfrac{1}{(\log n)^2}\right)\right)$$

LOCAL TIME

$$\times \sum_{k=k_1+1}^{k_2} \left(\frac{1}{k\pi} + O\left(\frac{1}{k^2}\right) - \frac{1}{k\pi} - \|x\|^2 O\left(\frac{1}{k^2}\right) \right)$$

$$+ \sum_{k=k_2+1}^{n} \left(\frac{1}{k\pi} + O\left(\frac{1}{k^2}\right) - \frac{1}{k\pi} - \|x\|^2 O\left(\frac{1}{k^2}\right) \right)$$

$$\leq \frac{\log \|x\|^2 + O(1)}{\log n}.$$

Similarly, for $1 < k_3 < n$

$$\mathbf{P}\{\xi(x, 2n) = 0\} - \gamma_{2n+2} \geq \gamma_{2n+2} \sum_{k=1}^{k_3} (p_{2k} - \mathbf{P}\{S_{2k} = x\}).$$

Take

$$k_3 = \frac{\|x\|^2}{\log_2 \|x\|^2}.$$

Then

$$\mathbf{P}\{\xi(x, 2n) = 0\} - \gamma_{2n+2} \geq \frac{\log \|x\|^2}{\log n} \left(1 - O\left(\frac{\log_3 \|x\|^2}{\log \|x\|^2}\right)\right),$$

hence we have Theorem 20.3 in the case $20 < \|x\| < n^{1/3}$. The case $n^{1/6} < \|x\| < n^{1/2}/20$ can be treated similarly.

As a trivial consequence of Theorem 20.3 we prove

LEMMA 20.2 *Let* $n_k = [\exp(e^{k \log k})]$. *Then for any k big enough*

$$\mathbf{P}\{\xi(0, n_k^{\log k}) - \xi(0, n_k) = 0 \mid S_j;\ j = 0, 1, 2, \ldots, n_k\} \leq \frac{3}{\log k}.$$

Proof. Since $\|S_{n_k}\| \leq n_k$, for any $x \in \mathbb{Z}^2$ with $\|x\| = n_k$ we have

$$\mathbf{P}\{\xi(0, n_k^{\log k}) - \xi(0, n_k) = 0 \mid S_j;\ j = 0, 1, 2, \ldots, n_k\}$$

$$\leq \mathbf{P}\{\xi(x, n_k^{\log k} - n_k) = 0\} \leq \frac{3}{\log k}.$$

The next theorem gives a complete description of the strong behaviour of $\xi(0, n)$.

THEOREM 20.4 (Erdős–Taylor, 1960/A). *Let $d = 2$ and let $f(x)$ (resp. $g(x)$) be a decreasing (resp. increasing) function for which*

$$f(x) \log x \nearrow \infty, \quad g(x)(\log x)^{-1} \searrow 0.$$

Then
$$\pi^{-1}g(n)\log n \in \text{UUC}(\xi(0,n)) \tag{20.9}$$
if and only if
$$\int_1^\infty \frac{g(x)}{x\log x} e^{-g(x)} dx < \infty, \tag{20.10}$$
and
$$f(n)\log n \in \text{LLC}(\xi(0,n)) \tag{20.11}$$
if and only if
$$\int_1^\infty \frac{f(x)}{x\log x} dx < \infty. \tag{20.12}$$

Remark 1. The function
$$g(x) = \log_3 x + 2\log_4 x + \log_5 x + \cdots + \log_k x + \tau \log_{k+1} x$$
satisfies (20.10) if and only if $\tau > 1$. The function
$$f(x) = (\log\log x)^{-1-\varepsilon}$$
satisfies (20.12) if and only if $\varepsilon > 0$.

Proof of (20.9). Instead of (20.9) we prove the somewhat weaker statement only: for any $\varepsilon > 0$
$$(1+\varepsilon)\pi^{-1}(\log n)\log_3 n \in \text{UUC}(\xi(0,n)) \tag{20.13}$$
and
$$(1-\varepsilon)\pi^{-1}(\log n)\log_3 n \in \text{ULC}(\xi(0,n)). \tag{20.14}$$

Let $n_k = [\exp((1+\varepsilon/3)^k)]$. Then by Theorem 20.1
$$\mathbf{P}\left\{\xi(0,n_k) \geq \left(1+\frac{\varepsilon}{3}\right)\pi^{-1}(\log n_k)\log_3 n_k\right\} = O(k^{-1-\varepsilon/3}),$$
and by Borel–Cantelli lemma
$$\xi(0,n_k) < \left(1+\frac{\varepsilon}{3}\right)\pi^{-1}(\log n_k)\log_3 n_k \quad \text{a.s.}$$
for all but finitely many k. Let $n_k \leq n \leq n_{k+1}$. Then
$$\xi(0,n) \leq \xi(0,n_{k+1}) \leq \left(1+\frac{\varepsilon}{3}\right)\pi^{-1}(\log n_{k+1})\log_3 n_{k+1}$$
$$\leq (1+\varepsilon)\pi^{-1}(\log n)\log_3 n$$

LOCAL TIME

if k is large enough. Hence we have (20.13).

Now we turn to the proof of (20.14). Let

$$n_k = [\exp(e^{k \log k})] \quad (k = 2, 3, \ldots),$$
$$E_k = \left\{ \frac{\pi \xi(0, n_k)}{\log n_k} > \log_3 n_k \right\}$$

and

$$F_k = \left\{ \frac{\pi(\xi(0, n_{k+1}) - \xi(0, n_k^{\log k}))}{\log n_{k+1}} > \log_3 n_{k+1}, \; \xi(0, n_k^{\log k}) - \xi(0, n_k) > 0 \right\}.$$

Then clearly $F_k \subset E_{k+1}$ and by Theorem 20.1 and Lemma 20.2

$$\mathbf{P}\{F_k\} \geq \left(1 - \frac{3}{\log k}\right) \frac{1}{k \log k}(1 + o(1)),$$

and similarly for any $j < k$

$$\mathbf{P}\{F_k \mid \bar{F}_j\} \geq \left(1 - \frac{3}{\log k}\right) \frac{1}{k \log k}(1 + o(1)).$$

Hence we obtain (20.14) by Borel–Cantelli lemma.

Proof of (20.11). Instead of (20.11) we prove the somewhat weaker statement only: for any $\varepsilon > 0$

$$(\log n)(\log \log n)^{-1} \in \mathrm{LUC}(\xi(0, n)) \tag{20.15}$$

and

$$(\log n)(\log \log n)^{-1-\varepsilon} \in \mathrm{LLC}(\xi(0, n)). \tag{20.16}$$

Let $n_k = [\exp(e^k)]$. Then by Theorem 20.1

$$\mathbf{P}\left\{\xi(0, n_k) < \frac{\log n_k}{(\log \log n_k)^{1+\varepsilon}}\right\} = O(k^{-1-\varepsilon}),$$

and by Borel–Cantelli lemma

$$\xi(0, n_k) \geq \frac{\log n_k}{(\log \log n_k)^{1+\varepsilon}} \quad \text{a.s.}$$

for all but finitely many k. Let $n_k \leq n \leq n_{k+1}$. Then

$$\xi(0, n) \geq \xi(0, n_k) \geq \frac{\log n_k}{(\log \log n_k)^{1+\varepsilon}} \geq \frac{\log n}{(\log \log n)^{1+2\varepsilon}}$$

if k is large enough. Hence we have (20.16).

Now we turn to the proof of (20.15). For any $r = 1, 2, \ldots$ we have

$$\mathbf{P}\{\rho_{2^r} - \rho_{2^{r-1}} < \exp(Cr2^r)\} = \mathbf{P}\{\rho_{2^{r-1}} < \exp(Cr2^r)\}$$
$$= \mathbf{P}\{\xi(0, \exp(Cr2^r)) \geq 2^{r-1}\} \approx \exp\left(-\frac{\pi}{2Cr}\right) \approx 1 - \frac{\pi + o(1)}{2Cr}.$$

Since the r.v.'s $\rho_{2^r} - \rho_{2^{r-1}}$ ($r = 1, 2, \ldots$) are independent we obtain

$$\rho_{2^r} > \rho_{2^r} - \rho_{2^{r-1}} \geq \exp(Cr2^r) \quad \text{i.o. a.s.}$$

and consequently

$$\xi(0, \exp(Cr2^r)) \leq 2^r \quad \text{i.o. a.s.}$$

Let $n = \exp(Cr2^r)$ with $C = \log 2$. We get

$$\xi(0, n) \leq \frac{\log n}{\log \log n} \quad \text{i.o. a.s.}$$

Hence the proof of Theorem 20.4 (in fact a slightly weaker version of it) is complete.

Note that Theorem 20.4 easily implies

THEOREM 20.5 *For any $\varepsilon > 0$*

$$\exp(n(\log n)^{1+\varepsilon}) \in \mathrm{UUC}(\rho_n),$$
$$\exp(n(\log n)^{1-\varepsilon}) \in \mathrm{ULC}(\rho_n),$$
$$\exp\left((1+\varepsilon)\frac{\pi n}{\log_2 n}\right) \in \mathrm{LUC}(\rho_n),$$
$$\exp\left((1-\varepsilon)\frac{\pi n}{\log_2 n}\right) \in \mathrm{LLC}(\rho_n).$$

20.2 $\xi(n)$ in $\mathbb{Z}^d$

As we have seen (Recurrence Theorem, Chapter 17)

$$\lim_{n \to \infty} \xi(0, n) < \infty \quad \text{a.s.} \quad \text{if} \quad d \geq 3.$$

Similarly for any fixed $x \in \mathbb{Z}^d$

$$\lim_{n \to \infty} \xi(x, n) < \infty \quad \text{a.s.} \quad \text{if} \quad d \geq 3.$$

However, it turns out that

THEOREM 20.6 *For any $d \geq 1$ we have*
$$\lim_{n\to\infty} \xi(n) = \lim_{n\to\infty} \sup_{x\in\mathbb{Z}^d} \xi(x,n) = \infty \quad a.s.$$

Proof. Theorem 7.1 told us that the length Z_n of the longest head run till n is a.s. larger than or equal to $(1-\varepsilon)\log n/\log 2$ for any $\varepsilon > 0$ if n is large enough. Similarly one can show that the sequence $X_1, X_2, \ldots, X_n$ contains a run $e_1, -e_1, e_1, -e_1, \ldots, e_1$ of size $(1-\varepsilon)\log n/\log 2d$. This implies that

$$\liminf_{n\to\infty} \frac{\xi(n)}{\log n} \geq \frac{(\log 2d)^{-1}}{2}$$

which, in turn, implies Theorem 20.6.

A more exact result was obtained by Erdős and Taylor (1960/A). They proved

THEOREM 20.7 *For any $d \geq 3$*
$$\lim_{n\to\infty} \frac{\xi(n)}{\log n} = \lambda_d \quad a.s. \tag{20.17}$$

where
$$\lambda_d = -(\log \mathbf{P}\{\lim_{n\to\infty} \xi(0,n) \geq 1\})^{-1} = -(\log(1-\gamma))^{-1}.$$

In the case $d = 2$
$$\frac{1}{4\pi} \leq \liminf_{n\to\infty} \frac{\xi(n)}{(\log n)^2} \leq \limsup_{n\to\infty} \frac{\xi(n)}{(\log n)^2} \leq \frac{1}{\pi} \quad a.s.$$

Erdős and Taylor (1960/A) also formulated the following:

Conjecture. For $d = 2$
$$\lim_{n\to\infty} \frac{\xi(n)}{(\log n)^2} = \frac{1}{\pi} \quad a.s.$$

This Conjecture was proved by Dembo–Peres–Rosen–Zeitouni (2001).

They also investigated the number of points which are visited nearly $(\log n)^2/\pi$ times up to n. Let

$$M(n,\alpha) = \#\{x: \, x \in \mathbb{Z}^2, \, \xi(x,n) \geq \alpha(\log n)^2\} \quad (0 < \alpha < 1).$$

They proved

THEOREM 20.8
$$\lim_{n\to\infty} \frac{\log M(n,\alpha)}{\log n} = 1 - \alpha\pi \quad a.s.$$

20.3 A few further results

First we give an analogue of Theorem 12.8.

THEOREM 20.9 (Erdős–Taylor, 1960/A).

$$\lim_{n\to\infty} \frac{1}{\log n} \sum_{k=1}^{n} \frac{1}{\log \rho_k} = \frac{1}{\pi} \quad a.s. \quad if \quad d=2.$$

The next theorem is an analogue of Theorem 12.1.

THEOREM 20.10 (Erdős–Taylor, 1960/A). *Let $d = 2$, $f(n) \uparrow \infty$ as $(n \to \infty)$ and E_n be the event that the random walk S_n does not return to the origin between n and $n^{f(n)}$. Then*

$$\mathbf{P}\{E_n \; i.o.\} = 0 \quad or \quad 1$$

depending on whether the series

$$\sum_{k=1}^{\infty} \frac{1}{f(2^{2^k})}$$

converges or diverges.

Now we turn to the analogue of Theorem 13.9.

THEOREM 20.11 (Erdős–Taylor, 1960/A, Flatto, 1976.) *Let $Q_k(n)$ be the number of points visited exactly k times ($k = 1, 2, \ldots$) up to n. i.e.*

$$Q_k(n) = \#\{j : 0 \leq j \leq n, \; \xi(S_j, n) = k\}.$$

Then

$$\lim_{n\to\infty} \frac{Q_k(n) \log^2 n}{\pi^2 n} = 1 \quad a.s. \quad if \quad d=2 \tag{20.18}$$

and

$$\lim_{n\to\infty} \frac{Q_k(n)}{n} = \gamma^2 (1-\gamma)^{k-1} \quad a.s. \quad if \quad d \geq 3, \; k = 1, 2, \ldots$$

where $\gamma = \gamma(d)$ is the probability that the path will never return to the origin.

Remark 2. Observe that in case $d = 2$ the limit properties of $Q_k(n)$ do not depend on k (cf. (20.18)). An explanation of this surprising fact can be found in Hamana (1997). Further properties of $Q_k(n)$ are studied by Pitt (1974) and Hamana (1995, 1997). For example it is proved that $Q_k(n)$ ($k = 1, 2, \ldots$) obeys the central limit theorem if $d \geq 5$.

Chapter 21

The Range

21.1 The strong law of large numbers

Let $R(n)$ be the number of different vectors among $S_1, S_2, \ldots, S_n$, i.e. $R(n)$ is the number of points visited by the particle during the first n steps. The r.v.

$$R(n) = \sum_{x \in \mathbb{Z}^d} f(\xi(x,n)), \quad f(i) = \begin{cases} 0 & \text{if } i = 0, \\ 1 & \text{if } i \geq 1 \end{cases}$$

will be called the *range* of $S_1, S_2, \ldots, S_n$. In the case $d = 1$ Theorem 5.7 essentially tells us that $R(n)$ is going to infinity like $n^{1/2}$. In the case $d = 2$ Theorems 20.1 and 20.4 suggest that $R(n) \sim n(\log n)^{-1}$. (Since any fixed point is visited $\log n$ times the number of points visited at all till n is $n(\log n)^{-1}$. Clearly it is not a proof since some points are visited more frequently (cf. Theorem 20.7) and some less frequently (cf. Theorem 20.11) among the points visited at all.) In fact we prove

THEOREM 21.1 (Dvoretzky–Erdős, 1950).

$$\mathbf{E}R(n) = \begin{cases} \dfrac{\pi n}{\log n} + O\left(\dfrac{n \log \log n}{(\log n)^2}\right) & \text{if } d = 2, \\ n\mathbf{P}\left\{\lim_{n \to \infty} \xi(0,n) = 0\right\} + O(n^{1/2}) & \text{if } d = 3, \\ n\mathbf{P}\left\{\lim_{n \to \infty} \xi(0,n) = 0\right\} + O(\log n) & \text{if } d = 4, \\ n\mathbf{P}\left\{\lim_{n \to \infty} \xi(0,n) = 0\right\} + \beta_d + O(n^{2-d/2}) & \text{if } d \geq 5 \end{cases} \quad (21.1)$$

where β_d ($d = 5, 6, \ldots$) are positive constants and

$$\mathbf{Var}R(n) = \mathbf{E}R^2(n) - (\mathbf{E}R(n))^2 \leq \begin{cases} O\left(\dfrac{n^2 \log \log n}{(\log n)^3}\right) & \text{if } d = 2, \\ O(n^{3/2}) & \text{if } d = 3, \\ O(n \log n) & \text{if } d = 4, \\ O(n) & \text{if } d \geq 5. \end{cases}$$
(21.2)

Further, the strong law of large numbers

$$\lim_{n\to\infty} \frac{R(n)}{\mathbf{E}R(n)} = 1 \quad a.s. \quad if \quad d \geq 2 \tag{21.3}$$

holds.

Remark 1. Theorem 5.7 implies that the range does not satisfy the strong law of large numbers in the case $d = 1$.

In case $d = 2$ resp. 3 (Jain–Pruitt, 1972/B resp. Bass–Kumagai, 2002) for $\operatorname{Var} R(n)$ the following stronger results are known:

$$\frac{c_1 n^2}{(\log n)^4} \leq \operatorname{Var} R(n) \leq \frac{c_2 n^2}{(\log n)^4} \quad if \quad d = 2,$$

$$c_1 n \log n \leq \operatorname{Var} R(n) \leq c_2 n \log n \quad if \quad d = 3$$

where $c_1 < c_2$ are positive constants.

The proof of (21.1) is based on the following:

LEMMA 21.1

$$\mathbf{P}\{S_n \neq S_i \text{ for } i = 1, 2, \ldots, n-1\} = \mathbf{P}\{\xi(0, n-1) = 0\} = \gamma_n. \tag{21.4}$$

Remark 2. The left hand side of (21.4) is the probability that the n-th step takes the path to a new point.

Proof of Lemma 21.1.

$$\mathbf{P}\{S_n \neq S_i \text{ for } i = 1, 2, \ldots, n-1\}$$
$$= \mathbf{P}\{X_n + X_{n-1} + \cdots + X_{i+1} \neq 0 \text{ for } i = 1, 2, \ldots, n-1\}$$
$$= \mathbf{P}\{X_1 + X_2 + \cdots + X_{n-i} \neq 0 \text{ for } i = 1, 2, \ldots, n-1\}$$
$$= \mathbf{P}\{S_j \neq 0 \text{ for } j = 1, 2, \ldots, n-1\} = \mathbf{P}\{\xi(0, n-1) = 0\}.$$

Hence we have (21.4).

Let

$$\psi_n = \begin{cases} 1 & \text{if } S_n \neq S_i \text{ for } i = 1, 2, \ldots, n-1, \\ 0 & \text{otherwise.} \end{cases}$$

Then

$$R(n) = \sum_{k=1}^{n} \psi_k.$$

THE RANGE

Consequently by Lemma 21.1

$$\mathbf{E}R(n) = \mathbf{E}\sum_{k=1}^{n}\psi_k = \sum_{k=1}^{n}\mathbf{P}\{\xi(0,k-1)=0\}.$$

Hence (21.1) in the case $d = 2$ follows from Lemma 20.1, and in the case $d \geq 3$ it follows from Lemma 17.11.

In order to prove (21.2) we present two lemmas.

LEMMA 21.2 *Let $1 \leq m \leq n$. Then*

$$\mathbf{E}\psi_m\psi_n \leq \mathbf{E}\psi_m\mathbf{E}\psi_{n-m+1}.$$

Proof.

$\mathbf{E}\psi_m\psi_n$
$= \mathbf{P}\{S_i \neq S_m, \ i = 1,\ldots, m-1; S_j \neq S_n, \ j = 1,\ldots, n-1\}$
$\leq \mathbf{P}\{S_i \neq S_m, \ i = 1,\ldots, m-1; S_j \neq S_n, \ j = m,\ldots, n-1\}$
$= \mathbf{P}\{S_i \neq S_m, \ i = 1,\ldots, m-1\}\mathbf{P}\{S_j \neq S_n, \ j = m,\ldots, n-1\}$
$= \mathbf{P}\{S_i \neq S_m, \ i = 1,\ldots, m-1\}\mathbf{P}\{S_j \neq S_{n-m+1}, \ j = 1,\ldots n-m\}$
$= \mathbf{E}\psi_m\mathbf{E}\psi_{n-m+1}.$

LEMMA 21.3

$$\mathrm{Var}R(n) \leq 2\mathbf{E}R(n)(\mathbf{E}R(n-[n/2]) - \mathbf{E}R(n) + \mathbf{E}R([n/2])).$$

Proof. By Lemma 21.2

$\mathrm{Var}R(n)$
$= \sum_{i,j=1}^{n}\mathbf{E}\psi_i\psi_j - \left(\sum_{i=1}^{n}\mathbf{E}\psi_i\right)^2 = \sum_{i,j=1}^{n}(\mathbf{E}\psi_i\psi_j - \mathbf{E}\psi_i\mathbf{E}\psi_j)$
$\leq 2\sum_{1\leq i\leq j\leq n}(\mathbf{E}\psi_i\psi_j - \mathbf{E}\psi_i\mathbf{E}\psi_j) \leq 2\sum_{1\leq i\leq j\leq n}(\mathbf{E}\psi_i\mathbf{E}\psi_{j-i+1} - \mathbf{E}\psi_i\mathbf{E}\psi_j)$
$\leq 2\sum_{i=1}^{n}\mathbf{E}\psi_i \max_{1\leq i\leq n}\left(\sum_{j=i}^{n}(\mathbf{E}\psi_{j-i+1} - \mathbf{E}\psi_j)\right).$

Since by Lemma 21.1 $\mathbf{E}\psi_j$ is nonincreasing, the max is attained for $i = [n/2] + 1$ and Lemma 21.3 is proved.

Then (21.2) follows from Lemma 21.3 and (21.1). (21.3) can be obtained by routine methods.

Donsker and Varadhan (1979) were interested in another property of $R(n)$. In fact they investigated the limit behaviour of $\mathbf{E}\exp(-\nu R(n))$ ($\nu > 0$, $n \to \infty$). They proved

THEOREM 21.2 *For any $\nu > 0$ and $d = 2, 3, \ldots$*

$$\lim_{n\to\infty} n^{-d/(d+2)} \log \mathbf{E}\exp(-\nu R(n)) = -k(\nu)$$

where

$$k(\nu) = \nu^{2/(d+2)} \left(\frac{d+2}{2}\right) \left(\frac{2\alpha_d}{d}\right)^{d/(d+2)}$$

and α_d is the lowest eigenvalue of $-1/2\Delta$ for the sphere of unit volume in $\mathbb{R}^d$ with zero boundary values.

Remark 3. In the case $d = 2$ Theorem 21.1 claims that $R(n)$ is typically $\pi n/\log n$. Hence we could expect that $\mathbf{E}\exp(-\nu R(n)) \sim \exp(-\nu\pi n/\log n)$. However, Theorem 21.2 claims that $\mathbf{E}\exp(-\nu R(n)) \sim \exp(-k(\nu)n^{1/2})$. Comparing these two results it turns out that in the asymptotic behaviour of $\mathbf{E}\exp(-\nu R(n))$ the very small values of $R(n)$ contribute most. This fact is explained by the following:

LEMMA 21.4 *For any $\nu > 0$ there exists a $C_\nu > 0$ such that*

$$\mathbf{E}\exp(-\nu R(n)) \geq \exp(-C_\nu n^{1/2}).$$

Proof.

$$\mathbf{E}\exp(-\nu R(n)) \geq \mathbf{E}(\exp(-\nu R(n)) \mid M_n \leq n^{1/4})\mathbf{P}(M_n \leq n^{1/4})$$
$$\geq \exp(-\nu\pi n^{1/2})\mathbf{P}(M_n \leq n^{1/4}).$$

By Theorem 18.5

$$\mathbf{P}\{M_n \leq n^{1/4}\} = \exp(-O(n^{1/2}))$$

and we have Lemma 21.4.

21.2 CLT, LIL and Invariance Principle

Having the strong law of large numbers (21.3) it looks natural to ask about the CLT, LIL and Invariance Principle.

The central limit theorem was proved by Jain and Pruitt (1971, 1974) in case $d \geq 3$ and by Le Gall (1986) in case $d = 2$. It turns out that in case $d = 2$ the limit distribution is not normal but it is exactly described.

The law of iterated logarithm for $d \geq 4$ was proved again by Jain and Pruitt (1972/A). The case $d = 2$ is settled by Bass and Kumagai (2002) who proved

THEOREM 21.3 *Let $d = 2$. Then*

$$\limsup_{n \to \infty} (\log n)^2 \sup_{j \leq n} \frac{(R(j) - \mathbf{E}R(j))}{n \log_3 n} = c \quad a.s.$$

for some $c > 0$.

The almost sure invariance principle for $d \geq 4$ was proved by Hamana (1998).

THEOREM 21.4 *Assuming that $d \geq 4$ and the probability space is rich enough one can find a Wiener process $W(\cdot)$ such that*

$$n^{1/2} \frac{R(n) - \mathbf{E}R(n)}{(\mathrm{Var} R(n))^{1/2}} - W(n) = O(n^{2/5+\varepsilon}) \quad a.s.$$

for any $\varepsilon > 0$.

The case $d = 3$ is harder. This problem was solved by Bass and Kumagai (2002). They proved the following surprising result.

THEOREM 21.5 *Assuming that $d = 3$ and the probability space is rich enough one can find a Wiener process $W(\cdot)$ such that*

$$2^{1/2} \pi \frac{R(n) - \mathbf{E}R(n)}{\gamma^2} - W(n \log n) = O(n^{1/2} (\log n)^q) \quad a.s.$$

where

$$\gamma = \mathbf{P}\{S_n \neq 0, \ n = 1, 2, \ldots\},$$
$$q = \frac{15}{32}.$$

Clearly this Theorem also implies the law of iterated logarithm for $d = 3$.

21.3 Wiener sausage

Let $W(t) \in \mathbb{R}^d$ be a Wiener process. Consider the random set

$$B_r(T) = \bigcup_{0 \leq t \leq T} (W(t) + K_r)$$
$$= \{x : x \in \mathbb{R}^d, x = W(t) + a \text{ for some } 0 \leq t \leq T \text{ and } a \in K_r\}$$

where

$$K_r = \{x : \|x\| \leq r\}.$$

$B_r(T)$ is called *Wiener sausage*. The most important results are summarized in

THEOREM 21.6 (cf. Le Gall, 1988). *For any $r > 0$ and $d = 2$ we have*

$$\lim_{T \to \infty} \frac{\log T}{T} \lambda(B_r(T)) = 2\pi \quad a.s. \qquad (21.5)$$

If $d \geq 3$ then

$$\lim_{T \to \infty} T^{-1} \lambda(B_r(T)) = c(r, d) \quad a.s. \qquad (21.6)$$

where $c(r, d)$ is a positive valued known function of r and d.
Further,

$$\lim_{T \to \infty} \mathbf{P}\{K_d(T)(\lambda(B_r(T)) - L_d(T)) < x\} = \Phi_d(\alpha x + \beta) \qquad (21.7)$$

where

$$L_d(T) = \begin{cases} 2\pi \dfrac{T}{\log T} & \text{if } d = 2, \\ \gamma(d, r)T & \text{if } d \geq 3, \end{cases}$$

$$K_d(T) = \begin{cases} \dfrac{(\log T)^2}{T} & \text{if } d = 2, \\ (T \log T)^{-1/2} & \text{if } d = 3, \\ T^{-1/2} & \text{if } d \geq 4, \end{cases}$$

Φ_d *is the normal law if $d \geq 3$ and it is non-normal if $d = 2$, and α, β, γ are known functions of r and d.*

Chapter 22

Heavy Points and Heavy Balls

22.1 The number of heavy points

Theorem 20.11 described the properties of the number $Q_k(n)$ of the points $x \in \mathbb{Z}^d$ ($d \geq 3$) visited exactly k times ($k = 1, 2, \ldots$) whenever k is a fixed positive integer. Now we wish to study the properties $Q_k(n)$ when $k = k(n)$ converges to infinity. By Theorem 20.7 $Q_k(n) = 0$ if $k = k(n) \geq (1+\varepsilon) \log n$ and n is large enough.

Introduce the following notations:

$$U(k,n) = \#\{j : 0 < j \leq n, \, \xi(S_j, \infty) = k, \, S_j \neq S_\ell \, (\ell = 1, \ldots, j-1)\},$$

$$R(k,n) = \sum_{j=k}^{\infty} Q_j(n) = \sum_{j=k}^{n} Q_j(n),$$

$$V(k,n) = \sum_{j=k}^{\infty} U(j,n).$$

THEOREM 22.1 (Csáki–Földes–Révész, 2005) *Let $d \geq 3$,*

$$\mu(t) = \gamma(1-\gamma)^{t-1}, \quad (t = 1, 2, \ldots)$$
$$t_n = t_n(B) = [\lambda \log n - \lambda B \log \log n], \quad (n = 3, 4, \ldots, \, B > 2)$$
$$\lambda = -(\log(1-\gamma))^{-1}$$

and $H(t,n)$ any of

$$\frac{U(t,n)}{n\gamma\mu(t)}, \quad \frac{Q(t,n)}{n\gamma\mu(t)}, \quad \frac{V(t,n)}{n\mu(t)}, \quad \frac{R(t,n)}{n\mu(t)}.$$

Then we have

$$\lim_{n \to \infty} \sup_{t \leq t_n} |H(t,n) - 1| = 0 \quad a.s.$$

In order to prove Theorem 22.1 we introduce a few notations and prove some Lemmas.

Let

$$X_i(t) = X_i$$
$$= \begin{cases} 1 & \text{if } S_j \neq S_i \ (j = 0, 1, 2, \ldots, i-1), \ \xi(S_i, \infty) \geq t, \\ 0 & \text{otherwise,} \end{cases}$$
$$Y_i(t, n) = Y_i$$
$$= \begin{cases} 1 & \text{if } S_j \neq S_i \ (j = 0, 1, 2, \ldots, i-1), \ \xi(S_i, n) \geq t, \\ 0 & \text{otherwise,} \end{cases}$$
$$\rho_i(t) = \rho_i = I(X_i = 1)(\min\{j : \xi(S_i, j) \geq t\} - i),$$
$$\mu_i(t) = \mu_i = f_i(1-\gamma)^{t-1},$$
$$\sigma_n^2 = \mathbf{E}\left(\sum_{i=1}^n X_i - n\mu\right)^2.$$

Clearly we have

$$V(t, n) = \sum_{i=1}^n X_i,$$
$$R(t, n) = \sum_{i=1}^n Y_i.$$

The next lemma can be easily obtained by Lemma 17.11.

LEMMA 22.1 *Let*

$$\mathbf{E} X_i = \mu_i,$$
$$n\mu \leq \mathbf{E}\sum_{i=1}^n X_i = \sum_{i=1}^n \mu_i \leq n\mu + \mu A_n O(1)$$

where

$$A_n = \sum_{i=1}^n \frac{1}{i^{d/2-1}} = \begin{cases} O(1) & \text{if} \quad d > 4, \\ O(1)\log n & \text{if} \quad d = 4, \\ O(1) n^{1/2} & \text{if} \quad d = 3. \end{cases}$$

LEMMA 22.2

$$\sigma_n^2 \leq n\mu + \mu A_n O(1) - n^2\mu^2 + 2(I + II + III)$$

where

$$I = \sum_{1 \leq i < j \leq n} \mathbf{P}\{X_i = 1,\ X_j = 1,\ \rho_i \geq n^\alpha\},$$

$$II = \sum_{1 \leq i < j \leq i+3n^\alpha \leq n} \mathbf{P}\{X_i = 1,\ X_j = 1,\ \rho_i < n^\alpha\},$$

$$III = \sum_{1 \leq i < i+3n^\alpha < j \leq n} \mathbf{P}\{X_i = 1,\ X_j = 1,\ \rho_i < n^\alpha\},$$

$$\alpha = 2/d.$$

Proof. Clearly we have

$$\sigma_n^2 = \mathbf{E}\left(\sum_{i=1}^n X_i\right)^2 + n^2\mu^2 - 2n\mu \mathbf{E}\left(\sum_{i=1}^n X_i\right)$$

$$= \mathbf{E}\sum_{i=1}^n X_i + 2 \sum_{1 \leq i < j \leq n} \mathbf{E} X_i X_j + n^2\mu^2 - 2n\mu \sum_{i=1}^n \mu_i \leq$$

$$\leq n\mu + \mu A_n O(1) + 2 \sum_{1 \leq i < j \leq n} \mathbf{E} X_i X_j - n^2\mu^2.$$

Further

$$\sum_{1 \leq i < j \leq n} \mathbf{E} X_i X_j = \sum_{1 \leq i < j \leq n} \mathbf{P}\{X_i = 1,\ X_j = 1\} = I + II + III.$$

Hence Lemma 22.2 is proved.

LEMMA 22.3

$$z_x + s_x \leq \frac{2(1-\gamma)}{2-\gamma} \qquad (22.1)$$

and

$$z_x + s_x \leq (1-\gamma)\left(1 + \frac{c}{(1-\gamma)R^{(d-2)}}\right) \qquad (22.2)$$

if $\|x\| = R$.

Proof. (22.1) follows from (17.6) and Lemma 17.13. Since

$$z_x + s_x = \mathbf{P}\{T < T_x\} + \mathbf{P}\{T_x < T\} \leq$$
$$\leq \mathbf{P}\{S_n = 0 \text{ for some } n\} + \mathbf{P}\{S_n = x \text{ for some } n\}.$$

By Lemma 17.8 we have (22.2).

LEMMA 22.4 *For $t \leq t_n$, any $\varepsilon > 0$ and large enough n we have*

$$I \leq O(1)n^{2/d+\varepsilon}\left(n + \left(\frac{2}{2-\gamma}\right)^{2t_n}\right)\mu^2. \tag{22.3}$$

Proof. Now we need to estimate the probability

$$\mathbf{P}\{X_i = 1,\ X_j = 1,\ \rho_i \geq n^\alpha\}.$$

Define the events B_k by

$$B_k = \{\xi(S_i, \infty) - \xi(S_i, i) + \xi(S_j, \infty) - \xi(S_j, i) = k\}$$

and consider the k time intervals between the consecutive visits of $\{S_i, S_j\}$. Then at least one of these intervals is larger than

$$\frac{\rho_i(t)}{k} \geq \frac{n^\alpha}{k} \tag{22.4}$$

(provided that $\{X_i = 1,\ X_j = 1,\ \rho_i \geq n^\alpha\}$). Denote this event by D_k. Then

$$\mathbf{P}\{X_i = 1,\ X_j = 1,\ \rho_i \geq n^\alpha\}$$
$$\leq \sum_{x \in \mathbb{Z}^d} \mathbf{P}\left\{S_j - S_i = x,\ \bigcup_{k \geq 2t-1} B_k D_k\right\}$$
$$\leq \sum_{x \in \mathbb{Z}^d} \mathbf{P}\{S_{j-i} = x\} \sum_{k \geq 2t-1} \mathbf{P}\{B_k D_k\}. \tag{22.5}$$

The event $B_k D_k$ means that placing a new origin at the point S_i and starting the time at i there are exactly k visits in the set $\{0, x\}$ and at least one time interval between consecutive visits is larger than n^α/k. Hence applying (17.7) of Lemma 17.14 and Lemma 17.11 we have

$$\mathbf{P}\{B_k D_k\} \leq O(1)k\left(\frac{k}{n^\alpha}\right)^{d/2-1}(1 - z_x - s_x)(z_x + s_x)^{k-1}$$
$$\leq O(1)k^{d/2}n^{2/d-1}(z_x + s_x)^{k-1}$$

where $O(1)$ is uniform in k and x, hence

$$\sum_{k \geq 2t-1} \mathbf{P}\{B_k D_k\} \leq O(1)n^{2/d-1} \sum_{k \geq 2t-1} k^{d/2}(z_x + s_x)^{k-1}$$
$$\leq O(1)n^{2/d-1}t^{d/2}(z_x + s_x)^{2t-2}. \tag{22.6}$$

By (22.5) and (22.6) we have

$$\mathbf{P}\{X_i = 1, \ X_j = 1, \ \rho_i \geq n^\alpha\}$$
$$\leq O(1) n^{2/d-1} t^{d/2} \sum_{x \in \mathbb{Z}^d} \mathbf{P}\{S_{j-i} = x\}(z_x + s_x)^{2t-2}$$
$$= O(1) n^{2/d-1} t^{d/2} \left[\sum_{\|x\| \leq R} + \sum_{\|x\| > R} \right] \quad (22.7)$$

where $R = t^{1/(d-2)}$.

By (22.1) and Lemma 17.5

$$\sum_{\|x\| \leq R} \leq O(1) \frac{R^d}{(j-i)^{d/2}} \left(\frac{2(1-\gamma)}{2-\gamma} \right)^{2t-2}$$
$$\leq O(1) \mu^2 \frac{t^{d/(d-1)}}{(j-i)^{d/2}} \left(\frac{2}{2-\gamma} \right)^{2t}. \quad (22.8)$$

By (22.2) we have

$$\sum_{\|x\| > R} \leq O(1) \sum_{\|x\| > R} \mathbf{P}\{S_{j-i} = x\} \left((1-\gamma) \left(1 + \frac{c}{(1-\gamma) R^{d-2}} \right) \right)^{2t-2}$$
$$\leq O(1) \mu^2 \sum_{\|x\| > R} \mathbf{P}\{S_{j-i} = x\} \left(1 + \frac{c}{(1-\gamma)t} \right)^{2t-2}$$
$$\leq O(1) \mu^2. \quad (22.9)$$

By (21.7), (21.8) and (21.9)

$$\mathbf{P}\{X_i = 1, \ X_j = 1, \ \rho_i \geq n^\alpha\}$$
$$\leq O(1) n^{2/d-1} t^{d/2} \mu^2 \left(1 + \frac{t^{d/(d-2)}}{(j-i)^{d/2}} \left(\frac{2}{2-\gamma} \right)^{2t} \right). \quad (22.10)$$

Hence

$$I \leq O(1) n^{2/d} t_n^{d/2} \mu^2 \left(n + t_n^{d/(d-2)} \left(\frac{2}{2-\gamma} \right)^{2t_n} \right).$$

Since $t_n \leq \lambda \log n$, we have Lemma 22.4.

LEMMA 22.5 *Let $i < j$. Then for $t \geq 1$ integer we have*

$$\mathbf{P}\{X_i = 1, \ X_j = 1\} \leq c\mu^2 \left(1 + \frac{t^{d/(d-2)}}{(j-i)^{d/2}} \left(\frac{2}{2-\gamma} \right)^{2t} \right)$$

where c is a constant, independent of i, j, t.

Proof.

$$\mathbf{P}\{X_i = 1,\ X_j = 1\} = \sum_{x \in \mathbb{Z}^d} \mathbf{P}\left\{S_j - S_i = x,\ \bigcup_{k \geq 2t-1} B_k\right\}$$

$$= \sum_{x \in \mathbb{Z}^d} \mathbf{P}\{S_{j-i} = x\}\mathbf{P}\{\xi(0,\infty) + \xi(x,\infty) \geq 2t - 1\}$$

$$= \sum_{x \in \mathbb{Z}^d} \mathbf{P}\{S_{j-i} = x\}(z_x + s_x)^{2t-1}.$$

By (22.8) and (22.9) we get Lemma 22.5.

LEMMA 22.6 *For $t \leq t_n$ any $\varepsilon > 0$ and large enough n we have*

$$II \leq O(1) n^{2/d+\varepsilon} \mu^2 \left(n + n^{1-2/d}\left(\frac{2}{2-\gamma}\right)^{2t_n}\right). \qquad (22.11)$$

Proof. It is a simple consequence of Lemma 22.5.

LEMMA 22.7

$$III \leq \frac{\mu^2 n^2}{2} + O(1) n^{3/2} \mu^2. \qquad (22.12)$$

Proof. Let

$$A = \{S_i \text{ is a new point i.e. } S_i \neq S_j\ j = 1, 2, \ldots, i-1\},$$
$$B = \{\xi(S_i, i + n^\alpha) - \xi(S_i, i) \geq t - 1\},$$
$$D = \{S_j \text{ is a new point}\},$$
$$E = \{\xi(S_j, \infty) - \xi(S_j, j) \geq t - 1\},$$
$$D \subset G = \left\{\xi(S_j, j) - \xi\left(S_j, i + \frac{2(j-i)}{3}\right) = 0\right\},$$
$$B \subset H = \{\xi(S_i, \infty) - \xi(S_i, i) \geq t - 1\}.$$

Let $j > i + 3n^\alpha$. Then

$$\mathbf{P}\{X_i = 1,\ X_j = 1,\ \rho_i < n^\alpha\}$$
$$\leq \mathbf{P}\{ABDE\} \leq \mathbf{P}\{ABGE\} = \mathbf{P}\{A\}\mathbf{P}\{B\}\mathbf{P}\{G\}\mathbf{P}\{E\}$$
$$\leq \mathbf{P}\{A\}\mathbf{P}\{H\}\mathbf{P}\{G\}\mathbf{P}\{E\} = \gamma(i+1)(1-\gamma)^{t-1}\gamma((j-i)/3)(1-\gamma)^{t-1}.$$

Clearly we have

$$III \leq \sum \gamma(i+1)(1-\gamma)^{t-1} \gamma((j-i)/3)(1-\gamma)^{t-1}$$

$$\leq \gamma^2(1-\gamma)^{2t-2} \sum \left(1 + \frac{O(1)}{(j-i)^{d/2-1}}\right)\left(1 + \frac{O(1)}{i^{d/2-1}}\right)$$

$$\leq \gamma^2(1-\gamma)^{2t-2}\left[\binom{n}{2} + O(1)(K+L+M)\right]$$

where

$$K = \sum \frac{1}{i^{d/2-1}} \leq nA_n,$$
$$L = \sum \frac{1}{(j-i)^{d/2-1}} \leq nA_n,$$
$$M = \sum \frac{1}{i^{d/2-1}} \frac{1}{(j-i)^{d/2-1}} \leq nA_n.$$

Hence we have Lemma 22.7.

LEMMA 22.8
$$\sigma_n^2 = O(1)(n\mu + \mu^2 n^{1.8}). \tag{22.13}$$

Proof is based on Lemmas 22.2, 22.4, 22.6 and 22.7. The numerical values of λ can be obtained by a result of Griffin (1990):

$$1 - \gamma_3 = 0.341,$$
$$1 - \gamma_4 = 0.193,$$
$$1 - \gamma_5 = 0.131,$$
$$1 - \gamma_6 = 0.104.$$

Consequently

$$\lambda_3 = 0.929,$$
$$\lambda_4 = 0.608,$$
$$\lambda_5 = 0.492,$$
$$\lambda_6 = 0.442.$$

By using $t_n < \lambda \log n$, one can verify (numerically)

$$\left(\frac{2}{2-\gamma}\right)^{2t_n} < n^{2\lambda \log(2/(2-\gamma))} < n^{0.75}$$

for $d = 3$ and hence also for all $d \geq 3$. By choosing an appropriate ε we can see that each term on the right-hand sides of (22.3), (22.11) and (22.12) is smaller than the right-hand side of (22.13), proving Lemma 22.8.

Lemma 22.8 implies

LEMMA 22.9 *For any $0 < C < B$*

$$\sigma_n (\log n)^{C/2} \leq O(1)((n\mu)^{1/2}(\log n)^{C/2} + \mu n^{0.9}(\log n)^{C/2}) = o(1)n\mu.$$

Proof of Theorem 22.1. Now we prove Theorem 22.1 in case

$$H(t,n) = \frac{V(t,n)}{n\mu(t)}. \tag{22.14}$$

By Markov's inequality for any $C > 0$ we have

$$\mathbf{P}\{|V(t,n) - n\mu(t)| \geq \sigma_n (\log n)^{C/2}\} \leq (\log n)^{-C}.$$

By Lemma 22.9, if $C < B$,

$$\mathbf{P}\{|V(t,n) - n\mu(t)| \geq o(1)n\mu(t)\} \leq (\log n)^{-C}.$$

Consequently, since $t_n < \lambda \log n$,

$$\mathbf{P}\left\{\sup_{t \leq t_n+1} \frac{|V(t,n) - n\mu(t)|}{n\mu(t)} \geq o(1)\right\} \leq O(1)(\log n)^{-C+1}. \tag{22.15}$$

Choose $C > 2$, $n(k) = \exp(k/\log k)$. (22.13) and Borel–Cantelli lemma imply

$$\lim_{k \to \infty} \sup_{t \leq t(n(k))+1} \left|\frac{V(t,n(k))}{n(k)\mu(t)} - 1\right| = 0 \quad \text{a.s.} \tag{22.16}$$

Let $n(k) \leq n < n(k+1)$. Then for $t \leq t_n$ we have

$$V(t, n(k)) \leq V(t,n) \leq V(t, n(k+1))$$

and

$$\lim_{k \to \infty} \frac{n(k+1)}{n(k)} = 1.$$

Hence for any $\varepsilon > 0$ and large enough n,

$$\frac{V(t,n)}{n\mu(t)} \leq \frac{V(t,n(k+1))}{n(k+1)\mu(t)} \frac{n(k+1)}{n} \leq (1+\varepsilon) \quad \text{a.s.},$$

since $t \leq t_n \leq t(n(k+1))$. Similarly,

$$\frac{V(t,n)}{n\mu(t)} \geq \frac{V(t,n(k))}{n(k)\mu(t)} \frac{n(k)}{n} \geq (1-\varepsilon) \quad \text{a.s.}$$

Hence we have Theorem 22.1 in case (22.14).

The other three statements of Theorem 22.1 can be obtained similarly.

Consequence 1. Apply Theorem 22.1 for
$$H(t,n) = \frac{R(t,n)}{n\mu(t)} \quad \text{and} \quad t = 1.$$

Since $R(1,n) = R(n)$ and $\mu(1) = \gamma$, we obtain (21.3) for $d \geq 3$ as a special case of Theorem 22.1.

Consequence 2. Apply Theorem 22.1 for
$$H(t,n) = \frac{Q(t,n)}{n\gamma\mu(t)}.$$

We obtain Theorem 20.11 for $d \geq 3$ as a special case of Theorem 22.1.

As we mentioned Hamana (1995) proved a CLT for $Q(t,n)$ whenever t is fixed. It looks interesting to try to prove the CLT in case $t = t_n \to \infty$.

Consequence 3. Apply Theorem 22.1 for
$$H(t,n) = \frac{R(t,n)}{n\mu(t)}, \quad t = [\lambda \log n - \lambda B \log \log n], \quad B > 2$$

we obtain
$$\xi(n) \geq \lambda \log n - \lambda B \log \log n \quad B > 0.$$

i.e.
$$\lambda \log n - \lambda B \log \log n \in \text{LLC}(\xi(n)).$$

A theorem on the upper classes of $\xi(n)$ is given in Révész (2004):
$$\lambda \log n + (1+\varepsilon)\lambda \log \log n \in \text{UUC}(\xi(n)),$$
$$\lambda \log n + (1-\alpha-\varepsilon)\lambda \log \log n \in \text{ULC}(\xi(n))$$

if
$$d \geq 5, \quad \alpha = \frac{2}{d-2}, \quad 0 < \varepsilon < 1-\alpha.$$

We do not have any non-trivial result about $\text{LUC}(\xi(n))$. For example the following question looks interesting: Is it true that
$$\xi(n) \leq \lambda \log n \quad \text{i.o. a.s.}$$

Remark 1. Without any new idea one can prove the following slightly stronger version of Theorem 22.1: there exists an $\varepsilon > 0$ such that
$$\lim_{n \to \infty} n^\varepsilon \sup_{t \leq t_n} |H(t,n) - 1| = 0 \quad \text{a.s.}$$

22.2 Heavy balls

Theorem 20.7 told us how heavy can be the heaviest lattice point. In fact we saw that
$$\lim_{n\to\infty} \frac{\xi(n)}{\lambda \log n} = 1 \quad \text{a.s.}$$
Similarly we might ask about the weight of the heaviest ball. Let
$$\mathcal{S}(1) = \{x : x \in \mathbb{Z}^d, |x| = 1\}$$
and for any $A \subset \mathbb{Z}^d$ let
$$\mu_n(A) = \sum_{z \in A} \xi(z, n).$$
Then we are interested in the limit properties of
$$\sup_{x \in \mathbb{Z}^d} \mu_n(x + \mathcal{S}(1)).$$
An answer of this question is

THEOREM 22.2 (Csáki–Földes–Révész–Rosen–Shi, 2005)
$$\lim_{n\to\infty} \sup_{x \in \mathbb{Z}^d} \frac{\mu_n(x + \mathcal{S}(1))}{\log n} = -\frac{1}{\log\left(1 - \frac{\gamma}{2d(1-\gamma)}\right)} \quad a.s.$$

In order to prove this Theorem we present two notations and one Lemma.

1. Let ϑ be the probability that starting from $\mathcal{S}(1)$ the particle returns to $\mathcal{S}(1)$ before visiting 0 i.e.
$$\vartheta = \mathbf{P}\{\min\{k : k > 1, |S_k| = 1\} < \min\{k : k > 1, S_k = 0\}\}.$$

2. Let τ be the number of outward excursions from $\mathcal{S}(1)$ to $\mathcal{S}(1)$, i.e.
$$\tau = \sum_{j=1}^{\infty} I\{S_j \in \mathcal{S}(1), |S_{j+1}| > 1\}.$$

3. Let $Z(A)$ denote the number of visits in the set A up to first return to 0, i.e.
$$Z(A) = \sum_{n=1}^{T} I\{S_n \in A\}.$$

LEMMA 22.10

$$\vartheta = 1 - \frac{1}{2d(1-\gamma)}, \quad (22.17)$$

$$\mathbf{P}\{\mu_\infty(\mathcal{S}(1)) = j\} = \left(1 - \vartheta - \frac{1}{2d}\right)\left(\vartheta + \frac{1}{2d}\right)^{j-1}, \quad (22.18)$$

$$\mathbf{P}\{\tau = M,\ \xi(0,\infty) = N\}$$
$$= \binom{N+M}{N}\left(1 - \vartheta - \frac{1}{2d}\right)\vartheta^M \left(\frac{1}{2d}\right)^N. \quad (22.19)$$

Proof. Clearly

$$\mathbf{P}\{Z(\mathcal{S}(1)) = j,\ T < \infty\} = \vartheta^{j-1}\frac{1}{2d}. \quad (22.20)$$

Summing (22.20) in j we get

$$1 - \gamma = \mathbf{P}\{T < \infty\} = \sum_{j=1}^{\infty} \vartheta^{j-1}\frac{1}{2d} = \frac{1}{2d(1-\vartheta)}$$

which, in turn, implies (22.17). The proof of (22.18) and (22.19) can be obtained similarly.

Proof of Theorem 22.2. First we prove the upper bound. Note that

$$\sup_{x \in \mathbb{Z}^d} \mu_n(x + \mathcal{S}(1)) \leq \sup_{x \in \mathbb{Z}^d} \mu_\infty(\mathcal{S}(1) + x)I(T_{\mathcal{S}(1)+x} \leq n)$$

and

$$\mathbf{P}\{\mu_\infty(\mathcal{S}(1) + x)I(T_{\mathcal{S}(1)+x} \leq n) \geq j\} \leq \mathbf{P}\{\mu_\infty(\mathcal{S}(1)) \geq j\}.$$

Then by (22.18) for some $\delta > 0$ we have

$$\mathbf{P}\{\mu_\infty(\mathcal{S}(1)) > c\log n\} \leq \left(\vartheta + \frac{1}{2d}\right)^{c\log n} = n^{-1}$$

where

$$c = \frac{1}{\log\left(1 - \frac{\gamma}{2d(1-\gamma)}\right)}.$$

Hence for any $\varepsilon > 0$ we have

$$\mathbf{P}\{\sup_{x \in \mathbb{Z}^d} \mu_\infty(\mathcal{S}(1) + x) I(T_{\mathcal{S}(1)+x} \leq n) > (c+\varepsilon)\log n\} \leq n^{-\varepsilon}.$$

Now the upper bound can be obtained by the usual application of the Borel–Cantelli lemma.

Now we turn to the lower bound. Let $h < c$ and $n_i = i(\log n)^3$ ($i = 0, 1, \ldots, [n(\log n)^{-3}]$). Since

$$\mathbf{P}\{\sup_x \mu_n(\mathcal{S}(1)) < h\log n\} \leq (\mathbf{P}\{\mu_{[\log n]^3}(\mathcal{S}(1)) < h\log n\})^{n(\log n)^{-3}},$$

applying again the Borel–Cantelli lemma we have the lower bound.

It is natural to ask how can the set $\mathcal{S}(1)$ be replaced in Theorem 22.2 by another bounded subset A of $\mathbb{Z}^d$. This question was studied by Csáki–Földes–Révész–Rosen–Shi (2005). They proved the following three Theorems.

THEOREM 22.3

$$\lim_{n \to \infty} \sup_{x \in \mathbb{Z}^d} \frac{\mu_n(A+x)}{\log n} = -\frac{1}{\log(1 - 1/\Lambda_A)} \quad a.s.$$

where A is a finite subset of $\mathbb{Z}^d$, Λ_A is the largest eigenvalue of the $|A| \times |A|$ matrix $G_A(x,y) = G(x-y)$ ($x,y \in A$) and $G(x) = \sum_{k=0}^\infty \mathbf{P}\{S_k = x\}$.

To get the exact value of Λ_A is very hard except for some simple A. For example we have

THEOREM 22.4

$$\Lambda_{\{0,y\}} = \frac{1 + t_y}{\gamma} \tag{22.21}$$

where $t_y = \mathbf{P}\{T_y < \infty\}$ and

$$\Lambda_{B(0,1)} = \frac{2}{2 - \vartheta - (\vartheta^2 + 2/d)^{1/2}} \tag{22.22}$$

where $B(0,1) = \{x : x \in \mathbb{Z}^d, |x| \leq 1\}$.

Remark 1. (22.21) shows that the maximal total time spent in a small neighbourhood of any point is less than the number of points in the neighbourhood times the maximal local time for points. To study this phenomenon further, consider the sets of two points not too far from each other.

THEOREM 22.5 *Put $r(n) = (\log n)^K$ with any $K > 0$. Then*

$$\lim_{n\to\infty} \max_{x,y\in D} \frac{\mu_n(\{x,y\})}{\log n} = \frac{1}{\log \frac{2-\gamma}{2(1-\gamma)}} < \frac{-2}{\log(1-\gamma)}$$

where

$$D = \{x,y : x,y \in \mathbb{Z}^d, |x-y| \leq r(n)\}.$$

This Theorem expresses the fact that any two points with local time up to time n, both close to the maximum, should be at a distance larger than any power of $\log n$.

22.3 Heavy balls around heavy points

Theorems 22.2 and 22.4 tell us that for any $\varepsilon > 0$ there exist sequences $\{u_n = u_n(\varepsilon) \in \mathbb{Z}^d, n = 1, 2, \ldots\}$ and $\{v_n = v_n(\varepsilon) \in \mathbb{Z}^d, n = 1, 2, \ldots\}$ such that

$$\mu_n(u_n + \mathcal{S}(1)) \geq -(1-\varepsilon)\log n \frac{1}{\log\left(1 - \frac{\gamma}{2d(1-\gamma)}\right)} \tag{22.23}$$

and

$$\mu_n(v_n + B(1)) \geq -(1-\varepsilon)\log n \frac{1}{\log\left(\frac{\vartheta + (\vartheta^2 + 2/d)^{1/2}}{2}\right)}. \tag{22.24}$$

We are interested in the properties of the sequences $\{u_n\}$, $\{v_n\}$. For example how big (or how small) $\xi(u_n, n)$ and $\xi(v_n, n)$ can be. It is not hard to prove that none of them can be very big. More precisely for any $0 < \varepsilon < 1$ there exist a $\lambda_1 < \lambda$ and a $\lambda_2 < \lambda$ such that (22.23) and (22.24) hold true, but

$$\xi(u_n, n) \leq \lambda_1 \log n$$

and

$$\xi(v_n, n) \leq \lambda_2 \log n.$$

Let

$$\mathcal{F}(\beta, n) = \{x : x \in \mathbb{Z}^d, \xi(x, n) = [\lambda\beta\log n]\},$$
$$(0 < \beta < 1; n = 2, 3, \ldots)$$
$$\xi(\beta, n, \mathcal{S}) = \sup_{x\in\mathcal{F}(\beta,n)} \sum_{|y-x|=1} \xi(y, n),$$
$$\xi(\beta, n, B) = \sup_{x\in\mathcal{F}(\beta,n)} \sum_{|y-x|\leq 1} \xi(y, n).$$

Conjecture. There exist functions $f(\beta, \mathcal{S})$ and $g(\beta, B)$ such that

$$\lim_{n \to \infty} \frac{\xi(\beta, n, \mathcal{S})}{\log n} = f(\beta, \mathcal{S}), \quad \text{a.s.}$$

$$\lim_{n \to \infty} \frac{\xi(\beta, n, B)}{\log n} = g(\beta, B), \quad \text{a.s.}$$

$$\sup_{0 < \beta < 1} f(\beta, \mathcal{S}) = -\frac{1}{\log\left(1 - \dfrac{\gamma}{2d(1-\gamma)}\right)},$$

$$\sup_{0 < \beta < 1} g(\beta, B) = -\frac{1}{\log\left(\dfrac{\vartheta + (\vartheta^2 + 2/d)^{1/2}}{2}\right)}.$$

22.4 Wiener process

The questions formulated for random walk in Sections 22.1 and 22.2 can be reformulated for Wiener process. Such problems were initiated and solved by Perkins–Taylor (1987) and Dembo–Peres–Rosen–Zeitouni (2000).

Let

$$\mu_T^W(A) = \int_0^T I_A(W(t)) dt$$

and let

$$B(x, r) = \{y : y \in \mathbb{R}^d, |x - y| \leq r\}.$$

THEOREM 22.6 *There exist absolute constants $0 < c_1 < c_2 < \infty$ such that*

$$\frac{c_1 \varepsilon^2}{|\log \varepsilon|} \leq \mu_T^W(B(x, \varepsilon)) \leq \frac{c_2 \varepsilon^2}{|\log \varepsilon|} \quad \text{a.s.} \tag{22.25}$$

for all $x \in \{W(t), 0 \leq t \leq T\}$ if ε is small enough.

This Theorem is a straight analogue of Theorems 22.2 and 22.3. It suggests the following two questions:

1. What can we say about c_1 and c_2?
2. What can we say about the number of heavy balls? More precisely we ask about the Hölder dimension of the set of the centers of the heavy balls.

The next Theorem claims that c_1 and c_2 are about $4q_d^{-2}$ where q_d is the first positive zero of the Bessel function $J_{d/2-2}(x)$.

THEOREM 22.7 *Let $d \geq 3$. Then for any $T > 0$ and all $0 < a < 4q_d^{-2}$*

$$\dim\left\{x : \limsup_{\varepsilon \to 0} \frac{\mu_T^W(B(x, \varepsilon))}{\varepsilon^2 |\log \varepsilon|} = a\right\} = 2 - \frac{a q_d^2}{2} \quad \text{a.s.}$$

Chapter 23

Crossing and Self-crossing

It is easy to see (and Theorem 21.1 also implies) that the path of a random walk crosses itself infinitely many times for any $d \geq 1$ with probability 1. We mean that there exists an infinite sequence $\{U_n, V_n\}$ of positive integer valued r.v.'s such that $S(U_n) = S(U_n + V_n)$, and $0 \leq U_1 < U_2 < \ldots, (n = 1, 2, \ldots)$. However, we ask the following question: will self-crossings occur after a long time? For example, we ask whether the crossing $S(U_n) = S(U_n + V_n)$ will occur for every $n = 1, 2, \ldots$ if we assume that V_n converges to infinity with a great speed and U_n converges to infinity much slower. In fact Erdős and Taylor (1960/B) proposed the following two problems.

Problem A. Let $f(n)$ be a positive integer valued function. What are the conditions on the rate of increase of $f(n)$ which are necessary and sufficient to ensure that the paths $\{S_0, S_1, \ldots, S_n\}$ and $\{S_{n+f(n)}, S_{n+f(n)+1}, \ldots\}$ have points in common for infinitely many values of n with probability 1?

Problem B. A point S_n of a path is said to be "good" if there are no points common to $\{S_0, S_1, \ldots, S_n\}$ and $\{S_{n+1}, S_{n+2}, \ldots\}$. For $d = 1$ or 2 there are no good points with probability 1. For $d \geq 3$ there might be some good points: how many are there?

As far as Problem A is concerned we have

THEOREM 23.1 (Erdős–Taylor, 1960/B). *Let $f(n) \uparrow \infty$ be a positive integer valued function and let E_n be the event that paths*

$$\{S_0, S_1, \ldots, S_n\} \quad \text{and} \quad \{S_{n+f(n)}, S_{n+f(n)+1}, \ldots\}$$

have points in common. Then

(i) *for $d = 3$, if $f(n) = n(\varphi(n))^2$ and $\varphi(n)$ is increasing, then*

$$\mathbf{P}\{E_n \text{ i.o. }\} = 0 \quad \text{or} \quad 1 \tag{23.1}$$

depending on whether $\sum_{k=1}^{\infty} (\varphi(2^k))^{-1}$ converges or diverges,

(ii) *for $d = 4$, if $f(n) = n\chi(n)$ and $\chi(n)$ is increasing, then we have (23.1) depending on whether $\sum_{k=1}^{\infty} (k\chi(2^k))^{-1}$ converges or diverges,*

(iii) *for $d \geq 5$, if*

$$\sup_{m \geq n} \frac{f(m)}{m} \geq C \frac{f(n)}{n}$$

(for some $C > 0$) then we have (23.1) depending on whether

$$\sum_{n=1}^{\infty}(f(n))^{(2-d)/2}$$

converges or diverges.

An answer of Problem B is

THEOREM 23.2 (Erdős–Taylor, 1960/B). *For $d \geq 3$ let $G^{(d)}(n)$ be the number of integers r $(1 \leq r \leq n)$ for which $(S_0, S_1, \ldots, S_r)$ and $(S_{r+1}, S_{r+2}, \ldots$ have no points in common. Then*

(i) *if $d = 3$, for any $\varepsilon > 0$*

$$\mathbf{P}\{G^{(3)}(n) > n^{1/2+\varepsilon} \quad i.o.\ \} = 0,$$

(ii) *if $d = 4$,*

$$\mathbf{P}\left\{0 = \liminf_{n \to \infty} \frac{G^{(4)}(n)\log n}{n} \leq \limsup_{n \to \infty} \frac{G^{(4)}(n)\log n}{n} \leq C\right\} = 1,$$

(iii) *if $d \geq 5$,*

$$\lim_{n \to \infty} \frac{G^{(d)}(n)}{n} = \tau_d \quad a.s.$$

where τ_d is an increasing sequence of positive numbers with $\tau_d \uparrow 1$ as $d \to \infty$.

Remark 1. Applying Theorem 23.1 for $d = 4$ and $f(n) = n - 1$ we find that for infinitely many n the paths $\{S_0, S_1, \ldots, S_n\}$ and $\{S_{2n}, S_{2n+1}, \ldots\}$ have a point in common. This statement is not true for $d \geq 5$.

Our next theorem is intuitively clear by Remark 1.

THEOREM 23.3 (Erdős–Taylor, 1960/B, Lawler, 1980). *For $d = 4$, two independent random walks which start from any two given fixed points have infinitely many common points with probability 1; whereas for $d \geq 5$, two independent random walks meet only finitely often, with probability 1.*

Remark 2. Theorem 23.3 tells us that the paths of two independent random walks in $\mathbb{Z}^d$ ($d \leq 4$) cross each other. It does not mean that the particles meet each other.

One can also investigate the self-crossing of a d-dimensional Wiener process. Dvoretzky–Erdős–Kakutani (1950) proved the following beautiful theorem:

THEOREM 23.4 *For $d \leq 3$ almost all paths of a Wiener process have double points (in fact they have infinitely many double points), i.e. there exist r.v.'s $0 \leq U < V \leq \infty$ with $W(U) = W(V)$. For $d \geq 4$ almost all paths of a Wiener process have no double points.*

Remark 3. Comparing Theorems 23.1 and 23.4 in the case $d = 4$ we obtain that for infinitely many n the paths $\{S_0, \ldots, S_n\}$ and $\{S_{2n}, S_{2n+1}, \ldots\}$ have a point in common, while for any $t > 0$ the paths $\{W(s);\ 0 \leq s \leq t\}$ and $\{W(s);\ 2t \leq s < \infty\}$ have no points in common with probability 1. This surprising fact is explained by Erdős and Taylor (1961) as follows: for $d = 4$ with probability 1 the paths $\{W(s);\ 0 \leq s \leq t\}$ and $\{W(s);\ 2t < s < \infty\}$ approach arbitrarily close to each other for arbitrarily large values of t. Thus they have infinitely many near misses, but fail to intersect. This explanation suggests the following:

Problem C. Characterize the set of those functions $f(\cdot)$ for which

$$\lim_{t \to \infty} f(t) \inf_{\substack{0 \leq s \leq t \\ 2t \leq u}} \|W(s) - W(u)\| = 0 \quad \text{a.s.} \quad (d = 4).$$

Since the paper of Erdős–Taylor (1960/B) a number of new results about the crossing of independent random walks appeared. Here we present some of them.

Let $\mathbf{S}(p, n) = \{S_1(n), S_2(n), \ldots, S_p(n)\}$ ($p \geq 2$, $n = 1, 2, \ldots$) where $S_1(n), S_2(n), \ldots, S_p(n)$ are independent random walks on $\mathbb{Z}^d$. Then

$$I_n = \sum_{k_1, k_2, \ldots, k_p = 1}^{n} I(S_1(k_1) = S_2(k_2) = \cdots = S_p(k_p))$$

is the intersection local time of $\mathbf{S}(p, n)$.

It is relatively easy to prove that

$$\lim_{n \to \infty} I_n = \infty \quad \text{a.s.} \tag{23.2}$$

if and only if $p(d - 2) \leq d$.

In case $d = 1$ the rate in (23.2) was studied by X. Chen and W. Li (2004). They proved

$$\limsup_{n \to \infty} n^{-(p+1)/2} (\log \log n)^{-(p-1)/2} I_n$$

$$= \left(\frac{4(p-1)}{p} \right)^{(p-1)/2} \left(\frac{p+1}{2} \right)^{(p-3)/2} \left(B\left(\frac{1}{p-1}, \frac{1}{2} \right) \right)^{-(p-1)} \quad \text{a.s.}$$

where $B(\cdot,\cdot)$ is the β-function.

The case $d=2$, $p \geq 2$ as well as the case $d=3$, $p=2$ was studied by X. Chen (2004). He proved

$$\limsup_{n\to\infty} \frac{I_n}{n(\log\log n)^{p-1}} = \left(\frac{2}{p}\right)^{p-1} \kappa(2,p)^{2p} \quad \text{a.s.}$$

if $d=2$, $p \geq 2$ and

$$\limsup_{n\to\infty} \frac{I_n}{(n(\log\log n)^3)^{1/2}} = \kappa(3,2) \quad \text{a.s.}$$

if $d=3$, $p=2$ where $\kappa(d,p)$ is known.

The case $d=4$, $p=2$ is settled by Marcus and Rosen (1997):

$$\limsup_{n\to\infty} \frac{I_n}{\log n \log\log\log n} = \frac{1}{2\pi^2} \quad \text{a.s.}$$

In case $d=p=3$ Rosen (1997) proved that

$$\limsup_{n\to\infty} \frac{I_n}{\log n \log\log\log n} = \frac{1}{\pi} \quad \text{a.s.}$$

The very interesting monograph of Lawler (1991) is proposed to the reader who is interested in a more detailed study of the subject of this Chapter.

Chapter 24

Large Covered Balls

24.1 Completely covered discs centered in the origin of $\mathbb{Z}^2$

We say that the disc

$$Q(r) = \{x \in \mathbb{Z}^2, \|x\| \leq r\}$$

is covered by the random walk $\{S_k\}$ in time n if

$$\xi(x, n) > 0 \text{ for every } x \in Q(r).$$

Let $R(n)$ be the largest integer for which $Q(R(n))$ is covered in time n. The Recurrence Theorem of Chapter 17 implies that

$$\lim_{n \to \infty} R(n) = \infty \quad \text{a.s.} \tag{24.1}$$

We are interested in the rate of convergence in (24.1). We prove

THEOREM 24.1 (Erdős–Révész, 1988, Révész, 1989/A, 1989/B, Auer, 1990). *For any $\varepsilon > 0$ and $C > 0$ we have*

$$\exp\left(2(\log n)^{1/2} \log_3 n\right) \in \text{UUC}(R(n)), \tag{24.2}$$

$$\exp\left(\frac{1-\varepsilon}{120^{1/2}} (\log n \log_3 n)^{1/2}\right) \in \text{ULC}(R(n)), \tag{24.3}$$

$$\exp\left(C(\log n)^{1/2}\right) \in \text{LUC}(R(n)), \tag{24.4}$$

$$\exp\left((\log n)^{1/2}(\log \log n)^{-1/2-\varepsilon}\right) \in \text{LLC}(R(n)). \tag{24.5}$$

A sharper statement than (24.2) and (24.3) was proved by Hough–Peres (2005). They proved that

$$\limsup_{n \to \infty} \frac{(\log R(n))^2}{\log n \log_3 n} = \frac{1}{4} \quad \text{a.s.} \tag{24.2-3*}$$

Here we prove only (24.2) – (24.5). In fact instead of
(24.3) the stronger Theorem 24.5,
(24.4) the stronger Theorem 24.7,
(24.5) the stronger Theorem 24.3
will be proved. The proof of (24.2) is given as stated above.

About the limit distribution of $R(n)$ we prove

THEOREM 24.2 (Révész, 1989/A, 1989/B). *For any $z > 0$*

$$\exp(-120z) \leq \liminf_{n\to\infty} \mathbf{P}\left\{\frac{(\log R(n))^2}{\log n} > z\right\}$$
$$\leq \limsup_{n\to\infty} \mathbf{P}\left\{\frac{(\log R(n))^2}{\log n} > z\right\} \leq \exp\left(-\frac{z}{4}\right). \quad (24.6)$$

A sharper statement than (24.6) was proved by Lawler (1993). He proved that

$$\exp(-4z) \leq \liminf_{n\to\infty} \mathbf{P}\left\{\frac{(\log R(n))^2}{\log n} > z\right\}$$
$$\leq \limsup_{n\to\infty} \mathbf{P}\left\{\frac{\log R(n))^2}{\log n} > z\right\} \leq \exp(-2z).$$

Note that even the proof of the existence of the limit distribution is hard.

The final result is due to Dembo, Peres, Rosen and Zeitouni (2005). They proved that

$$\lim_{n\to\infty} \mathbf{P}\left\{\frac{(\log R(n))^2}{\log n} > z\right\} = \exp(-4z). \quad (24.6^*)$$

Here we prove only (24.6). In fact instead of
the upper part of (24.6) the stronger Theorem 24.4,
the lower part of (24.6) the stronger Theorem 24.6
will be proved.

In order to prove Theorem 24.2 at first we introduce a few notations and prove some lemmas.

Let $\alpha(r)$ be the probability that the random walk $\{S_n\}$ hits the circle of radius r before returning to the point $0 = (0,0)$, i.e.

$$\alpha(r) = \mathbf{P}\{\inf\{n: \|S_n\| \geq r\} < \inf\{n: n \geq 1, S_n = 0\}\}.$$

Further, let

$$p(0 \rightsquigarrow x) = \mathbf{P}\{\inf\{n: n \geq 1, S_n = 0\} > \inf\{n: n \geq 1, S_n = x\}\}$$
$$= \mathbf{P}\{\{S_n\} \text{ reaches } x \text{ before returning to } 0\}.$$

LEMMA 24.1
$$\lim_{r\to\infty} \alpha(r)\log r = \pi/2. \tag{24.7}$$

Proof. Clearly we have

$$\{\inf\{n : \|S_n\| \geq r\} > \inf\{n : n \geq 1, S_n = 0\}\}$$
$$\subset \{\xi(0, r^2 \log^2 r) > 0\} + \{\max_{0\leq k\leq r^2 \log^2 r} \|S_k\| \leq r\}.$$

Since by Lemma 20.1

$$\mathbf{P}\{\xi(0, r^2 \log^2 r) = 0\} \approx \pi/2\log r$$

and by a trivial calculation (cf. Theorem 18.5)

$$\mathbf{P}\{\max_{0\leq k\leq r^2 \log^2 r} \|S_k\| \leq r\} = o(1/\log r),$$

we have

$$\alpha(r) \geq \frac{\pi + o(1)}{2\log r}.$$

Observe also

$$\alpha(r) \leq \mathbf{P}\{\max_{0\leq k\leq r^2 (\log r)^{-1}} \|S_k\| \geq r\} + \mathbf{P}\{\xi(0, r^2(\log r)^{-1}) = 0\}.$$

Since by Theorem 18.5

$$\mathbf{P}\{\max_{0\leq k\leq r^2(\log r)^{-1}} \|S_k\| \geq r\} = \exp(-O(\log r)),$$

applying again Lemma 20.1 we obtain (24.7).

Remark 1. Lemma 24.1 is closely related to Lemma 18.1.

LEMMA 24.2 *There exists a positive constant C such that*

$$p(0 \rightsquigarrow x) \geq \frac{C}{\log \|x\|}$$

for any $x \in \mathbb{Z}^2$ with $\|x\| \geq 2$. Further,

$$\frac{\pi}{12} \leq \liminf_{\|x\|\to\infty} p(0 \rightsquigarrow x)\log\|x\| \leq \limsup_{\|x\|\to\infty} p(0 \rightsquigarrow x)\log\|x\| \leq \frac{\pi}{2}.$$

Proof. Let $x = \|x\|e^{i\varphi}$. Then by Lemma 24.1 the probability that the particle crosses the arc $\|x\|e^{i\psi}$ ($\varphi - \pi/3 < \psi < \varphi + \pi/3$) before returning to 0 is larger than $(1-\varepsilon)\pi(6\log\|x\|)^{-1}$ (for any $\varepsilon > 0$ if $\|x\|$ is big enough). Since starting from any point of the arc $\|x\|e^{i\psi}$ ($\varphi - \pi/3 < \psi < \varphi + \pi/3$) the probability that the particle hits x before 0 is larger than $1/2$ we obtain the lower estimate of the liminf. The upper estimate is a trivial consequence of Lemma 24.1.

Spitzer (1964) obtained the exact order of $p(0 \rightsquigarrow x)$. He proved

LEMMA 24.3 (Spitzer, 1964, pp. 117, 124 and 125).

$$p(0 \rightsquigarrow x) = \frac{\pi + o(1)}{4\log\|x\|} \quad as \quad \|x\| \to \infty.$$

LEMMA 24.4 *Let*

$$Y_i = \xi(x, \rho_i) - \xi(x, \rho_{i-1}) - 1 \quad (i = 1, 2, \ldots)$$

and $Z_i = -Y_i$. Then there exists a positive constant C^* such that for any $\delta > 0$ and $f(n) \uparrow \infty$ for which $n/f(n) \to \infty$ we have

$$\mathbf{P}\{Y_1 + Y_2 + \cdots + Y_n \geq \delta\sigma n^{3/4}\} \leq \exp\left(-\frac{\delta^2 n^{1/2}}{2} + C^*\left(\frac{n}{f(n)}\right)^{1/2}\right) \quad (24.8)$$

and

$$\mathbf{P}\{Z_1 + Z_2 + \cdots + Z_n \geq \delta\sigma n^{3/4}\} \leq \exp\left(-\frac{\delta^2 n^{1/2}}{2} + C^*\left(\frac{n}{f(n)}\right)^{1/2}\right) \quad (24.9)$$

where

$$\sigma^2 = \mathbf{E}Y_1^2 = 2\frac{1 - p(0 \rightsquigarrow x)}{p(0 \rightsquigarrow x)}$$

and

$$\|x\| \leq \exp\left(\frac{n^{1/2}}{f(n)}\right). \quad (24.10)$$

(24.8) and (24.9) combined imply

$$\mathbf{P}\{|Y_1 + Y_2 + \cdots + Y_n| \geq \delta\sigma n^{3/4}\} \leq 2\exp\left(-\frac{\delta^2 n^{1/2}}{2} + C^*\left(\frac{n}{f(n)}\right)^{1/2}\right). \quad (24.11)$$

LARGE COVERED BALLS

Proof. By a simple combinatorial argument (cf. Theorem 9.7) we get

$$\mathbf{P}\{Y_1 = -1\} = 1 - p(0 \leadsto x),$$
$$\mathbf{P}\{Y_1 = k\} = (1 - p(0 \leadsto x))^k (p(0 \leadsto x))^2 \quad (k = 0, 1, 2, \ldots),$$
$$\mu_1 = \mathbf{E}Y_1 = \mathbf{E}Z_1 = 0,$$
$$\mu_2 = \mathbf{E}Y_1^2 = \mathbf{E}Z_1^2 = \sigma^2 = 2\frac{1 - p(0 \leadsto x)}{p(0 \leadsto x)}$$
$$\ldots \quad \ldots$$
$$\mu_m = \mathbf{E}Y_1^m = (-1)^m q + \sum_{k=0}^{\infty} k^m p^2 q^k$$
$$\leq (-1)^m q + \sum_{k=1}^{\infty} p^2 q^k k(k+1) \cdots (k+m-1)$$
$$= (-1)^m q + m! p^{1-m} q$$

where $p = p(0 \leadsto x)$ and $q = 1 - p$. Hence

$$|\mu_m| \leq q + m! p^{1-m} q \quad (m = 3, 4, \ldots).$$

Similarly

$$|\mu_m^*| = |\mathbf{E}Z_1^m| \leq q + m! p^{1-m} q \quad (m = 3, 4, \ldots).$$

Let

$$g(z) = \mathbf{E}\exp(zY_1) = qe^{-z} + \frac{p^2}{1 - qe^z},$$
$$\psi(z) = \log g(z)$$

and

$$s = s_n = \delta\sigma^{-1} n^{-1/4}.$$

Then by Lemma 24.2 and condition (24.10) we have

$$s = \delta \left(\frac{p}{2q}\right)^{1/2} n^{-1/4} = \frac{\delta p}{(2q)^{1/2} n^{1/4}} p^{-1/2} \leq \frac{\delta p}{(2q)^{1/2} n^{1/4}} \left(\frac{\log\|x\|}{C}\right)^{1/2}$$
$$\leq \frac{\delta p}{(2q)^{1/2} n^{1/4}} \frac{n^{1/4}}{(f(n)C)^{1/2}} = \frac{\delta}{(2qC)^{1/2}} \frac{1}{(f(n))^{1/2}} p \leq \frac{p}{2}$$

if n is big enough. Hence

$$\frac{|\mu_k| s^k}{k!} \leq q\frac{s^k}{k!} + pq\left(\frac{s}{p}\right)^k \leq q\frac{s^k}{k!} + pq\left(\frac{1}{2}\right)^k$$

($k = 3, 4, \ldots$) and

$$|\psi(s)| \leq \log\left(1 + \frac{\sigma^2 s^2}{2} + q\sum_{k=3}^{\infty} \frac{s^k}{k!} + pq\sum_{k=3}^{\infty} \left(\frac{s}{p}\right)^k\right)$$

$$\leq \log\left(1 + \frac{\sigma^2 s^2}{2} + qs^3 e^s + pq\left(\frac{s}{p}\right)^3 2\right) \leq \frac{\sigma^2 s^2}{2} + s^3 e^s + \frac{2s^3}{p^2}.$$

Similarly

$$|\log \mathbf{E} \exp(sZ_1)| = |\psi(-s)| \leq \frac{\sigma^2 s^2}{2} + s^3 e^s + \frac{2s^3}{p^2}.$$

Let

$$F_n(k) = \mathbf{P}\{Y_1 + Y_2 + \cdots + Y_n = k\}$$

and define a sequence $U_1, U_2, \ldots$ of i.i.d.r.v.'s with

$$\mathbf{P}\{U_1 = k\} = e^{-\psi(s)} e^{sk} \mathbf{P}\{Y_1 = k\}.$$

Then

$$\mathbf{P}\{Y_1 + Y_2 + \cdots + Y_n > \delta\sigma n^{3/4}\} = \sum_{k > \delta\sigma n^{3/4}} F_n(k) = e^{n\psi(s)} \sum_{k > \delta\sigma n^{3/4}} e^{-sk} V_n(k)$$

where

$$V_n(k) = \mathbf{P}\{U_1 + U_2 + \cdots + U_n = k\}.$$

Hence

$$\mathbf{P}\{Y_1 + Y_2 + \cdots + Y_n > \delta\sigma n^{3/4}\}$$

$$\leq \exp(n\psi(s) - s\delta\sigma n^{3/4}) \frac{1}{1 - e^{-s}}$$

$$\leq \exp\left(n\left(\frac{\sigma^2 s^2}{2} + s^3 e^s + \frac{2s^3}{p^2}\right) - s\delta\sigma n^{3/4}\right) \frac{1}{1 - e^{-s}}$$

$$\leq \exp\left(-\frac{n^{1/2} \delta^2}{2} + C'\left(\frac{n}{f(n)}\right)^{1/2}\right) \frac{1}{1 - e^{-s}}$$

$$\leq \exp\left(-\frac{n^{1/2} \delta^2}{2} + C^*\left(\frac{n}{f(n)}\right)^{1/2}\right)$$

and we have (24.8).

Note that the proof of (24.9) is going on the same line but instead of the sequence $\{U_n\}$ we have to use the sequence $\{U_n^*\}$ defined by

$$\mathbf{P}\{U_1^* = k\} = e^{-\psi(-s)} e^{sk} \mathbf{P}\{Z_1 = k\}$$

and we have the lemma.

Now we turn to the proof of (24.5). In fact we prove the much stronger

THEOREM 24.3 (Auer, 1990). *For any $\varepsilon > 0$ we have*

$$\lim_{n\to\infty} \sup_{\|x\|\leq r_n} \left|\frac{\xi(x,n)}{\xi(0,n)} - 1\right| = 0 \quad a.s. \tag{24.12}$$

where

$$r_n = \exp\left((\log n)^{1/2}(\log\log n)^{-1/2-\varepsilon}\right).$$

Remark 2. Note that (24.5) claims that the disc around the origin of radius r_n is covered in time n. The meaning of (24.12) is that the very same disc is "homogeneously" covered, i.e. every point of this disc will be visited about $\xi(0,n) \sim \log n$ times during the first n steps. For the one-dimensional analogue of this theorem, cf. Theorem 11.20.

Proof of Theorem 24.3. Clearly

$$\frac{\xi(x,\rho_n)}{\xi(0,\rho_n)} - 1 = \frac{Y_1 + Y_2 + \cdots + Y_n}{n}$$

and by Lemma 24.4 we have

$$\mathbf{P}\left\{\sup_{\|x\|\leq A_n}\left|\frac{\xi(x,\rho_n)}{\xi(0,\rho_n)} - 1\right| \geq \delta\sqrt{\frac{2}{C}}\frac{1}{\sqrt{f(n)}}\right\}$$

$$\leq A_n^2 \pi \sup_{\|x\|\leq A_n} \mathbf{P}\left\{|Y_1+Y_2+\cdots+Y_n| \geq \delta\sqrt{\frac{2}{C}}\frac{n}{\sqrt{f(n)}}\right\}$$

$$\leq A_n^2 \pi \sup_{\|x\|\leq A_n} \mathbf{P}\{|Y_1+Y_2+\cdots+Y_n| \geq \delta\sigma n^{3/4}\}$$

$$\leq 2\pi A_n^2 \exp\left(-\frac{\delta^2 n^{1/2}}{2} + C^*\left(\frac{n}{f(n)}\right)^{1/2}\right)$$

where $A_n = \exp(n^{1/2}/f(n))$. Let $f(n) = (\log n)^{\varepsilon}$ then by the Borel–Cantelli lemma we obtain

$$\lim_{n\to\infty} \sup_{\|x\|\leq A_n}\left|\frac{\xi(x,\rho_n)}{\xi(0,\rho_n)} - 1\right| = 0 \quad a.s.$$

which, in turn, implies

$$\lim_{n\to\infty} \sup_{\|x\|\leq A_n} \sup_{\rho_n \leq m \leq \rho_{n+1}}\left|\frac{\xi(x,m)}{\xi(0,\rho_n)} - 1\right| = 0 \quad a.s.$$

Then replacing n by $\xi(0, N)$ and recalling that

$$\xi(0, N) \geq \frac{\log N}{(\log \log N)^{1+\varepsilon}} \quad \text{a.s.}$$

for all but finitely many N (cf. Theorem 20.4) we obtain the theorem.

In order to prove (24.2) of Theorem 24.1 we introduce a few notations and we prove three lemmas. Let

$$I(x, n) = \begin{cases} 1 & \text{if } \xi(x, n) > 0, \\ 0 & \text{if } \xi(x, n) = 0, \end{cases}$$
$$m_k = m_k(x_1, x_2, \ldots, x_k; n) = \mathbf{E}(I(x_1, n)I(x_2, n)\ldots, I(x_k, n))$$
$$= \mathbf{P}\{\xi(x_1, n) > 0, \xi(x_2, n) > 0, \ldots, \xi(x_k, n) > 0\},$$
$$\nu(z) = \nu_z = \min\{k, k > 0, S_k = z\},$$
$$\mathcal{M}(k, n) = \mathcal{M}(n) = \sum_{i=1}^{k} m_{k-1}(x_1, x_2, \ldots, x_{i-1}, x_{i+1}, \ldots, x_k; n).$$

Then we have

LEMMA 24.5 *For any $0 < q < 1$ $k = 2, 3, \ldots$ we have*

$$m_1(u; n - qn)[\mathcal{M}(qn) + (1-k)m_k(x_1, x_2, \ldots, x_k; qn)]$$
$$\leq m_k(x_1, x_2, \ldots, x_k; n) \leq \frac{m_1(v; n)\mathcal{M}(n)}{1 + (k-1)m_1(v; n)}$$

where $u \in \mathbb{Z}^2$ and $v \in \mathbb{Z}^2$ are defined by

$$m_1(v, n) = \max_{1 \leq i \leq k} m_1(x_i, n) \quad \text{and} \quad m_1(u, n) = \min_{1 \leq i \leq k} m_1(x_i, n).$$

In order to present the proof in an intelligible form we prove Lemma 24.5 first in the case $k = 2$. That is, we prove

LEMMA 24.6 *For any $0 < q < 1$ we have*

$$m_1(x - y; n - qn)[m_1(x; qn) + m_1(y; qn) - m_2(x, y; qn)]$$
$$\leq m_2(x, y; n) \leq \frac{m_1(x - y; n)[m_1(x; n) + m_1(y; n)]}{1 + m_1(x - y; n)}.$$

Proof.

$$m_2(x, y; n) = \mathbf{P}\{I(x, n) = 1, I(y, n) = 1\}$$

LARGE COVERED BALLS 253

$$= \sum_{k=0}^{n} \mathbf{P}\{I(x,n) = 1, I(y,n) = 1 \mid \nu_x = k < \nu_y\}\mathbf{P}\{\nu_x = k < \nu_y\}$$

$$+ \sum_{k=0}^{n} \mathbf{P}\{I(x,n) = 1, I(y,n) = 1 \mid \nu_y = k < \nu_x\}\mathbf{P}\{\nu_y = k < \nu_x\}$$

$$= \sum_{k=0}^{n} \mathbf{P}\{I(y,n) = 1 \mid \nu_x = k < \nu_y\}\mathbf{P}\{\nu_x = k < \nu_y\}$$

$$+ \sum_{k=0}^{n} \mathbf{P}\{I(x,n) = 1 \mid \nu_y = k < \nu_x\}\mathbf{P}\nu_y = k < \nu_x\}$$

$$= \sum_{k=0}^{n} \mathbf{P}\{I(y-x, n-k) = 1\}\mathbf{P}\{\nu_x = k < \nu_y\}$$

$$+ \sum_{k=0}^{n} \mathbf{P}\{I(x-y, n-k) = 1\}\mathbf{P}\{\nu_y = k < \nu_x\}.$$

Consequently we have

$$m_2(x,y;n)$$
$$\leq \mathbf{P}\{I(x-y,n) = 1\}\mathbf{P}\left\{\sum_{k=0}^{n}((\nu_x = k < \nu_y) + (\nu_y = k < \nu_x))\right\}$$
$$= \mathbf{P}\{I(x-y,n) = 1\}\mathbf{P}\{I(x,n) = 1 \text{ or } I(y,n) = 1\}$$
$$= m_1(x-y;n)[m_1(x;n) + m_1(y,n) - m_2(x,y,n)]$$

which implies the upper part of Lemma 24.6.

We also have

$$m_2(x,y;n) \geq \sum_{k=0}^{qn} \mathbf{P}\{I(y-x, n-k) = 1\}\mathbf{P}\{\nu_x = k < \nu_y\}$$

$$+ \sum_{k=0}^{qn} \mathbf{P}\{I(x-y, n-k) = 1\}\mathbf{P}\{\nu_y = k < \nu_x\}$$

$$\geq \mathbf{P}\{I(x-y, n-qn) = 1\}\mathbf{P}\{I(x, qn) = 1 \text{ or } I(y, qn) = 1\}$$
$$= m_1(x-y; n-qn)[m_1(x;qn) + m_1(y;qn) - m_2(x,y;qn)].$$

Hence we have Lemma 24.6.

Proof of Lemma 24.5. Let P_k (resp. $P_k(r)$) be the set of permutations of the integers $1, 2, \ldots, k$ (resp. $1, 2, \ldots, r-1, r+1, \ldots, k$). Further, let $A = A(i_1, i_2, \ldots, i_k; j) = \{\nu(x_{i_1}) < \nu(x_{i_2}) < \ldots < \nu(x_{i_{k-1}}) < \nu(x_{i_k}) = j\}$.

Then we have

$$m_k(x_1, x_2, \ldots, x_k; n)$$
$$= \sum_{\substack{(i_1,\ldots,i_k) \in P_k \\ 1 \le j \le n}} \mathbf{P}\{I(x_1, n) = \ldots = I(x_k, n) = 1 \mid A\}\mathbf{P}\{A\}$$
$$= \sum \mathbf{P}\{I(x_{i_k}, n) = 1 \mid A\}\mathbf{P}\{A\}$$
$$= \sum \mathbf{P}\{I(x_{i_k} - x_{i_{k-1}}, n - j) = 1\}\mathbf{P}\{A\}.$$

Consequently

$$m_k \le \mathbf{P}\{I(v, n) = 1\} \sum \mathbf{P}\{A\}$$
$$= \mathbf{P}\{I(v, n) = 1\} \sum_{r=1}^{k} \sum_{\substack{(i_1,\ldots,i_{k-1}) \in P_k(r) \\ 1 \le j \le n, i_k = r}} \mathbf{P}\{A\} = \mathbf{P}\{I(v, n) = 1\}$$
$$\times \mathbf{P}\left\{\sum_{r=1}^{k} \{I(x_j, n) = 1; \ j = 1, \ldots, r-1, r+1, \ldots, k\}\right\}$$
$$= m_1(v, n)\left[\mathcal{M}(n) + m_k\left(-\binom{k}{2} + \binom{k}{3} - \cdots (-1)^{k+1}\binom{k}{k}\right)\right]$$
$$= m_1(v, n)[\mathcal{M}(n) + (1-k)m_k]$$

which implies the upper part of the inequality of Lemma 24.5.

We also have

$$m_k \ge \sum_{\substack{(i_1,\ldots,i_k) \in P_k \\ 1 \le j \le qn}} \mathbf{P}\{I(x_{i_k} - x_{i_{k-1}}, n - j) = 1\}\mathbf{P}\{A\}$$
$$\ge \mathbf{P}\{I(u, n - qn) = 1\}\mathbf{P}\left\{\sum_{r=1}^{k} \sum_{\substack{(i_1,\ldots,i_{k-1}) \in P_k(r) \\ 1 \le j \le qn, i_k = r}} A\right\}$$
$$= \mathbf{P}\{I(u, n - qn) = 1\}$$
$$\times \mathbf{P}\left\{\sum_{r=1}^{k} \{I(x_j, qn) = 1; \ j = 1, \ldots, r-1, r+1, \ldots, k\}\right\}$$
$$= m_1(u, n - qn)[\mathcal{M}(qn) + (1-k)m_k].$$

Hence we have Lemma 24.5.

Let $N = N(n) \nearrow \infty$ and $k = k(n) \nearrow \infty$ be sequences of positive integers with $N(n) < n^{1/3}$. Assume that there exists an $\varepsilon > 0$ for which

LARGE COVERED BALLS

$k(n) \leq (N(n))^\varepsilon$. Then for any $n = 1, 2, \ldots$ there exists a sequence $x_1 = x_1(n), x_2 = x_2(n), \ldots, x_k = x_k(n) \in \mathbb{Z}^2(k = k(n))$ such that

$$N - 1 \leq \|x_i\| \leq N \quad (i = 1, 2, \ldots, k),$$
$$N^{1-\varepsilon} \leq \|x_i - x_j\| \leq N \quad (1 \leq i < j \leq k).$$

Now we formulate our

LEMMA 24.7 *For the above defined $x_1, x_2, \ldots, x_k$ we have*

$$m_k(x_1, x_2, \ldots, x_k; n) \leq \exp\left(-2\frac{\log N^{1-\varepsilon}}{\log n}\left(1 + O\left(\frac{\log_3 N^{1-\varepsilon}}{\log N^{1-\varepsilon}}\right)\right) \log k\right). \tag{24.13}$$

Further, if

$$N(n) = \exp((\log n)^{1/2} \log_3 n) \quad \text{and} \quad k(n) = \exp((\log n)^{1/2}) \tag{24.14}$$

then

$$m_k(x_1, x_2, \ldots, x_k; n) \leq \exp(-(2 - 4\varepsilon) \log_3 n). \tag{24.15}$$

Proof. By Lemma 24.5 and Theorem 20.3 we have

$$m_k \leq \frac{\left(1 - \frac{2 \log N^{1-\varepsilon}}{\log n}\left(1 + O\left(\frac{\log_3 N^{1-\varepsilon}}{\log N^{1-\varepsilon}}\right)\right)\right) \mathcal{M}(n)}{1 + (k-1)\left(1 - \frac{2 \log N^{1-\varepsilon}}{\log n}\left(1 + O\left(\frac{\log_3 N^{1-\varepsilon}}{\log N^{1-\varepsilon}}\right)\right)\right)}$$

$$\leq \frac{1}{k}\left(1 - \frac{2}{k}\frac{\log N^{1-\varepsilon}}{\log n}\left(1 + O\left(\frac{\log_3 N^{1-\varepsilon}}{\log N^{1-\varepsilon}}\right)\right)\right) \mathcal{M}(n)$$

$$\leq \exp\left(-\frac{2}{k}\frac{\log N^{1-\varepsilon}}{\log n}\left(1 + O\left(\frac{\log_3 N^{1-\varepsilon}}{\log N^{1-\varepsilon}}\right)\right)\right) \frac{1}{k}\mathcal{M}(n) \leq \ldots$$

$$\leq \exp\left(-2\frac{\log N^{1-\varepsilon}}{\log n}\left(1 + O\left(\frac{\log_3 N^{1-\varepsilon}}{\log N^{1-\varepsilon}}\right)\right) \log k\right).$$

Hence we have (24.13). If (24.14) holds then

$$\exp\left(-2\frac{\log N^{1-\varepsilon}}{\log n}\left(1 + O\left(\frac{\log_3 N^{1-\varepsilon}}{\log n^{1-\varepsilon}}\right)\right) \log k\right)$$

$$\leq \exp\left(-(2 - 4\varepsilon)\frac{(\log n)^{1/2} \log_3 n}{\log n}(\log n)^{1/2}\right) = \exp(-(2 - 4\varepsilon) \log_3 n)$$

and we obtain (24.15).

Now we can present the

Proof of (24.2). Let $N(n)$ be defined by (24.14) and let

$$n_j = [\exp(e^j)],$$
$$\tilde{N}(n) = \exp(2(\log n)^{1/2} \log_3 n),$$
$$k_{j+1} = \exp((\log n_{j+1})^{1/2}).$$

Then

$$\mathbf{P}\{R(n_{j+1}) \geq \tilde{N}(n_j)\} \leq \mathbf{P}\{R(n_{j+1}) \geq N(n_{j+1})\}$$
$$\leq m_{k_{j+1}}(x_1, x_2, \ldots, x_{k_{j+1}}; n_{j+1}) \leq (j+1)^{-(2-4\varepsilon)}$$

where $x_i = x_i(n_{j+1})$ $(i = 1, 2, \ldots, k_{j+1})$. Hence by the Borel–Cantelli lemma

$$R(n_{j+1}) \leq \tilde{N}(n_j) \text{ a.s.}$$

for all but finitely many j. Let $n_j \leq n < n_{j+1}$. Then

$$R(n) \leq R(n_{j+1}) \leq \tilde{N}(n_j) \leq \tilde{N}(n)$$

which proves (24.2).

Now we turn to the proof of the upper inequality of (24.6). In fact we prove a bit more:

THEOREM 24.4 (Révész, 1990/A). *For any $z > 0$ and for any $n = 2, 3, \ldots$ we have*

$$\mathbf{P}\left\{\frac{(\log R(n))^2}{\log n} > z\right\} \leq \exp\left(-\frac{z}{4}\right). \qquad (24.16)$$

In order to prove Theorem 24.4 let

$$M = M(n) = \exp(C(\log n)^{1/2}) \quad (C > 0)$$

and

$$K = K(n) = C^* \exp\left(\frac{C}{4}(\log n)^{1/2}\right) \quad (C^* > 0).$$

Then for any $n = 1, 2, \ldots$ there exists a sequence $y_1 = y_1(n), y_2 = y_2(n), \ldots, y_K = y_K(n)$ such that

$$M - 1 \leq \|y_i\| \leq M \quad (i = 1, 2, \ldots, K),$$

$$M^{3/4} \leq \|y_i - y_j\| \leq M \quad (1 \leq i < j \leq K).$$

if C^* is small enough. Now we formulate our

LEMMA 24.8

$$m_K(y_1, y_2, \ldots, y_K; n) \leq \exp\left(-\frac{C^2}{4}\right).$$

Proof. In the same way as we proved Lemma 24.7 we obtain

$$m_K \leq \frac{1}{K}\left(1 - \frac{2}{K}\frac{\log M^{3/4}}{\log n}\left(1 + O\left(\frac{\log_3 M^{3/4}}{\log M^{3/4}}\right)\right)\right)\mathcal{M}(K,n) \leq \ldots$$

$$\leq \exp\left(-2\frac{\log M^{3/4}}{\log n}\left(1 + O\left(\frac{\log_3 M^{3/4}}{\log M^{3/4}}\right)\right)\log K\right)$$

$$\leq \exp\left(-\frac{C^2}{4}\right).$$

Hence we have Lemma 24.8.

Proof of Theorem 24.4. Clearly

$$\mathbf{P}\{R(n) \geq \exp(C(\log n)^{1/2})\} \leq m_K \leq \exp\left(-\frac{C^2}{4}\right)$$

which proves Theorem 24.4.

Instead of proving (24.3) we prove the following stronger

THEOREM 24.5 *For any* $0 < \Theta < (\pi/120)^{1/2}, 9/10 < \delta^2 < 1$ *and* $\varepsilon > 0$ *we have*

$$\inf_{\|x\| \leq h(n)} \xi(x,n) \geq \frac{(1-\delta)}{\sqrt{\pi}}(\log n \log_3 n)^{1/2} \quad i.o. \ a.s.$$

where

$$h(n) = \exp\left((1-\varepsilon)\frac{\Theta}{\sqrt{\pi}}(\log n \log_3 n)^{1/2}\right).$$

In order to prove Theorem 24.5 let $\rho_1(0 \leadsto x), \rho_2(0 \leadsto x), \ldots$ resp. $\rho_1(x \leadsto 0), \rho_2(x \leadsto 0), \ldots$ be the first, second, ... waiting times to reach x from 0, resp. to reach 0 from x, i.e.

$$\rho_1(0 \leadsto x) = \inf\{n : n \geq 1, S_n = x\},$$
$$\rho_1(x \leadsto 0) = \inf\{n : n \geq \rho_1(0 \leadsto x), S_n = 0\} - \rho_1(0 \leadsto x),$$
$$\rho_2(0 \leadsto x) = \inf\{n : n \geq \rho_1(0 \leadsto x) + \rho_1(x \leadsto 0), S_n = x\}$$
$$\quad - (\rho_1(0 \leadsto x) + \rho_1(x \leadsto 0)),$$
$$\rho_2(x \leadsto 0) = \inf\{n : n \geq \rho_1(0 \leadsto x) + \rho_1(x \leadsto 0) + \rho_2(0 \leadsto x), S_n = 0\}$$
$$\quad - (\rho_1(0 \leadsto x) + \rho_1(x \leadsto 0) + \rho_2(0 \leadsto x)), \ldots$$

Let $\tau(0 \leadsto x, n)$ be the number of $0 \leadsto x$ excursions completed before n, i.e.

$$\tau(0 \leadsto x, n) = \max\left\{i : \sum_{j=1}^{i-1}(\rho_j(0 \leadsto x) + \rho_j(x \leadsto 0)) + \rho_i(0 \leadsto x) \leq n\right\}.$$

In the proof of Theorem 24.5 the following lemma will be used.

LEMMA 24.9 *For any $0 < \Theta < (\pi/120)^{1/2}, 9/10 < \delta^2 < 1$ and n big enough we have*

$$\mathbf{P}\{n^{-1/2} \inf_{\|x\| \leq e^{\Theta\sqrt{n}}} \tau(0 \leadsto x, \rho_n) \leq 1 - \delta\} \leq \exp\left(-\frac{\pi}{60}\frac{n^{1/2}}{\Theta}\right). \quad (24.17)$$

(For ρ_n, see Notation 6.)

Proof. Let $q = 1 - p = 1 - p(0 \leadsto x)$. Then applying Bernstein inequality (Theorem 2.3) with $\varepsilon = \delta p$ and Lemma 24.2 we obtain

$$\mathbf{P}\left\{\left|\frac{\tau(0 \leadsto x, \rho_n)}{n} - p\right| \geq \delta p\right\} \leq 2\exp\left(-\frac{n\delta^2 p}{2q(1 + \frac{\delta}{2q})^2}\right)$$

$$\leq \begin{cases} \exp\left(-\dfrac{\pi}{60}\dfrac{n}{\log\|x\|}\right) & \text{if } \|x\| \geq 2, \\ \exp(-O(1)n) & \text{if } \|x\| = 1. \end{cases}$$

Hence

$$\mathbf{P}\{\inf_{\|x\| \leq e^{\Theta\sqrt{n}}} n^{-1/2}\tau(0 \leadsto x, \rho_n) \leq 1 - \delta\}$$

$$\leq \mathbf{P}\left\{\inf_{\|x\| \leq e^{\Theta\sqrt{n}}} \frac{\tau(0 \leadsto x, \rho_n)}{np} \leq 1 - \delta\right\}$$

$$\leq \mathbf{P}\left\{\sum_{\|x\| \leq e^{\Theta\sqrt{n}}} \left\{\frac{\tau(0 \leadsto x, \rho_n)}{np} \leq 1 - \delta\right\}\right\}$$

$$\leq e^{2\Theta\sqrt{n}}\pi\exp\left(-\frac{\pi}{60}\frac{n^{1/2}}{\Theta}\right) \quad (24.18)$$

which implies (24.17).

Proof of Theorem 24.5. (24.17) clearly implies that

$$\liminf_{n \to \infty} n^{-1/2} \inf_{\|x\| \leq e^{\Theta\sqrt{n}}} \tau(0 \leadsto x, \rho_n) \geq 1 - \delta$$

for any $0 < \Theta < (\pi/120)^{1/2}$ and $9/10 < \delta^2 < 1$.

Observe that Theorem 20.5 and (24.18) imply

$$n^{-1/2} \inf_{\|x\| \leq e^{\Theta\sqrt{n}}} \tau\left(0 \rightsquigarrow x, \exp\left(\frac{(1+\varepsilon)\pi n}{\log_2 n}\right)\right) \geq 1 - \delta \quad \text{i.o. a.s.,}$$

i.e.

$$\inf_{\|x\| \leq h(n)} \tau(0 \rightsquigarrow x, n) \geq \frac{1-\delta}{\sqrt{\pi}} (\log n \log_3 n)^{1/2} \quad \text{i.o. a.s.}$$

which in turn implies Theorem 24.5.

Now, we have to prove the lower inequality of (24.6). In fact instead of proving the lower part of (24.6) we prove the following stronger

THEOREM 24.6 *For any $\varepsilon > 0$ and $z > 0$ there exists a positive integer $N_0 = N_0(\varepsilon, z)$ such that*

$$\mathbf{P}\{\inf_{\|x\| \leq \exp(\Theta(z\log n)^{1/2})} \tau(0 \rightsquigarrow x, n) \geq (1-\delta)(z\log n)^{1/2}\} \geq \exp(-\pi z) - \varepsilon \quad (24.19)$$

if $n \geq N_0$, $0 < \Theta < (\pi/120)^{1/2}$ and $9/10 < \delta^2 < 1$.

Proof. Theorem 20.2 and (24.18) imply that for any $\varepsilon > 0$ and $z > 0$ there exists a positive integer $N_0 = N_0(\varepsilon, z)$ such that

$$\mathbf{P}\{n^{-1/2} \inf_{\|x\| \leq e^{\Theta\sqrt{n}}} \tau(0 \rightsquigarrow x, \rho_n) \leq 1 - \delta\} \leq \varepsilon$$

and

$$\mathbf{P}\left\{\rho_n < \exp\left(\frac{n}{z}\right)\right\} \geq \exp(-\pi z) - \varepsilon$$

if $n \geq N_0$. Consequently

$$\mathbf{P}\left\{\inf_{\|x\| \leq e^{\Theta\sqrt{n}}} \tau\left(0 \rightsquigarrow x, \exp\left(\frac{n}{z}\right)\right) \geq (1-\delta)n^{1/2}\right\}$$

$$\geq \mathbf{P}\left\{\inf_{\|x\| \leq e^{\Theta\sqrt{n}}} \tau\left(0 \rightsquigarrow x, \exp\left(\frac{n}{z}\right)\right) \geq (1-\delta)n^{1/2}, \rho_n < \exp\left(\frac{n}{z}\right)\right\}$$

$$\geq \mathbf{P}\left\{\inf_{\|x\| \leq e^{\Theta\sqrt{n}}} \tau(0 \rightsquigarrow x, \rho_n) \geq (1-\delta)n^{1/2}, \rho_n < \exp\left(\frac{n}{z}\right)\right\}$$

$$\geq \mathbf{P}\{\inf_{\|x\| \leq e^{\Theta\sqrt{n}}} \tau(0 \rightsquigarrow x, \rho_n) \geq (1-\delta)n^{1/2}\} - \mathbf{P}\left\{\rho_n \geq \exp\left(\frac{n}{z}\right)\right\}$$

$$\geq 1 - \varepsilon - (1 - \exp(-\pi z)) - \varepsilon = \exp(-\pi z) - 2\varepsilon$$

if $n \geq N_0$.

Hence we have Theorem 24.6.

Instead of proving (24.4) we prove the stronger

THEOREM 24.7

$$\exp\left(\frac{(\log n)^{1/2}}{\log_3 n}\right) \in \mathrm{LUC}(R(n)).$$

Proof. Let $R(n, N)$ be the largest integer ρ for which the disc

$$Q(\rho, n) = \{x \in \mathbb{Z}^2,\ |S_n - x| \leq \rho\}$$

is completely covered by the path $\{S_n, S_{n+1}, \ldots, S_N\}$ i.e. for each $x \in Q(\rho, n)$ there exists a $k \in [n, N]$ such that $\xi(x, N) - \xi(x, n) > 0$.

By (24.6)

$$\mathbf{P}\left\{\frac{(\log R(n, N))^2}{\log(N - n)} > z\right\} \leq \exp\left(-\frac{z}{4}\right).$$

Consequently

$$\mathbf{P}\{\log R(n, N) \leq (z \log(N - n))^{1/2}\} \geq 1 - \exp\left(-\frac{z}{4}\right). \qquad (24.20)$$

Let

$$n_k = \exp(\exp(\exp k))$$

and

$$z = z_k = \frac{1}{\log_3 n_k}.$$

Apply (24.20) with $n = n_k$, $N = n_{k+1}$. Then we get

$$\mathbf{P}\left\{\log R(n_k, n_{k+1}) \leq \frac{(\log(n_{k+1} - n_k))^{1/2}}{\log_3 n_k}\right\}$$

$$\geq 1 - \exp\left(\frac{-1}{4\log_3 n_k}\right) \sim \frac{1}{4\log_3 n_k} = \frac{1}{k}.$$

Clearly $R(n_k, n_{k+1})$ is a sequence of independent r.v.'s. Hence

$$\log R(n_k, n_{k+1}) \leq \frac{(\log(n_{k+1} - n_k))^{1/2}}{\log_3 n_k} \qquad \text{i.o. a.s.}$$

and

$$S_{n_k} \leq (2 n_k \log \log n_k)^{1/2} \qquad \text{a.s.}$$

Since

$$(2 n_k \log \log n_k)^{1/2} \ll \frac{(\log(n_{k+1} - n_k))^{1/2}}{\log_3 n_k},$$

if there were a ball around the origin having radius

$$\frac{(\log(n_{k+1} - n_k))^{1/2}}{\log_3 n_k}$$

completely covered at time n_{k+1} then there were a ball around S_{n_k} of radius

$$(1-\varepsilon)\frac{(\log(n_{k+1} - n_k))^{1/2}}{\log_3 n_k}$$

completely covered at time n_{k+1}. Hence Theorem 24.7 is proved.

Theorem 24.3 clearly implies that for any fixed $x \in \mathbb{Z}^2$

$$\lim_{n \to \infty} \frac{\xi(x,n)}{\xi(0,n)} = 1 \quad \text{a.s.} \tag{24.21}$$

It is worthwhile to note that for fixed x in (24.21) a rate of convergence can also be obtained. In fact we have

THEOREM 24.8

$$\limsup_{n \to \infty} \frac{|\xi(0,n) - \xi(x,n)|}{(2\xi(0,n) \log \log \xi(0,n))^{1/2}} = \sigma(x) \quad \text{a.s.}$$

where $\sigma(x)$ is a positive constant depending only on x.

Remark 3. Theorem 24.8 and Theorem 20.4 combined imply

$$\lim_{n \to \infty} \left(\frac{\log n}{(\log \log n)^{1+\varepsilon}}\right)^{1/2} \left|\frac{\xi(x,n)}{\xi(0,n)} - 1\right| = 0 \quad \text{a.s.}$$

Before presenting the proof of Theorem 24.8 introduce the following notations: let $\Xi_1, \Xi_2, \ldots$ be the local time of the walk at 0 during the first, second, $\ldots 0 \rightsquigarrow x$ excursions, i.e.

$$\Xi_1(x) = \Xi_1 = \xi(0, \rho_1(0 \rightsquigarrow x)),$$
$$\Xi_2(x) = \Xi_2 = \xi(0, \rho_1(0 \rightsquigarrow x) + \rho_1(x \rightsquigarrow 0) + \rho_2(0 \rightsquigarrow x)) - \Xi_1,$$
$$\Xi_3(x) = \Xi_3 = \xi(0, R_3) - (\Xi_1 + \Xi_2),$$
$$\ldots \quad \ldots$$

where

$$R_3 = \rho_1(0 \rightsquigarrow x) + \rho_1(x \rightsquigarrow 0) + \rho_2(0 \rightsquigarrow x) + \rho_2(x \rightsquigarrow 0) + \rho_3(0 \rightsquigarrow x).$$

Observe that $\Xi_1, \Xi_2, \ldots$ are i.i.d.r.v.'s with distribution

$$\mathbf{P}\{\Xi_1 = k\} = \mathbf{P}\{\xi(0, \rho_1(0 \leadsto x)) = k\}$$
$$= (1 - p(0 \leadsto x))^{k-1} p(0 \leadsto x) \quad (k = 1, 2, \ldots).$$

Consequently

$$\mathbf{E}\Xi_1 = (p(0 \leadsto x))^{-1},$$
$$\mathbf{E}(\Xi_1 - (p(0 \leadsto x))^{-1})^2 = (p(0 \leadsto x))^{-2}(1 - p(0 \leadsto x))$$

and

$$\limsup_{n \to \infty} \frac{\Xi_1 + \cdots + \Xi_n - (H_1 + \cdots + H_n)}{(2n \log \log n)^{1/2}} = 2^{1/2} \frac{(1 - p(0 \leadsto x))^{1/2}}{p(0 \leadsto x)}$$

where

$$H_1 = \xi(x, \rho_1(0 \leadsto x) + \rho_1(x \leadsto 0)),$$
$$H_2 = \xi(x, \rho_1(0 \leadsto x) + \rho_1(x \leadsto 0) + \rho_2(0 \leadsto x) + \rho_2(x \leadsto 0)) - H_1,$$
$$\ldots \quad \ldots \ldots$$

Since

$$\Xi_1 + \Xi_2 + \cdots + \Xi_{\tau(0 \leadsto x, n)} \leq \xi(0, n) \leq \Xi_1 + \Xi_2 + \cdots + \Xi_{\tau(0 \leadsto x, n)+1}$$

and the sequence $\tau(0 \leadsto x, n)$ takes every positive integer we have

$$\limsup_{n \to \infty} \frac{\xi(0, n) - \xi(x, n)}{(2\tau(0 \leadsto x, n) \log \log \tau(0 \leadsto x, n))^{1/2}} = 2^{1/2} \frac{(1 - p(0 \leadsto x))^{1/2}}{p(0 \leadsto x)} \quad \text{a.s.}$$

By the law of large numbers

$$\lim_{n \to \infty} \frac{\xi(0, n) - \tau(0 \leadsto x, n)(p(0 \leadsto x))^{-1}}{\tau(0 \leadsto x, n)} = 0 \quad \text{a.s.}$$

Hence

$$\limsup_{n \to \infty} \frac{\xi(0, n) - \xi(x, n)}{(2\xi(0, n) \log \log \xi(0, n))^{1/2}} = \left(\frac{2(1 - p(0 \leadsto x))}{p(0 \leadsto x)}\right)^{1/2} = \sigma(x) \quad \text{a.s.}$$

and we have Theorem 24.8.

Remark 4. Theorem 13.3 claimed that the favourite values of a random walk in $\mathbb{Z}^1$ converge to infinity. It is natural to ask the analogue question in higher dimension. In the case $d \geq 3$ the Pólya Recurrence Theorem implies that the favourite values are also going to infinity. Comparing Theorem 20.4 and Theorem 20.7 (in case $d = 2$) we find that in $\mathbb{Z}^2$ the favourite values are also going to infinity. Theorem 24.3 also says that the rate of convergence is not very slow. In fact we get

THEOREM 24.9 Let $d = 2$ and consider a sequence $\{x_n\}$ for which $\xi(x_n, n) = \xi(n)$. Then for any $\varepsilon > 0$ we have

$$\liminf_{n \to \infty} \frac{x_n}{\exp((\log n)^{1/2}(\log \log n)^{-1/2-\varepsilon})} = \infty \quad a.s.$$

Remark 5. Replacing the random walk in Theorems 24.1 and 24.2 by a Wiener sausage $B_r(T)$ one can ask whether these theorems remain valid. A harder question is to study the case where $r = r_T \downarrow 0$. In fact Theorem 18.4 implies that if $r_T \leq T^{-(\log T)^{\varepsilon}}$ (with some $\varepsilon > 0$) then the analogues of Theorems 24.1 and 24.2 cannot be true anymore.

24.2 Completely covered disc in $\mathbb{Z}^2$ with arbitrary centre

The results of Section 24.1 suggest the question whether the largest completely covered disc is located around the origin or somewhere else. Investigating this question it turns out that the largest covered disc is much-much larger than the one around the origin.

Formally speaking, let $u \in \mathbb{Z}^2$ and define

$$Q(u, N) = \{x \in \mathbb{Z}^2 : \|x - u\|^2 \leq N^2\}.$$

Let $r(n)$ be the largest integer for which there exists a random vector $u = u(n) \in \mathbb{Z}^2$ such that $Q(u, r(n))$ is covered by the random walk in time n, that is,

$$\xi(x, n) \geq 1 \text{ for every } x \in Q(u, r(n)).$$

Then we formulate the following theorem.

THEOREM 24.10 (Révész, 1993/A, 1993/B) We have

$$n^{1/50} \leq r(n) \leq n^{0.42} \quad a.s.$$

for all but finitely many n.

This Theorem suggests the following:

Conjecture 1. There exists a $1/50 \leq q_0 \leq 0.42$ such that

$$\lim_{n \to \infty} \frac{\log r(n)}{\log n} = q_0 \quad a.s.$$

Zhan Shi conjectured $q_0 = 1/4$. This conjecture was nearly proved by Dembo–Peres–Rosen–Zeitouni (2005/A). In fact they proved:

THEOREM 24.11

$$\lim_{n\to\infty} \frac{\log r(n)}{\log n} = \frac{1}{4} \quad \text{in probability.}$$

24.3 Almost covered discs centred in the origin of $\mathbb{Z}^2$

The results of Section 24.1 claimed that the radius of the largest covered disc is about $\exp((\log n)^{1/2})$. Now we are interested in the relative frequency of the visited points in a larger disc.

In order to formulate our results, introduce the following notations:

$$I(x,n) = \begin{cases} 1 & \text{if } \xi(x,n) > 0, \\ 0 & \text{if } \xi(x,n) = 0, \end{cases}$$

$$K(N,n) = \frac{1}{N^2\pi} \sum_{x \in Q(N)} I(x,n);$$

i.e. $K(N,n)$ is the density (relative frequency) of the points of $Q(N)$ covered by the random walk $\{S_k,\ 0 \leq k \leq n\}$.

Our first theorem claims that if we consider the disc around the origin of radius $\exp((\log n)^\alpha)$ ($\alpha < 1$) or even of radius $\exp(\log n (\log \log n)^{-2-\varepsilon})$ ($\varepsilon > 0$) then the density of the covered points converges to one 1 a.s. In fact we have

THEOREM 24.12 (Auer–Révész, 1990). *For any $\varepsilon > 0$*

$$\lim_{n\to\infty} K\left(\exp\left(\frac{\log n}{(\log\log n)^{2+\varepsilon}}\right), n\right) = 1 \quad a.s.$$

Proof. Consider

$$K(N_n, \rho_n) = \frac{1}{N^2\pi} \sum_{x \in Q(N)} I(x, \rho_n)$$

where

$$N = N_n = \exp\left(\frac{n}{(\log n)^{1+\varepsilon}}\right).$$

Then by Lemma 24.2 we have

$$\mathbf{E}(1 - I(x, \rho_n)) = (1 - p(0 \rightsquigarrow x))^n \leq \left(1 - \frac{C}{\log \|x\|}\right)^n$$

$$\leq \left(1 - \frac{C(\log n)^{1+\varepsilon}}{n}\right)^n \leq \exp(-C(\log n)^{1+\varepsilon})$$

provided that $\|x\| \le N$. Hence

$$\mathbf{E}(1 - K(N, \rho_n)) = \frac{1}{N^2 \pi} \sum_{x \in Q(N)} \mathbf{E}(1 - I(x, \rho_n)) \le \exp(-C(\log n)^{1+\varepsilon}),$$

and by the Markov inequality for any $\delta > 0$

$$\mathbf{P}\{1 - K(N, \rho_n) \ge \delta\} \le \delta^{-1} \exp(-C(\log n)^{1+\varepsilon})$$

which, in turn, by Borel–Cantelli lemma implies that

$$\lim_{n \to \infty} (1 - K(N, \rho_n)) = 0 \quad \text{a.s.} \tag{24.22}$$

Let $m = m_n = [\exp(n(\log n)^{1+\varepsilon})]$. Then by Theorem 20.5 $m_n \ge \rho_n$ a.s. for all but finitely many n. Hence (24.22) implies

$$\lim_{n \to \infty} (1 - K(N, m)) = 0 \quad \text{a.s.}$$

Observe that given the choice of m and N we have

$$\exp\left(\frac{\log m}{(\log \log m)^{2+\varepsilon}}\right) \ge N \ge \exp\left(\frac{\log m}{(\log \log m)^{2+3\varepsilon}}\right)$$

and we obtain

$$\lim_{n \to \infty} K\left(\exp\left(\frac{\log m}{(\log \log m)^{2+\varepsilon}}\right), m\right) = 1 \quad \text{a.s.}$$

Consequently we also have

$$\lim_{n \to \infty} K\left(\exp\left(\frac{\log m_{n+1}}{(\log \log m_{n+1})^{2+2\varepsilon}}\right), m_n\right) = 1 \quad \text{a.s.}$$

This proves the theorem.

24.4 Discs covered with positive density in $\mathbb{Z}^2$

Theorem 24.12 tells us that almost all points of the disc $Q(\exp((\log n)^\alpha))$ ($1/2 < \alpha < 1$) will be visited by the random walk $\{S_0, S_1, \ldots, S_n\}$. At the same time by Theorem 24.1 we know that some points of $Q(\exp((\log n)^\alpha))$ will surely not be visited. We can ask how many points of $Q(\exp((\log n)^\alpha))$ will not be visited, i.e. what is the rate of convergence in Theorem 24.12? However, it is more interesting to investigate the geometrical properties of the non-visited points. For example: what is the area of the largest non-visited disc within $Q(\exp((\log n)^\alpha))$? By non-visited disc we mean a disc having only non-visited points. The following theorem claims that with probability 1 there exists a non-visited disc of radius $\exp((\log n)^\beta)$ within the disc $Q(\exp((\log n)^\alpha))$ for every $\beta < \alpha$ provided that $\alpha > 1/2$.

Let
$$Q(u, r) = \{x \in \mathbb{Z}^2, \ \|x - u\| \leq r\}.$$

Then we have

THEOREM 24.13 (Auer–Révész, 1990). *Let*

$$1/2 < \alpha < 1 \quad \text{and} \quad \beta < \alpha.$$

Then there exists a sequence of random vectors $u_1, u_2, \ldots$ *such that*

$$Q(u_n, \exp((\log n)^\beta)) \subset Q(0, \exp((\log n)^\alpha)) = Q(\exp((\log n)^\alpha))$$

and

$$I(x, n) = 0 \quad \text{for all} \quad x \in Q(u_n, \exp((\log n)^\beta)).$$

In order to prove Theorem 24.13, first we introduce a notation and present two lemmas.

Let $N > 0$ and $u_1, u_2, \ldots, u_k \in \mathbb{Z}^2$ $(k = 1, 2, \ldots)$ be such points for which the discs

$$Q_i = Q(u_i, N) \quad (i = 1, 2, \ldots, k)$$

are disjoint. Denote by

$$m_k(Q_1, Q_2, \ldots, Q_k; n)$$
$$= \mathbf{P}\{\forall i = 1, 2, \ldots, k, \ \exists y_i \in Q_i \text{ such that } I(y_i, n) = 1\}$$

the probability that the discs $Q_1, Q_2, \ldots, Q_k$ are visited during the first n steps.

LEMMA 24.10 *Let*

$$\exp((\log n)^\alpha) \leq \|u\| < n^{1/3}, \quad N \leq \exp((\log n)^\beta)$$

and

$$0 < \beta < \alpha < 1.$$

Then

$$m_1(Q(u, N); n) \leq 1 - C(\log n)^{\alpha - 1}$$

for a suitable constant $C > 0$.

Proof. It is easy to see that

$$\mathbf{P}\{I(u,2n)=1\}$$
$$\geq m_1(Q(u,N);n) \min_{y \in Q(u,N)} \mathbf{P}\{\xi(u,2n)-\xi(u,n)>0 \mid S_n=y\}.$$

Hence by Theorem 20.3

$$m_1(Q(u,N);n) \leq \frac{1-2(\log n)^{\alpha-1}+O\left(\dfrac{\log_3 n}{\log n}\right)}{1-2(\log n)^{\beta-1}+O\left(\dfrac{\log_3 n}{\log n}\right)} \leq 1-C(\log n)^{\alpha-1}.$$

Hence Lemma 24.10 is proved.

LEMMA 24.11 *Let*

(i) $0 < \beta < \alpha < 1$,

(ii) $N = \exp((\log n)^\beta)$,

(iii) $u_1, u_2, \ldots, u_k \in \mathbb{Z}^2$ *be a sequence for which*

$$\|u_i\| < n^{1/3-\varepsilon} \quad (i=1,2,\ldots,k),$$
$$\|u_i - u_j\| \geq \exp((\log n)^{\alpha+\varepsilon}) \quad (1 \leq i < j \leq k).$$

Then
$$m_k(Q_1,Q_2,\ldots,Q_k;n) \leq \exp(-C(\log n)^{\alpha-1}\log k)$$
for a suitable $C > 0$ where $Q_i = Q(u_i,N)$ $(i=1,2,\ldots,k)$.

Proof. Clearly

$$m_k(Q_1,Q_2,\ldots,Q_k;n)$$
$$= \mathbf{P}\left\{\bigcup_{i=1}^{k}\{\forall Q_j \text{ are visited before } n \text{ and } Q_i \text{ is the last visited disc}\}\right\}$$
$$\leq \mathbf{P}\left\{\bigcup_{i=1}^{k}\{\text{the discs } Q_1,\ldots,Q_{i-1},Q_{i+1},\ldots,Q_k \text{ are visited before } n\}\right\}$$
$$\quad \times \max_{i \neq j} \max_{x \in Q_i} \mathbf{P}\{Q_j \text{ is visited before } 2n \mid S_n = x\}.$$

Hence by Lemma 24.10
$$m_k(Q_1,\ldots,Q_k;n)$$
$$\leq \left(\sum_{i=1}^{k} m_{k-1}(Q_1,\ldots,Q_{i-1},Q_{i+1},\ldots,Q_k;n) - (k-1)m_k(Q_1,\ldots,Q_k;n)\right)$$
$$\times (1 - C(\log n)^{\alpha-1})$$
and
$$m_k(Q_1,\ldots,Q_k;n)$$
$$\leq \frac{\left(1 - C(\log n)^{\alpha-1}\right)\sum_{i=1}^{k} m_{k-1}(Q_1,\ldots,Q_{i-1},Q_{i+1},\ldots,Q_k;n)}{1 + (k-1)(1 - C(\log n)^{\alpha-1})}.$$

Since
$$\frac{1-a}{1+(k-1)(1-a)} \leq \frac{1}{k}\left(1 - \frac{a}{k}\right) \leq \frac{1}{k}\exp\left(-\frac{a}{k}\right)$$
for any $0 \leq a \leq 1$ and $k \geq 1$ we have
$$m_k(Q_1,\ldots,Q_k;n)$$
$$\leq \frac{1}{k}\exp\left(-\frac{C(\log n)^{\alpha-1}}{k}\right)\sum_{i=1}^{k} m_{k-1}(Q_1,\ldots,Q_{i-1},Q_{i+1},\ldots,Q_k;n),$$
by induction
$$m_k(Q_1,\ldots,Q_k;n)$$
$$\leq \exp\left(-C(\log n)^{\alpha-1}\sum_{i=2}^{k}\frac{1}{i}\right) \leq \exp(-C(\log n)^{\alpha-1}\log k)$$
and we have Lemma 24.11.

Remark 1. Lemma 24.11 is a natural analogue of Lemmas 24.5 and 24.7.

Proof of Theorem 24.13. Let
$$1/2 < \alpha < \alpha + \varepsilon < 1, \quad \beta < \alpha$$
and
$$N = \exp((\log n)^\beta).$$
Then there exist $k = k(n) = \exp((\log n)^\alpha)$ points $u_1, u_2, \ldots, u_k$ such that
$$\|u_i\| \leq \exp((\log n)^{\alpha+\varepsilon}) \quad (i = 1, 2, \ldots, k),$$

and
$$\|x_i - x_j\| \geq \exp((\log n)^{\alpha+\varepsilon/2}) \quad (i,j = 1, 2, \ldots, k;\ i \neq j).$$

Then by Lemma 24.11

$$m_k(Q_1, Q_2, \ldots, Q_k; n) \leq \exp(-C(\log n)^{2\alpha-1})$$

where $Q_i = Q(u_i, N)$. Choosing $n = n_j = e^j$ the Borel–Cantelli lemma implies that with probability 1 at least one Q_i ($i = 1, 2, \ldots, k(n_j)$) is not visited till n_j for all but finitely many j.

Let $n_j \leq n < n_{j+1}$. Then with probability 1 there exists a u with $\|u\| \leq \exp((j+1)^{\alpha+\varepsilon})$ such that the disc $Q(u, N)$ ($N = N_j = \exp((\log n_j)^\beta)$) is not visited before n if j is large enough. Consequently for all but finitely many n there exists a $u_0 = u_0(n, \omega) \in Q(\exp((\log n)^{\alpha+2\varepsilon}))$ such that $Q(u_0, N) \subset Q(\exp((\log n)^{\alpha+3\varepsilon}))$ is not visited before n. This proves Theorem 24.13.

Now we consider the density $K(N, n)$ for even larger N. The case $N = n^\alpha$ will be investigated and for any $0 < \alpha < 1/2$ we prove that $K(n^\alpha, n)$ has a limit distribution. In fact we have

THEOREM 24.14 (Révész, 1993/B). *For any $0 < \alpha < 1/2$ we have*

$$\lim_{n\to\infty} \mathbf{P}\{K([n^\alpha], n) < x\} = 1 - (1-x)^{2\alpha/(1-2\alpha)} \quad (0 \leq x \leq 1).$$

At first we present a few lemmas.

LEMMA 24.12 *Let*

$$n^\alpha (\log n)^{-\beta} \leq \|x\| \leq Cn^\alpha \quad (0 < \alpha < 1/2,\ \beta \geq 0,\ C \geq 1).$$

Then

$$\mathbf{P}\{I(x, n) = 1\} = 1 - 2\alpha + O\left(\frac{\log\log n}{\log n}\right).$$

Proof. It is a trivial consequence of Theorem 20.3.

LEMMA 24.13 *Let x and y be two points of $\mathbb{Z}^2$ such that*

$$n^\alpha (\log n)^{-\beta} \leq \|x\|, \|y\|, \|x-y\| \leq Cn^\alpha \quad (0 < \alpha < 1/2,\ \beta \geq 0,\ C \geq 1).$$

Then

$$\lim_{n\to\infty} m_2(x, y; n) = \frac{(1-2\alpha)^2}{1-\alpha}. \tag{24.23}$$

Proof. Lemmas 24.6 and 24.12 imply

$$m_2(x,y;n) \leq \frac{m_1(x-y;n)[m_1(x;n) + m_1(y;n)]}{1+m_1(x-y;n)}$$

$$= \frac{2\left(1-2\alpha+O\left(\frac{\log\log n}{\log n}\right)\right)^2}{2-2\alpha+O\left(\frac{\log\log n}{\log n}\right)}$$

$$= \frac{(1-2\alpha)^2}{1-\alpha} + O\left(\frac{\log\log n}{\log n}\right). \tag{24.24}$$

Similarly

$$m_2(x,y;n)$$
$$\geq m_1(x-y;n-qn)[m_1(x;qn) + m_1(y;qn) - m_2(x,y;qn)]$$
$$\geq \left(1-2\alpha+O\left(\frac{\log\log n}{\log n}\right)\right)\left[2(1-2\alpha)+O\left(\frac{\log\log n}{\log n}\right) - m_2(x,y;n)\right].$$

Consequently

$$m_2(x,y;n) \geq \frac{(1-2\alpha)^2}{1-\alpha} - O\left(\frac{\log\log n}{\log n}\right). \tag{24.25}$$

(24.24) and (24.25) together imply (24.23).

The next lemma is an extension of Lemma 24.13.

LEMMA 24.14 *Let $x_1, x_2, \ldots, x_k$ ($k = 1, 2, \ldots$) be a sequence in $\mathbb{Z}^2$ such that*

$$n^\alpha(\log n)^{-\beta} \leq \|x_i - x_j\|, \|x_i\| \leq Cn^\alpha$$

where $0 < \alpha < 1/2$, $\beta \geq 0$, $C \geq 1$, $1 \leq i < j \leq k$. Then

$$\lim_{n\to\infty} m_k(x_1,x_2,\ldots,x_k;n) = (1-2\alpha)^k \prod_{j=2}^{k}\left(1-\left(1-\frac{1}{j}\right)2\alpha\right)^{-1}.$$

Proof. Lemmas 24.5 and 24.12 imply

$$m_k \leq \frac{1-2\alpha+O\left(\frac{\log\log n}{\log n}\right)}{1+(k-1)\left(1-2\alpha+O\left(\frac{\log\log n}{\log n}\right)\right)}\mathcal{M}(n)$$

$$= \frac{1-2\alpha+O\left(\frac{\log\log n}{\log n}\right)}{1-\left(1-\frac{1}{k}\right)\left(2\alpha-O\left(\frac{\log\log n}{\log n}\right)\right)}\frac{1}{k}\mathcal{M}(n).$$

By induction we obtain

$$m_k \le \left(1 - 2\alpha + O\left(\frac{\log\log n}{n}\right)\right)^k$$
$$\times \prod_{j=2}^{k}\left(1 - \left(1 - \frac{1}{j}\right)\left(2\alpha - O\left(\frac{\log\log n}{\log n}\right)\right)\right)^{-1}. \quad (24.26)$$

Similarly

$$m_k \ge \frac{\left(1 - 2\alpha + O\left(\frac{\log\log n}{\log n}\right)\right)\mathcal{M}(qn)}{1 + (k-1)\left(1 - 2\alpha + O\left(\frac{\log\log n}{\log n}\right)\right)}.$$

Consequently

$$m_k \ge \frac{1 - 2\alpha + O\left(\frac{\log\log n}{\log n}\right)}{1 - \left(1 - \frac{1}{k}\right)\left(2\alpha - O\left(\frac{\log\log n}{\log n}\right)\right)} \frac{1}{k}\mathcal{M}(qn)$$

and by induction we obtain Lemma 24.14.

Proof of Theorem 24.14. Let $A(s,n)$ be the set of all possible s-tuples $(x_1, x_2, \ldots, x_s)$ of $Q(n^\alpha)$ with the property

$$\|x_i\|, \|x_i - x_j\| \ge \frac{n^\alpha}{\log n} \quad (i \ne j).$$

Then

$$\mathbf{E}((K(n^\alpha, n))^s) = \left(\frac{1}{n^{2\alpha}\pi}\right)^s \mathbf{E}\left(\sum_{x \in Q(n^\alpha)} I(x,n)\right)^s$$
$$\approx \left(\frac{1}{n^{2\alpha}\pi}\right)^s \sum_{(x_1,x_2,\ldots,x_s)\in A(s,n)} \mathbf{E}\left(\prod_{j=1}^{s} I(x_j,n)\right)$$

and by Lemma 24.14 we obtain

$$\lim_{n\to\infty} \mathbf{E}((K(n^\alpha,n))^s) = (1-2\alpha)^s \prod_{j=1}^{s}\left(1 - \left(1-\frac{1}{j}\right)2\alpha\right)^{-1}.$$

Consequently we have Theorem 24.14 with a distribution $G_\alpha(\cdot)$ satisfying

$$\int_0^1 x^s dG_\alpha(x) = (1-2\alpha)^s \prod_{j=1}^s \left(1 - \left(1 - \frac{1}{j}\right) 2\alpha\right)^{-1}$$

$$= \prod_{j=1}^s \left(1 + \frac{2\alpha}{1-2\alpha}\frac{1}{j}\right)^{-1} = O\left(s^{-2\alpha/(1-2\alpha)}\right). \quad (24.27)$$

It is easy to see that the only distribution function which satisfies the moment condition (24.27) is

$$G_\alpha(x) = 1 - (1-x)^{2\alpha/(1-2\alpha)}.$$

Hence Theorem 24.14 is proved.

24.5 Completely covered balls in $\mathbb{Z}^d$

Theorem 24.1 describes the area of the largest disc around the origin of $\mathbb{Z}^2$ covered by the random walk $\{S_k, k \leq n\}$. In $\mathbb{Z}^d$ ($d \geq 3$) the analogous problem is clearly meaningless since the largest covered ball around the origin is finite with probability 1. However, one can investigate in any dimension the radius of the largest ball (not surely around the origin) covered by the random walk in time n. Formally speaking let

$$Q(u, N) = \{x : x \in \mathbb{Z}^d, \|x - u\| \leq N\}$$

and $R^*(n) = R^*(n,d)$ be the largest integer for which there exists a random vector $u = u(n) \in \mathbb{Z}^d$ such that $Q(u, R^*(n))$ is covered by the random walk at time n, i.e.

$$\xi(x,n) \geq 1 \text{ for any } x \in Q(u, R^*(n)).$$

Then we formulate our

THEOREM 24.15 (Révész, 1990/B, 1993/B, Erdős–Révész, 1991) *Let $d \geq 3$. Then*

$$\lim_{n\to\infty} \frac{\log R^*(n)}{\log \log n} = \frac{1}{d-2} \quad a.s.$$

Before the proof we present a few lemmas.

LEMMA 24.15 *For any $0 < \alpha < 1$ and $L > 0$ there exists a sequence $x_1, x_2, \ldots, x_T$ of the points of $\mathbb{Z}^d$ such that*

$$L \leq \|x_i\| < L+1 \quad (i = 1, 2, \ldots, T),$$

$$\|x_i - x_j\| \geq L^\alpha \quad (i,j = 1,2,\ldots,T;\ i \neq j),$$
$$T = KL^{(1-\alpha)(d-1)}$$

where $K = K(d)$ is a positive constant depending on d only.

Proof is trivial.

LEMMA 24.16 *Let*

$$\alpha = \alpha_1 = \alpha_1(k) = \frac{D^{k+1} - D^k}{D^{k+1} - 1} + \varepsilon$$

where

$$0 < \varepsilon < \frac{D^k - 1}{D^{k+1} - 1}, \quad D = \frac{d-1}{d-2}, \quad k = 1,2,\ldots$$

and define $T = T_k$ and $x_1, x_2, \ldots, x_T$ as in Lemma 24.15. Then for any L big enough we have

$$\mathbf{P}\{Q(0,L) \text{ is covered eventually}\}$$
$$\leq \mathbf{P}\{x_1, x_2, \ldots, x_T \text{ are covered eventually}\} \leq e^{-(T-1)}.$$

Proof. Define the sequence $\alpha_1, \alpha_2, \ldots, \alpha_k$ ($k = 1, 2, \ldots$) by

$$\alpha_i = \frac{D^{k+1} - D^{k+1-i}}{D^{k+1} - 1} \quad (i = 2,3,\ldots,k).$$

Assuming that

$$0 < \varepsilon < \frac{D^k - D^{k-1}}{D^{k+1} - 1} < \frac{D^k - 1}{D^{k+1} - 1},$$

we have

$$0 < \alpha_1 < \alpha_2 < \ldots < \alpha_k < 1 \quad (k = 1,2,\ldots)$$

and

$$(\alpha_{i+1} - \alpha_1)(d-1) < \alpha_i(d-2) \quad (i = 1,2,\ldots,k)$$

where $\alpha_{k+1} = 1$.

Let $x_{i_1}, \ldots, x_{i_T}$ be an arbitrary permutation of the sequence $x_1, \ldots, x_T$. Consider the consecutive distances

$$\|x_{i_2} - x_{i_1}\|, \|x_{i_3} - x_{i_2}\|, \ldots, \|x_{i_T} - x_{i_{T-1}}\|.$$

Assume that among these distances l_1 (resp. l_2, resp. ,..., l_k) are lying between L^{α_1} and L^{α_2} (resp. L^{α_2} and L^{α_3}, resp. ,..., L^{α_k}) and $2L$.

Then (by Lemma 17.8) the probability that the random walk visits the points $x_{i_1}, x_{i_2}, \ldots, x_{i_T}$ in this given order is less than or equal to

$$\left(\frac{C_d + o(1)}{L^{\alpha_1(d-2)}}\right)^{l_1} \left(\frac{C_d + o(1)}{L^{\alpha_2(d-2)}}\right)^{l_2} \cdots \left(\frac{C_d + o(1)}{L^{\alpha_k(d-2)}}\right)^{l_k}.$$

Taking into consideration that the number of those j's $(1 \leq j \leq T)$ for which $L^{\alpha_i} \leq \|x_j - x_s\| < L^{\alpha_i+1}$ (where s is a fixed element of the sequence $(1, 2, \ldots, T))$ is less than or equal to

$$KL^{(\alpha_{i+1}-\alpha_1)(d-1)} \quad (i = 1, 2, \ldots, k),$$

we get

$\mathbf{P}\{x_1, x_2, \ldots, x_T \text{ are covered eventually}\}$

$$\leq \sum_{l_1+l_2+\cdots+l_k=T-1} \frac{(T-1)!}{l_1! \, l_2! \ldots l_k!} \prod_{i=1}^{k} \left(\frac{(C_d + o(1))KL^{(\alpha_{i+1}-\alpha_1)(d-1)}}{L^{\alpha_i(d-2)}}\right)^{l_i}$$

$$= \left[\sum_{i=1}^{k} \frac{(C_d + o(1))KL^{(\alpha_{i+1}-\alpha_1)(d-1)}}{L^{\alpha_i(d-2)}}\right]^{T-1} \leq e^{-(T-1)}$$

if L is big enough and we have Lemma 24.16.

In the same way as we proved Lemma 24.16 we can prove the following:

LEMMA 24.17 *For any L big enough and $u \in \mathbb{Z}^d$ we have*

$$\mathbf{P}\{Q(u, L) \text{ is covered eventually}\} \leq C^* L^d e^{-(T-1)}$$

where

$$T = KL^{(1-\alpha)(d-1)}, \quad \alpha = \frac{D^{k+1} - D^k}{D^{k+1} - 1} + \varepsilon$$

$$0 < \varepsilon < \frac{D^k - 1}{D^{k+1} - 1}, \quad D = \frac{d-1}{d-2},$$

k is an arbitrary positive integer and $C^* = C_d^*$ is a positive constant.

Proof of the upper part of Theorem 24.15. Let $L = [(\log n)^\theta]$ with

$$\theta = \theta_k = \frac{1}{(d-1)(1-\alpha)} + \varepsilon = \frac{1}{d-1}\left(\frac{D^k - 1}{D^{k+1} - 1} - \varepsilon\right)^{-1} + \varepsilon.$$

Then $T \geq (\log n)^\psi$ with some $\psi > 1$ and we obtain our statement observing that

$$\lim_{k \to \infty} \theta_k = \frac{1}{(d-2) - \varepsilon(d-1)} + \varepsilon.$$

LARGE COVERED BALLS 275

Proof of the lower part of Theorem 24.15.
At first we prove a few lemmas.

LEMMA 24.18 *There exists a constant $K > 0$ such that*

$$\mathbf{P}\{\xi(x,n) > 0\} \geq \left(\frac{C_d}{2}\right) R^{2-d} \qquad R = \|x\|$$

if $n \geq KR^2$ where C_d is the constant of Lemma 17.8.

Proof is essentially the same as that of Lemma 17.8.
Let

$$L = L(n) = [(\log n)^{(d-2)/(d-1)}]$$

and define

$$\tau_1 = n + [(\log n)^{2/(d-1)}],$$
$$\psi_1 = \inf\{k:\ k > \tau_1,\ S_n = S_k\} - \tau_1,$$
$$\tau_2 = \tau_1 + \psi_1 + [(\log n)^{2/(d-1)}],$$
$$\psi_2 = \inf\{k:\ k > \tau_2,\ S_n = S_k\} - \tau_2, \ldots$$

Clearly, with a positive probability (depending on n) ψ_1 is infinite. However, we have

LEMMA 24.19 *For any $\delta > 0$ there exists a constant $M = M(\delta) > 0$ such that*

$$\mathbf{P}\{\max_{1 \leq i \leq L} \psi_i \leq M(\log n)^{2/(d-2)}\} \geq n^{-\delta}. \qquad (24.28)$$

Proof. Clearly for any $N > 0$ there exists a constant $0 < p = p(N) < 1$ such that

$$\mathbf{P}\{\|S_{\tau_1} - S_n\| \leq N(\log n)^{1/(d-2)}\} \geq p > 0.$$

Observe that by Lemma 24.18 we have

$$\mathbf{P}\{\psi_1 \leq M(\log n)^{2/(d-2)}\}$$
$$\geq \mathbf{P}\{\psi_1 \leq M(\log n)^{2/(d-2)} \mid \|S_{\tau_1} - S_n\| \leq N(\log n)^{1/(d-2)}\}$$
$$\times \mathbf{P}\{\|S_{\tau_1} - S_n\| \leq N(\log n)^{1/(d-2)}\}$$
$$\geq \frac{C_d}{2}\left(N(\log n)^{1/(d-2)}\right)^{2-d} p = \frac{C_d p}{2N^{d-2}} \frac{1}{\log n}$$

provided that $M > KN^2$ where C_d and K are the constants of Lemma 24.18. Since $\psi_1, \psi_2, \ldots$ are i.i.d.r.v.'s we get

$$\mathbf{P}\{\max_{1 \leq i \leq L} \psi_i \leq K(\log n)^{2/(d-2)}\} \geq \left(\frac{C_d p}{2N^{d-2} \log n}\right)^{(\log n)^{1-\varepsilon}} \geq n^{-\delta}$$

which implies (24.28).

Let
$$A_n = A(n) = \{\max_{1 \leq i \leq L} \psi_i \leq M(\log n)^{2/(d-2)}\}$$

and x be an arbitrary element of $\mathbb{Z}^d$ for which

$$\|x - S_n\| \leq (\log n)^{1/(d-2)-2\varepsilon}. \tag{24.29}$$

Then applying again Lemma 24.18 we have

$$\mathbf{P}\{\xi(x, n + [(\log n)^{2/(d-2)}]) - \xi(x,n) = 0\} \leq 1 - \frac{C_d}{2((\log n)^{1/(d-2)-2\varepsilon})^{d-2}}$$
$$= 1 - \frac{C_d}{2(\log n)^{1-2\varepsilon(d-2)}}.$$

Hence the conditional probability (given $A(n)$) that x is not covered is less than or equal to

$$\left(1 - \frac{C_d}{2(\log n)^{1-2\varepsilon(d-2)}}\right)^{(\log n)^{1-\varepsilon}} \leq \exp\left(-\frac{C_d}{2}(\log n)^{2\varepsilon(d-2)-\varepsilon}\right).$$

Consequently the conditional probability that there exists a point for which (24.29) is satisfied and which is not covered is less than or equal to

$$O\left((\log n)^{d/(d-2)-2\varepsilon d}\right) \exp\left(-\frac{C_d}{2}(\log n)^{2\varepsilon(d-2)-\varepsilon}\right).$$

Let $\alpha > 1$ then by Lemma 24.19

$$\mathbf{P}\left\{\left(\sum_{k=n^\alpha}^{(n+1)^\alpha} A_k\right)^c\right\} \leq \left(1 - \frac{1}{n^\delta}\right)^{T(n)} \leq \exp\left(-\frac{T(n)}{n^\delta}\right) \leq \exp(-n^{\alpha-1-\delta-\varepsilon})$$

for any $\varepsilon > 0$ where $T(n) = n^{\alpha-1}(\log n)^{-(2/(d-2)+1-\varepsilon)}$.

Consequently with probability 1 for all but finitely many k there exists an n between 2^k and 2^{k+1} for which A_n holds. Given this n the conditional probability that $R_d(n) \leq (\log n)^{1/(d-2)-2\varepsilon}$ is less than or equal to

$$O\left((\log n)^{d/(d-2)-2\varepsilon d}\right) \exp\left(-\frac{C_d}{2}(\log n)^{2\varepsilon(d-2)-\varepsilon}\right).$$

Hence among these n's there are only finitely many for which $R_d(n) \leq (\log n)^{1/(d-2)-2\varepsilon}$, i.e. between 2^k and 2^{k+1} there exists an n (if k is large enough) for which $R_d(n) > (\log n)^{1/(d-2)-2\varepsilon}$. This implies the lower part of Theorem 24.15.

LARGE COVERED BALLS

Let $V_n(d)$ be the number of steps after the n-th step until the random walk $\{S_n\}$ on $\mathbb{Z}^d$ visits a previously unvisited site. Clearly $V_n(d) = 1$ i.o. a.s. At the same time Theorem 24.11 suggests that

$$\limsup_{n \to \infty} \frac{\log V_n(2)}{\log n} = \frac{1}{2} \quad \text{a.s.}$$

This conjecture was proved by Dembo et al. (2005/A).

Theorem 24.15 suggests the following

Conjecture 1.

$$\limsup_{n \to \infty} \frac{\log V_n(d)}{\log \log n} = \frac{2}{d-2} \quad \text{a.s.} \quad (d \geq 3).$$

Theorem 24.15 tells us that the path of the random walk in its first n steps covers relatively big balls. It is natural to ask where these big, covered balls are located in $\mathbb{Z}^d$. For example we might ask how close they can be to the origin. In fact we want to prove that if $Q(u, r_n)$ ($r_n = (\log n)^{1/(d-2)-\varepsilon}$) is covered at time n then u is big.

THEOREM 24.16 *Assume that $Q(u, r_n)$ is covered at time n. Then $\|u\| \geq \exp(\log n)^{1/2}$.*

Proof. By Theorem 24.15 the radius of the largest covered ball at time $\exp(3(\log n)^{1/2})$ is smaller than

$$(3(\log n)^{1/2})^{1/(d-2)+\varepsilon} < (\log n)^{1/(d-2)-\varepsilon}.$$

By Theorem 19.5 $S_k \notin Q(0, \exp((\log n)^{1/2}))$ if $k \geq \exp(3(\log n)^{1/2})$. Hence if $\|u\| \leq \exp((\log n)^{1/2})$ then after the time $\exp(3(\log n)^{1/2})$ the random walk cannot visit the ball $Q(u, r_n)$ at all, before this time it cannot cover the ball.

It is easy to get a much better result than the above one. However, to get the best possible result seems to be hard.

Question. Let $Q(u(n), R^*(n))$ be the largest covered ball and let $x_n \in \mathbb{Z}^d$ be a favourite value, i.e. $\xi(x_n, n) = \xi(n)$. Is it true that

$$x_n \in Q(u(n), R^*(n)) \quad \text{i.o. a.s.?}$$

24.6 Large empty balls

The previous Sections of this Chapter gave a description of the size of the covered or nearly covered balls. This Section is devoted to study the size of the large empty balls (left empty by a Wiener process).

Let $W(t) \in \mathbb{R}^d$ ($t \geq 0$, $d \geq 3$) be a Wiener process. We say that the ball
$$Q(x,r) = \{y: \ y \in \mathbb{R}^d, \ \|y - x\| \leq r\}$$
is left empty by $W(\cdot)$ forever if
$$\mathcal{D}(x,r) := Q(x,r) \cap \{W(t), \ t \geq 0\} = \emptyset.$$
Let
$$\rho(R) = \max\{r: \ \exists x \in \mathbb{R}^d \text{ such that } Q(x,r) \subset Q(0,R) \text{ and } \mathcal{D}(x,r) = \emptyset\}$$
be the radius of the largest empty ball in $Q(0, R)$. We are interested in studying the properties of the process $\{\rho(R), \ R \geq 0\}$. Since $W(0) = 0$, clearly
$$\rho(R) \leq \frac{R}{2}. \tag{24.30}$$

First we give a sharper upper bound than the trivial one of (24.30). The next four theorems are due to Erdős–Révész (1997).

THEOREM 24.17 *For any $\varepsilon > 0$*
$$\rho(R) \leq \frac{R}{2} - R^{1-\varepsilon} \qquad a.s. \tag{24.31}$$
if R is big enough.

Our next theorem tells us that the upper bound of (24.31) is not very far from the best possible result.

THEOREM 24.18
$$\rho(R) \geq \frac{R}{2} - \frac{R}{4 \log R} \qquad i.o. \ a.s. \tag{24.32}$$

Theorem 24.18 tells us that for some R the $\rho(R)$ will be very big. The next theorem tells that for some R the $\rho(R)$ will be much smaller.

THEOREM 24.19 *For any $\varepsilon > 0$ we have*
$$\rho(R) \leq \frac{R}{(\log \log R)^{1/d - \varepsilon}} \qquad i.o. \ a.s. \tag{24.33}$$

Now we show that the upper bound of (24.33) is close to the best possible result.

LARGE COVERED BALLS

THEOREM 24.20 *For any $\varepsilon > 0$*

$$\rho(R) \geq \frac{R}{(\log R)^{(1+\varepsilon)/(d-3)}} \quad a.s. \tag{24.34}$$

if R is big enough and $d \geq 4$. Further

$$\rho(R) \geq R(\log R)^{-(1+\varepsilon)} \quad a.s.$$

if R is big enough and $d = 3$.

The proof of Theorem 24.17 is based on the following simple Theorem 24.21. In order to formulate it we introduce a few notations.

For any $x \in \mathbb{R}^d$ with $\|x\| = 1$ and $0 < \vartheta < 1$ define the cone $\mathcal{K}(x, \vartheta)$ as follows:

$$\mathcal{K}(x, \vartheta) = \left\{ y : y \in \mathbb{R}^d, \left(\frac{y}{\|y\|}, x \right) \geq 1 - \vartheta \right\}.$$

Clearly for any $0 < \vartheta < 1$ there exists a positive integer $K = K(\vartheta)$ and a sequence $x_1, x_2, \ldots, x_K$ such that

$$x_i \in \mathbb{R}^d, \qquad \|x_i\| = 1, \qquad (i = 1, 2, \ldots, K)$$

$$\bigcup_{i=1}^K \mathcal{K}(x_i, \vartheta) = \mathbb{R}^d, \qquad K \leq L(1 - (1 - \vartheta)^2)^{-(d-1)/2}$$

where L is an absolute positive constant.

Let

$$\mathcal{L}_i = \mathcal{L}_i(R) = \mathcal{L}_i(R, \varepsilon, \vartheta) = \{ y : y \in \mathcal{K}(x_i, \vartheta), \ R^\varepsilon \leq (y, x_i) \leq R^{1-\varepsilon} \}$$

where

$$i = 1, 2, \ldots, K, \qquad 0 < \varepsilon < 1/2, \qquad R > 0.$$

Now we have

THEOREM 24.21 *For any $0 < \varepsilon < 1/2$, $1/2 < \vartheta < 1$,*

$$\mathbf{P}\left\{ \limsup_{R \to \infty} \bigcup_{i=1}^K \{ \mathcal{L} \cap \{ W(t, \omega), \ t \geq 0 \} = \emptyset \} \right\} = 0.$$

Note that Theorem 24.21 tells us that for any r big enough $W(t)$ meets all frustum of cones $\mathcal{L}_i(R)$ ($i = 1, 2, \ldots, K$).

In case $d = 3$ a much stronger Theorem was proved by Adelman–Burdzy–Pemantle (1998). Let f be a strictly positive increasing function on $\mathbb{R}^+$ and let $\mathcal{C}_f$ be the thorn

$$\{(x, y, z) \in \mathbb{R}^3 : x^2 + y^2 + z^2 \geq 1 \quad \text{and} \quad (x^2 + y^2)^{1/2} \leq f(|z|)\}.$$

Say that the Wiener process $W(\cdot)$ avoids f-thorns if there is with probability 1 a random set congruent to $\mathcal{C}_f$ avoided by W.

THEOREM 24.22 *If $f(z) = z\exp(-c(\log z)^{1/2})$ for $c > 0$ sufficiently small, then $W(\cdot)$ does not avoid f-thorns.*

24.7 Summary of Chapter 24

In this Section we summarize the most important results of this Chapter in an inaccurate form.

Let
$$Q(u,r) = \{x \in \mathbb{Z}^d,\ \|x - u\| \leq r\},$$
$$K(u,r,n) = \frac{\sum_{x \in Q(u,r)} I(\xi(x,n))}{r^d \omega_d}$$

where
$$I(z) = \begin{cases} 0 & \text{if } z = 0, \\ 1 & \text{if } z > 0 \end{cases}$$

and ω_d is the volume of the unit ball of $\mathbb{R}^d$.

Let $d = 2$. Then

(i)
$$K(0,r,n) = 1 \quad \text{if} \quad r \leq r_n^{(1)} := \exp((\log n)^{1/2})$$

i.e. $Q(0,r)$ is completely covered if $r \leq r_n^{(1)}$ (the exact results are (24.2-3*), Theorem 24.7, (24.5), (24.6*)),

(ii)
$$\lim_{n \to \infty} K(0,r,n) = 1 \quad \text{if} \quad r \leq r_n^{(2)} := \exp\left(\frac{\log n}{(\log \log n)^{2+\varepsilon}}\right)$$

i.e. $Q(0,r)$ is "almost" covered if $r \leq r_n^{(2)}$ (the exact result is Theorem 24.12),

(iii) $K(0,r,n)$ has a nondegenerated limit distribution if $r \leq r_n^{(3)} := n^\alpha$ ($0 < \alpha < 1/2$) (the exact result is Theorem 24.14),

(iv)
$$\sup_{u \in \mathbb{Z}^2} K(u,r,n) = 1 \quad \text{if} \quad r \leq r_n^{(4)} := n^{1/4}$$

(the exact result is Theorem 24.10).

Let $d \geq 3$. Then
$$\sup_{u \in \mathbb{Z}^d} K(u,r,n) = 1 \quad \text{if} \quad r \leq r_n^{(5)} := (\log n)^{1/(d-2)}$$

(the exact result is Theorem 24.15).

Chapter 25

Long Excursions

25.1 Long excursions in $\mathbb{Z}^2$

In Section 12.1 we have seen that the length of the longest excursion up to n in $\mathbb{Z}^1$ for some n can be nearly as big as n (cf. Theorem 12.1). At the same time for some other n it will be about $n/\log\log n$ only but it cannot be smaller than this (cf. Theorem 12.3). We also studied the length of the second (third, fourth,...) longest excursion and we have seen that if the length of the longest excursion is $n/\log\log n$ only then the length of the second, third,... longest excursion is about the same (cf. Theorem 12.4). As a consequence of this we obtained that the sum of the length of the $\log\log n$ longest excursions is nearly n (cf. Theorem 12.5).

Now we intend to investigate the analogue questions in $\mathbb{Z}^2$.

Let $\rho_1, \rho_2, \ldots$ be the consecutive return times of the planar random walk to the origin (cf. Notation 6). Put $\tau_k = \rho_k - \rho_{k-1}$ (the length of the k-th excursion). Now let $\Psi(n)$ be the last return to the origin before time n i.e.

$$\Psi(n) = \max\{j : j \leq n, \ S_j = 0\} = \rho_{\xi(0,n)}.$$

Denote by
$$M_n^{(1)} \geq M_n^{(2)} \geq \ldots \geq M_n^{(\xi(0,n)+1)}$$
the order statistics of the sequence

$$\tau_1, \tau_2, \ldots, \tau_{\xi(0,n)}, n - \Psi(n). \tag{25.1}$$

THEOREM 25.1 (Csáki–Révész–Rosen, 1998).

$$\lim_{n \to \infty} \frac{M_n^{(1)} + M_n^{(2)}}{n} = 1 \quad a.s.$$

Proof. Let
$$g(n) = \sum_{k=0}^{n} \mathbf{P}\{S_k = 0\} \sim \sum_{k=1}^{n} \frac{1}{\pi k} \sim \frac{\log n}{\pi}$$

(cf. Lemma 17.2). Define n_j by $g(n_j) = j$. Then clearly

$$0 \leq j - g\left(\frac{n_j}{j^2}\right) \leq C \log j$$

and by Lemma 17.11

$$\phi(j) := \mathbf{P}\{n_j(g(n_j))^{-2} < \rho_1 \leq n_{j+1}\} \leq Cj^{-2} \log j$$

if j is large enough. Let

$$N(n) = [g(n) \log g(n)],$$
$$\kappa(n) = \#\{i : i \leq N(n), \, n(g(n))^{-2} < \tau_i \leq n\},$$
$$\kappa^*(n_j) = \#\{i : i \leq N(n_{j+1}), \, n_j(g(n_j))^{-2} < \tau_i \leq n_{j+1}\}.$$

Then, since $N(n_j) = j \log j$, we have

$$\mathbf{P}\{\kappa^*(n_j) > 1\}$$
$$= 1 - (1 - \phi(j))^{N(n_{j+1})} - N(n_{j+1})(1 - \phi(j))^{N(n_{j+1})-1}\phi(j)$$
$$\sim \frac{(N(n_{j+1})\phi(j))^2}{2} \leq C\frac{(\log j)^4}{j^2}.$$

Thus $\kappa^*(n_j) \leq 1$ a.s. for all but finitely many j. Now take $n_j \leq n \leq n_{j+1}$. Since

$$N(n) \leq N(n_{j+1})$$

and

$$n_j(g(n_j))^{-2} \leq n(g(n))^{-2} < n \leq n_{j+1}$$

we obtain that

$$\kappa(n) \leq \kappa^*(n_j) \leq 1.$$

Since $\xi(0, n) \leq N(n)$ (cf. Theorem 20.4) there are no more than $N(n)$ excursions before time n. Thus the sum of those elements of the sequence (25.1) which are no larger than $n(g(n))^{-2}$ is bounded by

$$N(n)n(g(n))^{-2} = \frac{n \log g(n)}{g(n)} = o(n).$$

Hence the fact that $\kappa(n) \leq 1$ implies the Theorem.

Remark 1. As we told in case $d = 1$ the number of excursions required to nearly cover the time interval $[0, n]$ is between 1 and $\log \log n$ (cf. Theorems 12.1 and 12.5). In case $d = 2$ Theorem 25.1 tells us that the same number is 1 or 2. However Theorem 25.1 does not imply straight that for some n one excursion is enough to nearly cover the time interval $[0, n]$. This follows from:

THEOREM 25.2 Csáki–Révész–Rosen, 1998). *Let*

$$h(m_1, m_2, \tau; x) = \begin{cases} m_1 & if \quad 0 \leq x < \tau, \\ m_2 & if \quad \tau \leq x \leq 1, \end{cases}$$

$$\mathcal{H} = \{h(x) = h(m_1, m_2, \tau; x) : \ 0 \leq m_1 < m_2 \leq 1, \ 0 \leq \tau \leq 1\}.$$

Then the set of the limit points of the sequence

$$L_n(x) = \frac{\xi(0, xn)}{g(n) \log \log g(n)} \qquad (0 \leq x \leq 1)$$

is equal to $\mathcal{H}$ a.s.

Theorem 25.1 easily implies the following

Consequence 1. $\rho_{n+1} - \rho_n$ is either much smaller or much bigger than ρ_n.

In order to give a more accurate form of Consequence 1, let $\{\alpha_n\}$ and $\{\beta_n\}$ be sequences of positive numbers with

$$\sum_{n=1}^{\infty} \frac{1}{n\alpha_n} < \infty, \qquad \alpha_n \geq \log_2 n,$$

$$\sum_{n=1}^{\infty} \frac{1}{n\beta_n} = \infty, \qquad \inf_{n \geq 1} \beta_n > 0.$$

Further let

$$a_n = \exp\left(\frac{n}{\alpha_n}\right), \quad b_n = \exp\left(\frac{n}{\beta_n}\right), \quad c_n = \exp\left(\frac{n}{(1+\varepsilon)\beta_n}\right) \quad (\varepsilon > 0).$$

Finally let

$$T_n = \frac{\rho_{n+1} - \rho_n}{\rho_n}.$$

Then we have

THEOREM 25.3 (Csáki–Révész–Shi, 2001/B).

(i) $\quad T_n \notin (a_n^{-1}, a_n) \quad$ *a.s. for all but finitely many n,*
(ii) $\quad T_n \in (b_n^{-1}, c_n^{-1}) \quad$ *i.o. a.s.*
(iii) $\quad T_n \in (c_n, b_n) \quad$ *i.o. a.s.*

Example 1.

$$\alpha_n = (\log n)^{1+\delta} \quad (\delta > 0), \qquad \beta_n = \log n$$

i.e.

$$a_n = \exp\left(\frac{n}{(\log n)^{1+\delta}}\right), \quad b_n = \exp\left(\frac{n}{\log n}\right), \quad c_n = \exp\left(\frac{n}{(1+\varepsilon)\log n}\right)$$

satisfy the above conditions.

The next two Theorems give the limit distribution of T_n.

THEOREM 25.4 (Révész, 2000).

$$\mathbf{P}\left\{T_n < \exp\left(-\frac{n}{z}\right)\right\} = 1 - e^{-\pi z}(1 + O(n^{-1/7}))$$

uniformly for $0 < z < n^{3/7}$.

Theorem 25.4 tells us that T_n, with a big probability, is very small. However, Theorem 25.3 claims that T_n occasionally is very large. In our next theorem the limit distribution of T_n is evaluated when T_n is large.

THEOREM 25.5 (Révész, 2000). *For any $0 < \varepsilon < 1/7$ we have*

$$\mathbf{P}\left\{T_n > \exp\left(\frac{n}{z}\right) \mid T_n > 1\right\} = \pi^2 \int_0^\infty \frac{uz}{u+z} e^{-\pi u} du + O(n^{\varepsilon - 1/7})$$

uniformly for $0 \le z < n^{3/7}$.

25.2 Long excursions in high dimension

Since in $\mathbb{Z}^d$ ($d \ge 3$) a random walk returns to the origin only finitely many times, Theorems 25.1, 25.2 and 25.3 cannot hold true in higher dimension. However if we consider the longest excursion away from some $x \in \mathbb{Z}^d$ completed by the time n, then it can be long. For $i \ge 0$, define the random variable $\chi(i)$ by

$$S_{i+j} \ne S_i, \qquad j = 1, 2, \ldots, \chi(i) - 1, \qquad S_{i+\chi(i)} = S_i.$$

(If such $\chi(i)$ does not exist, we set $\chi(i) := \infty$.) Let

$$R(n) := \max\{\chi(i) : i + \chi(i) \le n\}$$

which in words denotes the length of the longest completed excursion (away from any point) at time n.

THEOREM 25.6 (Csáki–Révész–Shi, 2001/B) *Let $d \geq 3$. With probability one,*
$$\lim_{n \to \infty} \frac{\log R(n)}{\log n} = \begin{cases} 1 & if \quad d = 3, 4, \\ \dfrac{2}{d-2} & if \quad d \geq 5. \end{cases}$$

On the proof, here we only mention that its main ingredient is Theorem 23.1.

Chapter 26
Speed of Escape

Theorem 19.5 on the rate of escape suggests that the sphere $\{x : \|x\| = R\}$ is crossed about R times by the random walk if R is big enough and $d \geq 3$. In fact if $\|S_n\| = \sqrt{n}$ for every n then $\sum_{x \in Z(R)} \xi(x, \infty) = O(R)$ where

$$Z(R) = \{x : x \in \mathbb{Z}^d, \|\|x\| - R\| \leq 1\}.$$

Introduce the following notations:

$$J(x) = \begin{cases} 0 & \text{if } \xi(x, n) = 0 \text{ for every } n = 0, 1, 2, \ldots, \\ 1 & \text{otherwise} \end{cases}$$

and

$$\tilde{\theta}(R) = \sum_{x \in Z(R)} J(x),$$

i.e. $J(x) = 1$ if $x \in \mathbb{Z}^d$ is visited by the random walk and $\tilde{\theta}(R)$ is the number of points of $Z(R)$ visited eventually by the random walk. On the behaviour of $\tilde{\theta}(R)$ we have the following:

Conjecture 1. For any $d \geq 3$ there exists a distribution function $H(x) = H_d(x)$ for which $H(0) = 0$ and

$$\lim_{R \to \infty} \mathbf{P}\left\{\frac{\tilde{\theta}(R)}{R} < x\right\} = H(x) \quad (-\infty < x < \infty).$$

Unfortunately we cannot settle this conjecture but the analogous question for a Wiener process $W(t) = \{W_1(t), W_2(t), \ldots, W_d(t)\}$ ($d \geq 3$) can be solved. In order to present the corresponding theorem we introduce the following

Definition. $W(t)$ is crossing the sphere $\{x : \|x\| = R\}$ $\theta = \theta(R)$ times if $\theta(R)$ is the largest integer for which there exists a random sequence $0 < \alpha_1 = \alpha_1(R) < \beta_1 = \beta_1(R) < \alpha_2 = \alpha_2(R) < \beta_2 = \beta_2(R) \ldots < \alpha_\theta = \alpha_\theta(R) < \beta_\theta = \beta_\theta(R) < \infty$ such that

$\|W(t)\| < R$ if $t < \alpha_1$,
$\|W(\alpha_1)\| = R$ and $R - 1 < \|W(t)\| < R + 1$ if $\alpha_1 \leq t < \beta_1$,
$\|W(\beta_1)\| = R - 1$ or $R + 1$ and $\|W(t)\| \neq R$ if $\beta_1 \leq t < \alpha_2$,

$\|W(\alpha_2)\| = R$ and $R - 1 < \|W(t)\| < R + 1$ if $\alpha_2 \leq t < \beta_2$,
$\|W(\beta_2)\| = R - 1$ or $R + 1$ and $\|W(t)\| \neq R$ if $\beta_2 \leq t < \alpha_3, \ldots$
$\|W(\alpha_\theta)\| = R$ and $R - 1 < \|W(t)\| < R + 1$ if $\alpha_\theta \leq t < \beta_\theta$,
$\|W(\beta_\theta)\| = R + 1$ and $\|W(t)\| > R$ if $t \geq \beta_\theta$.

$(\theta(R))^{-1}$ will be called the *speed of escape* in R.

THEOREM 26.1

$$\lim_{R \to \infty} \mathbf{P}\left\{\frac{(d-2)\theta(R)}{2R} < t\right\} = 1 - e^{-t} \quad (t \geq 0).$$

The proof is based on the following:

LEMMA 26.1

$$\mathbf{P}\{\theta(R) = k\} = \lambda(1 - \lambda)^{k-1} \quad (k = 1, 2, \ldots)$$

where

$$\lambda = p(R)\left(1 - \left(\frac{R}{R+1}\right)^{d-2}\right),$$

and

$$p(R) = \mathbf{P}\{B(R, t) \mid \|W(t)\| = R\} = 1 - \frac{(R+1)^{2-d} - R^{2-d}}{(R+1)^{2-d} - (R-1)^{2-d}}$$

where

$B(R, t)$
$= \{\inf\{s : s > t, \|W(s)\| = R - 1\} > \inf\{s : s > t, \|W(s)\| = R + 1\}$.

Remark 1. Note that the last formula for $p(R)$ comes from Lemma 18.1. By (18.4)

$$\left(\frac{R}{R+1}\right)^{d-2} = \mathbf{P}\{\|W(t+s)\| \leq R \text{ for some } s > 0 \mid \|W(t)\| = R + 1\}.$$

Proof. Clearly we have

$$\mathbf{P}\{\theta(R) = 1\} = \lambda,$$

SPEED OF ESCAPE

$$\mathbf{P}\{\theta(R) = 2\}$$
$$= q(R)p(R)\left(1 - \left(\frac{R}{R+1}\right)^{d-2}\right)$$
$$+ p(R)\left(\frac{R}{R+1}\right)^{d-2} p(R)\left(1 - \left(\frac{R}{R+1}\right)^{d-2}\right)$$
$$= p(R)\left(1 - \left(\frac{R}{R+1}\right)^{d-2}\right)\left[q(R) + p(R)\left(\frac{R}{R+1}\right)^{d-2}\right] = \lambda(1-\lambda)$$

where $q(R) = 1 - p(R)$. Similarly

$$\mathbf{P}\{\theta(r) = k\}$$
$$= \sum_{j=0}^{k-1} \binom{k-1}{j} (q(R))^j \left(p(R)\left(\frac{R}{R+1}\right)^{d-2}\right)^{k-1-j}$$
$$\times p(R)\left(1 - \left(\frac{R}{R+1}\right)^{d-2}\right)$$
$$= \lambda\left(q(R) + p(R)\left(\frac{R}{R+1}\right)^{d-2}\right)^{k-1} = \lambda(1-\lambda)^{k-1}.$$

Hence we have Lemma 26.1.

Observe that

$$\mathbf{E}\theta(R) = \frac{1}{\lambda} = \frac{1}{p(R)} \frac{1}{1 - \left(1 - \frac{1}{R+1}\right)^{d-2}} \approx \frac{R+1}{p(R)(d-2)}.$$

Since $p(R) \to 1/2$ $(R \to \infty)$ we have

$$\lim_{R\to\infty} \frac{\mathbf{E}\theta(R)}{R} = \frac{2}{d-2}. \tag{26.1}$$

Lemma 26.1 together with (26.1) easily implies Theorem 26.1.

Studying the properties of the process $\{\theta(R),\ R > 0\}$ the following question naturally arises: does a sequence $0 < R_1 < R_2 < \ldots$ exist for which
$$\lim_{n\to\infty} R_n = \infty \quad \text{and} \quad \theta(R_i) = 1 \quad i = 1, 2, \ldots?$$

The answer to this question is affirmative. In fact we prove a much stronger theorem. In order to formulate this theorem we introduce the following

Definition. Let $\psi(R)$ be the largest integer for which there exists a positive integer $u = u(R) \leq R$ such that

$$\theta(k) = 1 \quad \text{for any } u \leq k \leq u + \psi(R).$$

It is natural to say that the speed of escape in the interval $(u, u + \psi(R))$ is maximal.

THEOREM 26.2

$$\psi(R) \geq \frac{\log\log R}{\log 2} \quad i.o. \ a.s.$$

Proof. Let

$$f(R) = \frac{\log\log R}{\log 2},$$
$$A(R) = \{\theta(k) = 1 \text{ for every } R \leq k \leq R + f(R)\}$$

and

$$C(R, t) = \{\|W(t)\| = R + f(R) + 1\}.$$

Then

$$\mathbf{P}\{A(R)\} = \prod_{j=0}^{f(R)} \mathbf{P}\{B(R+j, t) \mid \|W(t)\| = R + j\}$$
$$\times (1 - \mathbf{P}\{\|W(t+s)\| \leq R + f(R) \text{ for some } s > 0 \mid C(R, t)\})$$
$$\approx \frac{1}{\log R} \frac{d-2}{2R}$$

and

$$\mathbf{P}\{A(R)A(R+S)\} \approx \frac{1}{\log(R+S) - \frac{\log\log R}{\log 2}} \frac{1}{\log R} \frac{d-2}{2R}$$

if $\log\log R / \log 2 < S = o(R)$. In the case $S \geq O(R)$ the events $A(R)$ and $A(R+S)$ are asymptotically independent. Hence

$$\sum_{R=1}^{n} \mathbf{P}\{A(R)\} \approx \frac{d-2}{2} \log\log n,$$

and for any $\varepsilon > 0$ if n is big enough we have

$$\sum_{R=1}^{n} \sum_{S=[\log\log R/\log 2]}^{n} \mathbf{P}\{A(R)A(R+S)\} \leq \left(\frac{d-2}{2}\right)^2 (\log\log n)^2 (1+\varepsilon)$$

which implies Theorem 26.2 by Borel–Cantelli lemma.

Conjecture 2.
$$\lim_{R \to \infty} \frac{\psi(R)}{\log \log R} = \frac{1}{\log 2} \quad \text{a.s.}$$

Theorem 26.2 clearly implies that $\theta(R) = 1$ i.o. a.s. It is natural to ask: how big can $\theta(R)$ be? An answer to this question is

THEOREM 26.3 *For any $\varepsilon > 0$ we have*
$$\theta(R) \leq 2(d-2)^{-1}(1+\varepsilon) R \log R \quad \text{a.s.}$$
if R is big enough and
$$\theta(R) \geq (1-\varepsilon) R \log \log \log R \quad \text{i.o. a.s.}$$

Since this result is far from the best possible one and the proof is trivial we omit it.

Remark 2. Conjecture 1 suggests that $\tilde{\theta}(R) \sim R$. Instead of investigating the path up to ∞ consider it only up to $\rho_1 = \min\{k : k > 0, S_k = 0\}$. Taking into account that $\mathbf{P}\{\rho_1 = \infty\} > 0$ if $d \geq 3$ we obtain $\sum_{x \in Z(R)} \xi(x, \rho_1) \sim R$ with positive probability. Investigating the case $d = 1$ by Theorem 9.7 we get $\mathbf{E} \sum_{x \in Z(R)} \xi(x, \rho_1) = \mathbf{E}(\xi(R, \rho_1) + \xi(-R, \rho_1)) = 2$ for any $R \in \mathbb{Z}^1$. We may ask about the analogous question in the case $d = 2$. By Lemma 24.1 we obtain

$$\mathbf{P}\left\{\sum_{x \in Z(R)} \xi(x, \rho_1) > 0\right\} \approx \frac{\pi}{2 \log R} \quad (R \to \infty).$$

Lemma 18.1 suggests that the probability of returning to the origin from $Z(R-1)$ before visiting $Z(R)$ is $O(R^{-1}(\log R)^{-1})$. Hence we conjecture that $\sum_{x \in Z(R)} \xi(x, \rho_1) \sim O(R)$; for example,

$$\mathbf{E}\left(\sum_{x \in Z(R)} \xi(x, \rho_1)\right) = O(R) \quad \text{as} \quad R \to \infty$$

for any $d \geq 2$.

Chapter 27

A Few Further Problems

27.1 On the Dirichlet problem

Let U be an open, convex domain in $\mathbb{R}^2$ which is bounded by a simple closed curve Δ. Suppose that a continuous real function f is given on Δ. Then the Dirichlet problem requires us to find a function $u = u(x, y)$ which

(i) is continuous on $U + \Delta$,

(ii) agrees with f on Δ,

(iii) satisfies the Laplace equation

$$\frac{\partial^2 u}{\partial x^2} + \frac{\partial^2 u}{\partial y^2} = 0.$$

A probabilistic solution of this problem is the following. Let $\{W(t), t \geq 0\}$ be a Wiener process on $\mathbb{R}^2$ and for any $z \in U$ define $W_z(t) = W(t) + z$. Further, let σ_z be the first exit time of $W_z(t)$ from U, i.e.

$$\sigma_z = \min\{t : W_z(t) \in \Delta\} \quad (z \in U).$$

Then we have

THEOREM 27.1 *The function*

$$u(z) = \mathbf{E} f(\sigma_z)$$

is the solution of the Dirichlet problem.

The proof is very simple and is omitted. The reader can find a very nice presentation in Lamperti (1977), Chapter 9.6.

Here we present a discrete analogue of Theorem 27.1. Instead of an open, convex domain U we consider a sequence U_r $(r = 1, 2, \ldots)$ of domains defined as follows.

Consider the following sequences of integers

$$a_1 < a_2 < \ldots < a_{n_r},$$

$$b_1 < c_1, b_2 < c_2, \ldots, b_{n_r-1} < c_{n_r-1}$$

satisfying the conditions

(1) $b_{i+1} < c_i$, $c_{i+1} > b_i$ $(i = 1, 2, \ldots, n_r - 2)$,
(2) $a_{i+1} - a_i > \alpha r$, $c_i - b_i > \alpha r$,

with some $\alpha > 0$ and $n_r = 2, 3, \ldots$. Now let

$$U_r = \sum_{i=1}^{n_r-1} (a_i, a_{i+1}] \times (b_i, c_i].$$

Condition (1) implies that U_r is connected. Condition (2) has only some minor technical meaning. Let Δ_r be the boundary of U_r and define a "continuous" function $f_r(\cdot)$ on the integer grid of Δ_r, where by continuity we mean:

For any $\varepsilon > 0$ there exists a $\delta > 0$ such that $|f(z_1) - f(z_2)| \leq \varepsilon$ if $\|z_1 - z_2\| \leq \delta r$ where $z_1, z_2 \in \Delta_r \mathbb{Z}^2$.

Now we consider a random walk $\{S_n; n = 0, 1, 2, \ldots\}$ on $\mathbb{Z}^2$ and for any $z \in (U_r + \Delta_r)\mathbb{Z}^2$ we define

$$S_n^{(z)} = S_n + z \quad (n = 0, 1, 2, \ldots).$$

Let σ_z be the first exit time of $S_n^{(z)}$ from U_r, i.e.

$$\sigma_z = \min\{n : S_n^{(z)} \in \Delta_r\}.$$

We wish to prove that

$$u(z) = \mathbf{E}f(S_{\sigma_z}^{(z)})$$

is the solution of the discrete Dirichlet problem, meaning that

(i) u is "continuous" on $(U_r + \Delta_r)\mathbb{Z}^2$, i.e. for any $\varepsilon > 0$ there exists a $\delta > 0$ such that if $z_1, z_2 \in (U_r + \Delta_r)\mathbb{Z}^2$ and $\|z_1 - z_2\| \leq \delta r$, then $|u(z_1) - u(z_2)| \leq \varepsilon$,

(ii) u agrees with f on $\Delta_r \mathbb{Z}^2$,

(iii) u satisfies the Laplace equation, i.e.

$$(u(x+1, y) - 2u(x, y) + u(x-1, y)) \\ + (u(x, y+1) - 2u(x, y) + u(x, y-1)) = 0$$

whenever $(x, y), (x+1, y), (x, y+1), (x-1, y), (x, y-1) \in (U_r + \Delta_r)\mathbb{Z}^2$.

(ii) is trivial. (iii) follows from the trivial observation that

$$u(x,y) = \frac{1}{4}(u(x+1,y) + u(x-1,y) + u(x,y+1) + u(x,y-1))$$

if (x,y) satisfies the condition of (iii).

To see (i) we present a simple

LEMMA 27.1 *For any $\varepsilon > 0$ there exists a $\delta > 0$ such that if*

$$z \in U_r \mathbb{Z}^2, \quad q \in \Delta_r \mathbb{Z}^2, \quad \|z - q\| \leq \delta r$$

then

$$\mathbf{P}\{\|S^{(z)}(\sigma_z) - q\| \leq \varepsilon r\} \geq 1 - \varepsilon.$$

Consequently

$$|\mathbf{E}f(S^{(z)}(\sigma_z)) - f(q)| \leq \varepsilon^*.$$

Proof is simple and is omitted.

In order to prove (i) we have to investigate two cases:

(α) $z_1, z_2 \in U_r \mathbb{Z}^2$,

(β) one of z_1, z_2 is an element of $\Delta_r \mathbb{Z}^2$ and the other one is an element of $U_r \mathbb{Z}^2$.

In case (β) our statements immediately follow from Lemma 27.1. In case (α) assume that $\sigma_{z_1} \leq \sigma_{z_2}$ and observe that

$$\|S^{(z_1)}(\sigma_{z_1}) - S^{(z_2)}(\sigma_{z_1})\| = \|z_2 - z_1\| \leq \delta r.$$

Since $S^{(z_2)}(\sigma_{z_2}) = S^{(S^{(z_2)}(\sigma_{z_1}))}(\sigma_{S^{(z_2)}(\sigma_{z_1})})$ applying Lemma 27.1 with $q = S^{(z_1)}(\sigma_{z_1})$ and $z = S^{(z_2)}(\sigma_{z_1})$ we obtain (i).

Remark 1. Having the above result on the solution of the discrete Dirichlet problem, one can get a concrete solution by Monte Carlo method. In fact to get the value of $u(\cdot, \cdot)$ in a point $z_0 = (x_0, y_0) \in U_r$ observe the random walk starting in z_0 till the exit time σ_{z_0} and repeat this experiment n times. Then by the law of large numbers

$$\lim_{n \to \infty} n^{-1} \sum_{i=1}^{n} f(S_i^{(z_0)}(\sigma_{z_0})) = u(x_0, y_0) \text{ a.s.} \qquad (27.1)$$

where $S_1, S_2, \ldots$ are independent copies of a random walk. Hence the average in (27.1) is a good approximation of the discrete Dirichlet problem if n is big enough. A solution of the continuous Dirichlet problem in some z_0 or in a few fix points can be obtained by choosing r big enough and the length of the steps of the random walk small enough comparing to the underlying domain.

27.2 DLA model

Let $A_1 \subset A_2 \subset \ldots$ be a sequence of random subsets of $\mathbb{Z}^2$ defined as follows:

A_1 consists of the origin, i.e. $A_1 = \{0\}$,

$A_2 = A_1 + y_2$ where y_2 is an element of the boundary of A_1

obtained by the following chance mechanism. A particle is released at ∞ and performs a random walk on $\mathbb{Z}^2$. Then y_2 is the position where the random walk first hits the boundary of A_1.

The boundary of a set $A \subset \mathbb{Z}^2$ is defined as

$$\partial A = \{y : y \in \mathbb{Z}^2 \text{ and } y \text{ is adjacent to some site in } A, \text{ but } y \notin A\}.$$

For example, $\partial A_1 = \{(0,1),(1,0),(-1,0),(0,-1)\}$.

Having defined A_n, A_{n+1} is defined as $A_{n+1} = A_n + y_{n+1}$ where y_{n+1} is the position where the random walk starting from ∞ first hits ∂A_n.

In the above definition the meaning of "released at ∞" is not very clear. Instead we can say: let

$$R_n = \inf\{r : r > 0, A_n \subset Q(r) = \{x : \|x\| \leq r\}\}.$$

Then instead of starting from infinity the particle might start its random walk from $(R_n^6, 0)$ (say). It is easy to see that the particle goes round the origin before it hits A_n (a.s. for all but finitely many n). This means that the distribution of the hitting point will be the same as in case of a particle released at ∞.

Many papers are devoted to studying this model, called *Diffusion Limited Aggregation* (DLA). The reason for the interest in this model can be explained by the fact that simulations show that it mimicks several physical phenomena well.

The most interesting concrete problem is to investigate the behaviour of the "radius"

$$r_n = \max\{\|x\| : x \in A_n\}.$$

Trivially $r_n \geq (n/\pi)^{1/2}$ and it is very likely that r_n is much bigger than this trivial lower bound. Only a negative result is known saying that r_n is not very big. In fact we have

THEOREM 27.2 (Kesten, 1987). *There exists a constant $C > 0$ such that*

$$\limsup_{n \to \infty} n^{-2/3} r_n \leq C \quad a.s.$$

The proof of Kesten is based on estimates of the hitting probability of ∂A_n. He proved that there exists a $C > 0$ such that for any $y \in \partial A_n$ we have
$$\mathbf{P}\{y_{n+1} = y\} \leq Cr_n^{-1/2}. \tag{27.2}$$

In order to get a lower estimate of r_n we should get a lower estimate of the probability in (27.2) at least for some $y \in \partial A_n$. Auer (1989) studied the question of how one can get the lower bounds of the hitting probabilities of some points of the boundaries of certain sets (not necessarily formed by a DLA model). He investigated the following sets:

$$B_1 = \{(-r,0), (-r+1,0), \ldots, (r-1,0), (r,0)\},$$
$$B_2 = B_1 + \{(0,-r), (0,-r+1), \ldots, (0,r-1), (0,r)\},$$
$$B_3 = \{x = (x_1, x_2) : |x_1| + |x_2| = r\}.$$

Consider the point $y = (r, 0)$. Then the probability that the particle coming from infinity first hits y among the points of $\partial B_i (i = 1, 2, 3)$ is larger than or equal to
$$Cr^{-1/2} \quad \text{if} \quad i = 1, 2$$
and
$$Cr^{-2/3}(\log r)^{-1/3} \quad \text{if} \quad i = 3$$
with some $C > 0$.

27.3 Percolation

Consider $\mathbb{Z}^2$ and assume that each bond (edge) is "open" with probability p and "closed" with probability $1 - p$. All bonds are independent of each other. An open path is a path on $\mathbb{Z}^2$ all of whose edges are open.

One of the main problems of the percolation theory is to find the probability $\Theta(p)$ of the existence of an infinite open path. Kesten (1980) proved that
$$\Theta(p) \begin{cases} = 0 & \text{if} \quad p \leq 1/2, \\ > 0 & \text{if} \quad p > 1/2. \end{cases}$$

The value $1/2$ is called the critical value of the bond percolation in $\mathbb{Z}^2$.

An analogous problem is the so-called site percolation. In site percolation the sites of $\mathbb{Z}^2$ are independently open with probability p and closed with probability $q = 1 - p$. Similarly as in the case of the bond percolation a path of $\mathbb{Z}^2$ is called open if all its sites are open and we ask the probability $\Theta^*(p)$ of the existence of an infinite open path. The critical value of the site percolation in $\mathbb{Z}^2$ is unknown, but Tóth (1985) proved

$$\Theta^*(p) = 0 \quad \text{if} \quad p \leq x_0^4 \sim 0,503478$$

where x_0 is the root lying between 0 and 1 of the polynomial

$$3x^8 - 8x^7 + 6x^6 + x^4 - 1.$$

We call the attention of the reader to the survey of Kesten (1988) on percolation theory.

> And God said, "Let there be lights in the firmament of the heavens to separate the day from the night; and let them be for signs and for seasons and for days and years."
>
> The First Book of Moses

III. RANDOM WALK IN RANDOM ENVIRONMENT

10. RANDOM WALK IN RANDOM ENVIRONMENT

Notations

1. $\mathcal{E} = \{\ldots, E_{-2}, E_{-1}, E_0, E_1, E_2, \ldots\}$ is a sequence of i.i.d.r.v.'s satisfying $\beta < E_i < 1 - \beta$ with some $0 < \beta < 1/2$ called environment.

2. $\{\Omega_1, \mathcal{F}_1, \mathbf{P}_1\}, \{\Omega_2, \mathcal{F}_2, \mathbf{P}_\mathcal{E}\}, \{\Omega, \mathcal{F}, \mathbf{P}\}$ (see Introduction).

3.
$$U_j = \frac{1 - E_j}{E_j}, \quad V_j = \log U_j \quad (j = 0, \pm 1, \pm 2, \ldots).$$

4. $T_0 = 0$, $T_n = V_1 + V_2 + \cdots + V_n$, $T_{-n} = V_{-1} + V_{-2} + \cdots + V_{-n}$ $(n = 1, 2, \ldots)$.

5.
$$D(a,b) = \begin{cases} 0 & \text{if } b = a, \\ 1 & \text{if } b = a + 1, \\ 1 + \sum_{j=1}^{b-a-1} \prod_{i=1}^{j} U_{a+i} & \text{if } b \geq a + 2, \end{cases}$$

$$D(b) = D(0,b) = 1 + U_1 + U_1 U_2 + \cdots + U_1 U_2 \ldots U_{b-1}$$
$$= e^0 + e^{T_1} + e^{T_2} + \cdots + e^{T_{b-1}} \quad (b = 1, 2, \ldots).$$

6.
$$\frac{1}{D^*(n)} = 1 - \frac{D(0, n-1)}{D(0,n)} = 1 - \frac{1 + U_1 + U_1 U_2 + \cdots + U_1 U_2 \ldots U_{n-2}}{1 + U_1 + U_1 U_2 + \cdots + U_1 U_2 \ldots U_{n-1}}$$
$$= (1 + U_{n-1}^{-1} + (U_{n-1} U_{n-2})^{-1} + \cdots + (U_{n-1} U_{n-2} \ldots U_1)^{-1})^{-1},$$

i.e.
$$D^*(n) = e^0 + \exp(-(T_{n-1} - T_{n-2})) + \exp(-(T_{n-1} - T_{n-3})) + \cdots$$
$$+ \exp(-(T_{n-1} - T_0)) = D(n)e^{-T_{n-1}} \quad (n = 1, 2, \ldots).$$

7.
$$\frac{1}{D(-n)} = 1 - \frac{D(-n,-1)}{D(-n,0)} = (1 + (U_{-1})^{-1} + (U_{-1} U_{-2})^{-1} + \cdots$$
$$+ (U_{-1} U_{-2} \cdots U_{-n+1})^{-1})^{-1}$$

$(n = 1, 2, \ldots)$. Caution: $D(n) = D(0,n)$, however, $D(-n) \neq D(-n, 0)$.

8. $I(t)$ is the inverse function of $D(n)$, i.e.

$$I(t) = k \text{ if } D(k) \leq t < D(k+1),$$
$$I(-t) = k \text{ if } D(-k) \leq t < D(-k-1) \ (t \geq 1, \ k = 1, 2, \ldots).$$

9. $R_0, R_1, \ldots$ is a random walk in random environment (RWIRE) (see Introduction).

10. $p(a, b, c)$ (see Lemma 30.1).

Chapter 28

Introduction

The sequence $\{S_n\}$ of Part I was considered as a mathematical model of the linear Brownian motion. In fact it is a model of the linear Brownian motion in a homogeneous (non-random) environment.

We meet new difficulties when the environment is non-homogeneous. It is the case, for example, when the motion of a particle in a magnetic field is investigated. In this case we consider a random environment instead of a deterministic one. This situation can be described by different mathematical models.

At first we formulate only a special case of our model. It is given in the following two steps:

Step 1. (The Lord creates the Universe). The Lord visits all integers of the real line and tosses a coin when visiting i ($i = 0, \pm 1, \pm 2, \ldots$). During the first six days He creates a random sequence

$$\mathcal{E} = \{\ldots, E_{-2}, E_{-1}, E_0, E_1, E_2, \ldots\}$$

where E_i is head or tail according the result of the experiment made in i.

Step 2. (The life of the Universe after the Sixth Day). Having the sequence $\{\ldots, E_{-2}, E_{-1}, E_0, E_1, E_2, \ldots\}$ the Lord puts a particle in the origin and gives the command: if you are located in i and E_i is head then go to the left with probability 3/4 and to the right with probability 1/4, if E_i is tail then go to the left with probability 1/4 and to the right with probability 3/4. Creating the Universe and giving this order to the particle "God rested from all his work which he had done in creation" forever.

The general form of our above, special model can be described as follows:

Step 1. (The Lord creates the Universe). Having a sequence $\mathcal{E} = \{\ldots, E_{-2}, E_{-1}, E_0, E_1, E_2, \ldots\}$ of i.i.d.r.v.'s with distribution

$$\mathbf{P}\{E_0 < x\} = F(x), \quad F(0) = 0, \quad F(1) = 1,$$

the Lord creates a realization $\mathcal{E}$ of the above sequence. (The random sequence $\{\ldots, E_{-2}, E_{-1}, E_0, E_1, E_2, \ldots\}$ and a realization of it will be denoted by the same letter $\mathcal{E}$.) This realization is called a random environment (RE).

Step 2. (The life of the Universe after the Sixth Day). Having an RE $\mathcal{E}$ the Lord lets a particle make a random walk starting from the origin and going one step to the right resp. to the left with probability E_0 resp. $1 - E_0$. If the particle is located at $x = i$ (after n steps) then the particle moves one step to the right (resp. to the left) with probability E_i (resp. $1 - E_i$). That is, we define the random walk $\{R_0, R_1, \ldots\}$ by $R_0 = 0$ and

$$\mathbf{P}_{\mathcal{E}}\{R_{n+1} = i+1 \mid R_n = i, R_{n-1}, R_{n-2} \ldots, R_1\}$$
$$= 1 - \mathbf{P}_{\mathcal{E}}\{R_{n+1} = i-1 \mid R_n = i, R_{n-1}, R_{n-2}, \ldots, R_1\} = E_i. \qquad (28.1)$$

The sequence $\{R_n\}$ is called a random walk in RE (RWIRE).

Now we give a more mathematical description of this model as follows. Let $\{\Omega_1, \mathcal{F}_1, \mathbf{P}_1\}$ be a probability space and let

$$\mathcal{E} = \mathcal{E}(\omega_1)$$
$$= \{\ldots, E_{-1} = E_{-1}(\omega_1), E_0 = E_0(\omega_1), E_1 = E_1(\omega_1), \ldots\}$$

$(\omega_1 \in \Omega_1)$ be a sequence of i.i.d.r.v.'s with $\mathbf{P}_1\{E_1 < x\} = F(x)$ ($F(0) = 1 - F(1) = 0$).

Further, let $\{\Omega_2, \mathcal{F}_2\}$ be the measurable space of the sequences $\omega_2 = \{\varepsilon_1, \varepsilon_2, \ldots\}$ where $\varepsilon_i = 1$ or $\varepsilon_i = -1$ ($i = 1, 2, \ldots$) and $\mathcal{F}_2$ is the natural σ-algebra. Define the r.v.'s $Y_1, Y_2, \ldots$ on Ω_2 by $Y_i(\omega_2) = \varepsilon_i$ ($i = 1, 2, \ldots$) and let $R_0 = 0$, $R_n = Y_1 + Y_2 + \cdots + Y_n$ ($n = 1, 2, \ldots$). Then we construct a probability measure $\mathbf{P}$ on the measurable space $\{\Omega = \Omega_1 \times \Omega_2, \mathcal{F} = \mathcal{F}_1 \times \mathcal{F}_2\}$ as follows: for any given $\omega_1 \in \Omega_1$ we define a measure $\mathbf{P}_{\omega_1} = \mathbf{P}_{\mathcal{E}(\omega_1)} = \mathbf{P}_{\mathcal{E}}$ on $\mathcal{F}_2$ satisfying (28.1). (Clearly (28.1) uniquely defines $\mathbf{P}_{\mathcal{E}}$ on $\mathcal{F}_2$.) Having the measures $\mathbf{P}_{\mathcal{E}(\omega_1)}$ ($\omega_1 \in \Omega_1$) and $\mathbf{P}_1$ one can define the measure $\mathbf{P}$ on $\mathcal{F}$ the natural way.

Our aim is to study the properties of the sequence $\{R_n\}$. In this study we meet two types of questions.

(i) Question of the Lord. The Lord knows ω_1, i.e. the sequence $\mathcal{E}$; or in other words, He knows the measure $\mathbf{P}_{\mathcal{E}}$ and asks about the behaviour of the particle in the future, i.e., He asks about the properties of the sequence $\{R_n\}$ given $\mathcal{E}$.

(ii) Question of the physicist. The physicist does not know ω_1. Perhaps he has some information on F, i.e. he knows something on $\mathbf{P}_1$. He also wants to predict the location of the particle after n steps, i.e. also wants to describe the properties of the sequence $\{R_n\}$.

A typical answer to the first type of question is a theorem of the following type:

THEOREM 28.1 *There exist two sequences of $\mathcal{F}_1$-measurable functions $f_n^{(1)} = f_n^{(1)}(\mathcal{E}) \leq f_n^{(2)} = f_n^{(2)}(\mathcal{E})$ such that*

$$f_n^{(1)} \leq \max_{0 \leq k \leq n} |R_k| \leq f_n^{(2)} \quad a.s. \quad (\mathbf{P}_{\mathcal{E}}) \qquad (28.2)$$

INTRODUCTION

for all but finitely many n, i.e.

$$\mathbf{P}_{\mathcal{E}}\{f_n^{(1)}(\mathcal{E}) \leq \max_{0 \leq k \leq n} |R_k| \leq f_n^{(2)}(\mathcal{E}) \text{ for all but finitely many } n\} = 1.$$

Since the physicist does not know the environment $\mathcal{E}$ he will not be satisfied with an inequality like (28.2). However, he wants to prove an inequality like

THEOREM 28.2 *There exist two deterministic sequences* $\alpha_n^{(1)} \leq \alpha_n^{(2)}$ *such that*

$$\alpha_n^{(1)} \leq f_n^{(1)} \leq f_n^{(2)} \leq \alpha_n^{(2)} \quad a.s. \quad (\mathbf{P}_1) \tag{28.3}$$

for all but finitely many n.

Having inequalities (28.2) and (28.3) the physicist gets the following answer to his question:

THEOREM 28.3 *There exist two deterministic sequences* $\alpha_n^{(1)} \leq \alpha_n^{(2)}$ *such that*

$$\alpha_n^{(1)} \leq \max_{0 \leq k \leq n} |R_k| \leq \alpha_n^{(2)} \quad a.s. \quad (\mathbf{P}) \tag{28.4}$$

for all but finitely many n. Equivalently

$$\mathbf{P}\{\alpha_n^{(1)} \leq \max_{0 \leq k \leq n} |R_k| \leq \alpha_n^{(2)} \text{ for all but finitely many } n\}$$

$$= \mathbf{P}_1\{\mathbf{P}_{\mathcal{E}}\{\alpha_n^{(1)} \leq \max_{0 \leq k \leq n} |R_k| \leq \alpha_n^{(2)} \text{ for all but finitely many } n\} = 1\} = 1.$$

Remark 1. The exact forms of Theorems 28.1, 28.2 and 28.3 are given in Theorems 30.6, 30.8 and 30.9 where the exact forms of $\alpha_n^{(1)}, f_n^{(1)}, \alpha_n^{(2)}, f_n^{(2)}$ are given.

Remark 2. In the special case when

$$\mathbf{P}_1\{E_0 = 1/2\} = F(1/2 + 0) - F(1/2) = 1,$$

the RWIRE problem reduces to the simple symmetric random walk problem.

Chapter 29

In the First Six Days

In this chapter we study what might have happened during the creation of the Universe, i.e. the possible properties of the sequence $\mathcal{E}$ are investigated.

The following conditions will be assumed:

(C.1) there exists a $0 < \beta < 1/2$ such that $\mathbf{P}\{\beta < E_0 < 1 - \beta\} = 1$,

(C.2)
$$\mathbf{E}_1 V_0 = \int_{-\infty}^{\infty} x d\mathbf{P}_1\{V_0 < x\} = \int_{\beta}^{1-\beta} \log \frac{1-x}{x} dF(x) = 0,$$

where $F(x) = \mathbf{P}_1\{E_0 < x\}$, $V_0 = \log U_0$ and $U_0 = (1 - E_0)/E_0$,

(C.3)
$$0 < \sigma^2 = \mathbf{E}_1 V_0^2 = \int_{\beta}^{1-\beta} \left(\log \frac{1-x}{x}\right)^2 dF(x) < \infty.$$

Remark 1. For a simple symmetric random walk (i.e. $\mathbf{P}_1\{E_0 = 1/2\} = 1$) we have $\mathbf{P}_1\{U_0 = 1\} = \mathbf{P}_1\{V_0 = 0\} = 1$ and consequently (C.1) and (C.2) are satisfied; however, (C.3) is not satisfied since $\mathbf{E}_1 V_0^2 = \sigma^2 = 0$.

We also mention that if (C.1) and (C.2) hold and $\mathbf{E}_1 V_0^2 = \sigma^2 = 0$ then $\mathbf{P}_1\{E_0 = 1/2\} = 1$.

Remark 2. Most of the following lemmas remain true replacing (C.1) by a much weaker condition or omitting it. Here we are not interested in this type of generalizations.

LEMMA 29.1

$$\limsup_{n\to\infty} T_n = \limsup_{n\to\infty} T_{-n} = -\liminf_{n\to\infty} T_n = -\liminf_{n\to\infty} T_{-n} = \infty \quad a.s. \quad (\mathbf{P}_1). \tag{29.1}$$

If we assume (C.1) and (C.3) but instead of (C.2) we assume that $\mathbf{E}_1 V_0 = m \neq 0$. Then

$$\lim_{n\to\infty} T_n = \lim_{n\to\infty} T_{-n} = (\text{sign } m)\infty \quad a.s. \quad (\mathbf{P}_1) \tag{29.2}$$

where $T_n = V_1 + \cdots + V_n$, $T_{-n} = V_{-1} + \cdots + V_{-n}$, $V_j = \log U_j$, $U_j = (1 - E_j)/E_j$ and $T_0 = 0$.

Proof. (29.1) is a trivial consequence of the LIL of Hartmann and Wintner (cf. Section 4.4), (29.2) follows from the strong law of large numbers.

LEMMA 29.2
$$\lim_{n\to\infty} D(n) = \infty \quad a.s. \quad (\mathbf{P}_1) \qquad (29.3)$$

(*cf. Notation 5*).

Proof. Since
$$D(n) = 1 + U_1 + U_1 U_2 + \cdots + U_1 U_2 \ldots U_{n-1} = e^0 + e^{T_1} + e^{T_2} + \cdots + e^{T_{n-1}}, \qquad (29.4)$$

(29.3) follows from Lemma 29.1.

By (29.4) we have
$$\exp(\max_{0\leq k\leq n-1} T_k) \leq D(n) \leq n\exp(\max_{0\leq k\leq n-1} T_k) \qquad (29.5)$$

and the LIL implies

LEMMA 29.3 *For any $\varepsilon > 0$ and for any $p = 1, 2, \ldots$ we have*

$$\max_{1\leq k\leq n} T_k \leq (1+\varepsilon)\sigma(2n\log\log n)^{1/2}$$
$$a.s. \quad (\mathbf{P}_1) \text{ for all but finitely many } n, \qquad (29.6)$$

$$\max_{1\leq k\leq n} T_k \geq (1-\varepsilon)\sigma(2n\log\log n)^{1/2} \quad i.o. \; a.s. \quad (\mathbf{P}_1), \qquad (29.7)$$

$$\max_{1\leq k\leq n} T_k \leq n^{1/2}(\log n\log\log n \cdots \log_p n)^{-1} \quad i.o. \; a.s. \quad (\mathbf{P}_1), \qquad (29.8)$$

$$\max_{1\leq k\leq n} T_k \geq n^{1/2}(\log n\log\log n \cdots (\log_p n)^{1+\varepsilon})^{-1}$$
$$a.s. \quad (\mathbf{P}_1) \text{ for all but finitely many } n. \qquad (29.9)$$

By (29.5) we also get

$$D(n) \leq \exp\{(1+\varepsilon)\sigma(2n\log\log n)^{1/2}\}$$
$$a.s. \quad (\mathbf{P}_1) \text{ for all but finitely many } n, \qquad (29.10)$$

$$D(n) \geq \exp\{(1-\varepsilon)\sigma(2n\log\log n)^{1/2}\} \quad i.o. \; a.s. \quad (\mathbf{P}_1), \qquad (29.11)$$

$$D(n) \leq \exp\{n^{1/2}(\log n\log\log n \cdots \log_p n)^{-1}\} \quad i.o. \; a.s. \quad (\mathbf{P}_1), \qquad (29.12)$$

$$D(n) \geq \exp\{n^{1/2}(\log n\log\log n \cdots (\log_p n)^{1+\varepsilon})^{-1}\}$$
$$a.s. \quad (\mathbf{P}_1) \text{ for all but finitely many } n. \qquad (29.13)$$

Replacing the $\max_{1\leq k\leq n}$ *by* $\max_{-n\leq k\leq -1}$ *the inequalities* (29.6) – (29.9) *remain true. Replacing* $D(n)$ *by* $D(-n)$ *in* (29.10) – (29.13) *they remain true.*

$$\limsup_{n\to\infty} \frac{\log D^*(n)}{\sqrt{2n\log\log n}} = \sigma \quad a.s. \quad (\mathbf{P}_1), \qquad (29.14)$$

$$\liminf_{n\to\infty} \max_{0\leq k\leq n} \frac{\log D^*(k)}{\sqrt{n}}\sqrt{\log\log n} = \sigma\pi/\sqrt{8} \quad a.s. \quad (\mathbf{P}_1), \qquad (29.15)$$

$$D^*(n) \geq 1, \qquad (29.16)$$

$$D^*(n) \leq n \quad i.o.\ a.s. \quad (\mathbf{P}_1). \qquad (29.17)$$

Proof. Inequalities (29.6)–(29.13) are clear as they are. The following simple analogue of (29.5),

$$\exp(-\min_{0\leq k\leq n-1}(T_{n-1}-T_k)) \leq D^*(n) \leq n\exp(-\min_{0\leq k\leq n-1}(T_{n-1}-T_k)), \qquad (29.18)$$

implies (29.16) and (29.17).

In order to get (29.14) and (29.15) approximate the process $\{T_k,\ k\geq 0\}$ by a Wiener process $\{\sigma W(t),\ 0\leq t<\infty\}$. By Theorem 10.2 the process

$$-\min_{0\leq t\leq T}(W(T)-W(t))$$

is identical in distribution to the process $\{|W(t)|,\ t\geq 0\}$. Hence the LIL and the Other LIL imply (29.14) and (29.15).

LEMMA 29.4 *For any $\varepsilon > 0$ and for any $p = 1, 2, \ldots$ we have*

$$I(t) \leq (\log|t|\log\log|t|\cdots\log_{p-1}|t|(\log_p|t|)^{1+\varepsilon})^2$$
$$a.s. \quad (\mathbf{P}_1)\ if\ |t|\ is\ big\ enough, \qquad (29.19)$$

$$I(t) \geq (\log|t|\log\log|t|\cdots\log_p|t|)^2 \quad i.o.\ a.s.\quad (\mathbf{P}_1), \qquad (29.20)$$

$$I(t) \leq \frac{1+\varepsilon}{2\sigma^2}\frac{\log^2|t|}{\log_3|t|} \quad i.o.\ a.s.\quad (\mathbf{P}_1), \qquad (29.21)$$

$$I(t) \geq \frac{1-\varepsilon}{2\sigma^2}\frac{\log^2|t|}{\log_3|t|} \quad a.s.\quad (\mathbf{P}_1)\ if\ |t|\ is\ big\ enough. \qquad (29.22)$$

Proof. (29.19), (29.20), (29.21) (resp. (29.22)) follows from (29.13), (29.12), (29.11) (resp. (29.10)).

LEMMA 29.5

$$D(I(t)) \leq t < D(I(t)+1) \quad (t\geq 1), \qquad (29.23)$$

$$D(-I(-t)) \leq t < D(-I(-t)-1) \quad (t\geq 1), \qquad (29.24)$$

$$D(n+1) = D(n)+U_1U_2\cdots U_n = D(n)+U_n\frac{D(n)}{D^*(n)} \quad (n=1,2,\ldots), \qquad (29.25)$$

$$t\left(1+\frac{U_{I(t)}}{D^*(I(t))}\right)^{-1} \leq D(I(t)) \leq t, \qquad (29.26)$$

$$D(n+1) \leq \frac{1}{\beta}D(n)+1, \qquad (29.27)$$

$$I(\lambda t+1) \geq I(t)+1 \quad \text{if} \quad \lambda > \frac{1}{\beta}, \qquad (29.28)$$

where β is the constant of (C.1).

Proof. (29.23), (29.24), (29.25) follow immediately from the definitions. (29.23) and (29.25) combined imply

$$t < D(I(t)+1) = D(I(t)) + U_{I(t)}\frac{D(I(t))}{D^*(I(t))}.$$

This, in turn, implies (29.26). (29.27) follows from (C.1). (29.28) follows from (29.23) and (29.27).

Chapter 30

After the Sixth Day

30.1 The recurrence theorem of Solomon

THEOREM 30.1 (Solomon, 1975). *Assuming conditions* (C.1), (C.2), (C.3) *we have*

$$\mathbf{P}\{R_n = 0 \ i.o.\} = \mathbf{P}_1\{\mathbf{P}_{\mathcal{E}}\{R_n = 0 \ i.o.\} = 1\} = 1.$$

Assuming (C.1), (C.3) *and* $\mathbf{E}_1 V_0 \neq 0$ *we have*

$$\mathbf{P}_1\{\mathbf{P}_{\mathcal{E}}\{R_n = 0 \ i.o.\} > 0\} = 0.$$

Remark 1. The statement of the above Theorem can be formulated as follows: with probability 1 ($\mathbf{P}_1$) the Lord creates such an environment in which the recurrence theorem is true, i.e. the particle returns to the origin i.o. with probability 1 ($\mathbf{P}_{\mathcal{E}}$). Before the proof of Theorem 30.1 we present an analogue of Lemma 3.1.

LEMMA 30.1 *Let*

$$p(a,b,c)$$
$$= \mathbf{P}_{\mathcal{E}}\{\min\{j: \ j > m, \ R_j = a\} < \min\{j: \ j > m, \ R_j = c\} \mid S_m = b\}$$

$(a \leq b \leq c)$, *i.e.* $p(a,b,c) = p(a,b,c,\mathcal{E})$ *is the probability that a particle starting from b hits a before c given the environment* $\mathcal{E}$. *Then*

$$p(a,b,c) = 1 - \frac{D(a,b)}{D(a,c)}.$$

Especially

$$p(0,1,n) = 1 - \frac{1}{D(n)} \quad \text{and} \quad p(0,n-1,n) = \frac{1}{D^*(n)}.$$

Proof. Clearly, we have

$$p(a,a,c) = 1, \quad p(a,c,c) = 0,$$

$$p(a,b,c) = E_b p(a,b+1,c) + (1-E_b)p(a,b-1,c).$$

Consequently,

$$p(a,b+1,c) - p(a,b,c) = \frac{1-E_b}{E_b}(p(a,b,c) - p(a,b-1,c)).$$

By iteration we get

$$p(a,b+1,c) - p(a,b,c) = U_b U_{b-1} \cdots U_{a+1}(p(a,a+1,c) - p(a,a,c))$$
$$= U_b U_{b-1} \cdots U_{a+1}(p(a,a+1,c) - 1). \qquad (30.1)$$

Adding the above equations for $b = a, a+1, \ldots, c-1$ we get

$$-1 = p(a,c,c) - p(a,a,c) = D(a,c)(p(a,a+1,c) - 1),$$

i.e.

$$p(a,a+1,c) = 1 - \frac{1}{D(a,c)}. \qquad (30.2)$$

Hence (30.1) and (30.2) imply

$$p(a,b+1,c) - p(a,b,c) = -\frac{1}{D(a,c)} U_b U_{b-1} \cdots U_{a+1}.$$

Adding these equations we obtain

$$p(a,b+1,c) - 1 = p(a,b+1,c) - p(a,a,c)$$
$$= \frac{-1}{D(a,c)}(1 + U_{a+1} + U_{a+1}U_{a+2} + \cdots + U_{a+1}U_{a+2}\cdots U_b)$$
$$= -\frac{D(a,b+1)}{D(a,c)}.$$

Hence we have the Lemma.

Consequence 1.

$$\mathbf{P}\left\{\lim_{n\to\infty} p(0,1,n;\mathcal{E}) = \lim_{n\to\infty}\left(1 - \frac{1}{D(n)}\right) = 1\right\} = 1, \qquad (30.3)$$

$$\mathbf{P}\{\lim_{n\to\infty} p(-n,-1,0;\mathcal{E}) = 0\} = 1. \qquad (30.4)$$

(30.3) follows from Lemma 30.1 and (29.3). In order to see (30.4) observe

$$p(-n,-1,0) = 1 - \frac{D(-n,-1)}{D(-n,0)} = \frac{1}{D(-n)}$$

and apply (29.13) for $D(-n)$.

The following lemma is a trivial analogue of Lemma 3.2, the proof will be omitted.

AFTER THE SIXTH DAY

LEMMA 30.2 *For any $-\infty < a \leq b < \infty$ we have*

$$\mathbf{P}\{\liminf_{n \to \infty} R_n = a\} = \mathbf{P}\{\limsup_{n \to \infty} R_n = b\} = 0.$$

Proof of Theorem 30.1. Assume that $R_1 = 1$, say. Then by Lemma 30.2 the particle returns to 0 or it is going to $+\infty$ before returning. However, by (30.3) for any $\varepsilon > 0$ there exists an $n_0 = n_0(\varepsilon, \mathcal{E})$ such that $p(0, 1, n) = 1 - 1/D(n) \geq 1 - \varepsilon$ if $n \geq n_0$. Consequently the probability that the particle returns to 0 is larger than $1 - \varepsilon$ for any $\varepsilon > 0$ which proves the Theorem.

30.2 Guess how far the particle is going away in an RE

Introduce the following notations:

$$M^+(n) = \max_{0 \leq k \leq n} R_k,$$
$$M^-(n) = -\min_{0 \leq k \leq n} R_k,$$
$$M(n) = \max\{M^+(n), M^-(n)\} = \max_{0 \leq k \leq n} |R_k|,$$
$$\rho_0 = 0,$$
$$\rho_1 = \min\{k : \ k > 0, \ R_k = 0\},$$
$$\ldots \quad \ldots\ldots$$
$$\rho_{j+1} = \min\{k : \ k > \rho_j, \ R_k = 0\},$$
$$\xi(k, n) = \#\{l : 0 < l \leq n, R_l = k\},$$
$$\xi(n) = \max_k \xi(k, n),$$
$$\nu(n) = \#\{i : \ 0 \leq i \leq n - 1, \ R_{\rho_i + 1} = 1\}.$$

Observe that $\xi(0, \rho_n) = n$.

Our aim is to study the behaviour of $M(n)$. Especially in this section a reasonable guess will be given.

Consider the simple environment when

$$\mathbf{P}_1\left\{U_i = \frac{3}{4}\right\} = \mathbf{P}_1\left\{U_i = \frac{1}{4}\right\} = \frac{1}{2}.$$

Note that conditions (C.1), (C.2) and (C.3) are satisfied. Note also that in the environment $\mathcal{E} = \{\ldots, 3/4, 1/4, 3/4, 1/4, 3/4, \ldots\}$ the behaviour of the random walk is the same as that of the simple symmetric random walk. For example, it is trivial to prove that

$$\limsup_{n\to\infty} b_n M(n) = 1 \quad \text{a.s.}$$

if $\mathcal{E}$ is the given environment and $b_n = (2n \log\log n)^{-1/2}$.

One can guess that since environment $\{\ldots, 3/4, 1/4, 3/4, 1/4, \ldots\}$ is nearly the typical one, $M(n)$ will be practically $n^{1/2}$ in most environments. This way of thinking is not correct because we know that in a typical environment there are long blocks containing mostly 3/4's and long blocks containing mostly 1/4's.

Assume that in our environment

$$\max_{1\leq k\leq n} T_k = n^{1/2} \quad \text{and} \quad -\min_{1\leq k\leq n} T_{-k} = n^{1/2}$$

which is a typical situation. Then by (29.10), (29.11) and Lemma 30.1 we have

$$p(0,1,n) = 1 - \frac{1}{D(n)} \sim 1 - \exp(-n^{1/2})$$

and

$$p(-n,-1,0) = \frac{1}{D(-n)} \sim \exp(-n^{-1/2}).$$

This means that the particle will return to the origin $\exp(n^{1/2})$ times before arriving n or $-n$. Hence to arrive n requires at least $\exp(n^{1/2})$ steps. Conversely, in n steps the particle cannot go farther than $(\log n)^2$.

This way of thinking is due to Sinai (1982). He was the first one who realized that having high peaks and deep valleys in the environment, for the particle it takes a long time to go through. Clearly high peak means that $T(k)$ is a big positive number for $k > 0$ resp. it is a big negative number for $k < 0$ while the meaning of the deep valley is just the opposite.

30.3 A prediction of the Lord

LEMMA 30.3 *For any environment $\mathcal{E}$ we have*

$$\mathbf{P}_{\mathcal{E}}\{\nu(n) = k\} = \binom{n}{k} E_0^k (1 - E_0)^{n-k}, \tag{30.5}$$

$$\limsup_{n\to\infty} \frac{|\nu(n) - nE_0|}{(2nE_0(1-E_0)\log\log n)^{1/2}} = 1 \quad \text{a.s.} \quad (\mathbf{P}_{\mathcal{E}}) \tag{30.6}$$

where ν_n is defined in Section 30.2.

AFTER THE SIXTH DAY

Proof is trivial.

LEMMA 30.4 *For any environment $\mathcal{E}$ and $k = 1, 2, \ldots$ we have*

$$\mathbf{P}_{\mathcal{E}}\{M^+(\rho_n) < k \mid \nu_n\} = (p(0,1,k))^{\nu_n} = \left(1 - \frac{1}{D(k)}\right)^{\nu_n}, \qquad (30.7)$$

$$\begin{aligned}
\mathbf{P}_{\mathcal{E}}&\{M^+(\rho_n) < k\} \\
&= \sum_{l=0}^{n} \left(1 - \frac{1}{D(k)}\right)^l \binom{n}{l} E_0^l (1-E_0)^{n-l} \\
&= \left(E_0\left(1 - \frac{1}{D(k)}\right) + 1 - E_0\right)^n = \left(1 - \frac{E_0}{D(k)}\right)^n. \qquad (30.8)
\end{aligned}$$

Proof is trivial.

Now we prove our

THEOREM 30.2 *For any environment $\mathcal{E}$ we have*

$$I(n(\log n)^{-1-\varepsilon}) \leq M^+(\rho_n) \leq I(n(\log n)^{1+\varepsilon}) \quad a.s. \quad (\mathbf{P}_{\mathcal{E}}), \qquad (30.9)$$
$$I(-n(\log n)^{-1-\varepsilon}) \leq M^-(\rho_n) \leq I(-n(\log n)^{1+\varepsilon}) \quad a.s. \quad (\mathbf{P}_{\mathcal{E}}), \qquad (30.10)$$

$$\begin{aligned}
\max\{I(n(\log n)^{-1-\varepsilon}), I(-n(\log n)^{-1-\varepsilon})\} &\leq M(\rho_n) \\
\leq \max\{I(n(\log n)^{1+\varepsilon}), I(-n(\log n)^{1+\varepsilon})\} &\quad a.s. \quad (\mathbf{P}_{\mathcal{E}})
\end{aligned} \qquad (30.11)$$

for all but finitely many n.

Proof. By Lemma 30.4 and (29.26) we have

$$\mathbf{P}_{\mathcal{E}}\left\{M^+(\rho_n) \geq I\left(\frac{1}{3}n(\log n)^{1+\varepsilon}\right)\right\}$$

$$= 1 - \left(1 - \frac{E_0}{D\left(I\left(\frac{1}{3}n(\log n)^{1+\varepsilon}\right)\right)}\right)^n \leq 1 - \left(1 - \frac{E_0 Q_n}{\frac{1}{3}n(\log n)^{1+\varepsilon}}\right)^n$$

$$\approx \frac{E_0 Q_n}{\frac{1}{3}(\log n)^{1+\varepsilon}}$$

where

$$Q_n = 1 + \frac{U_N}{D^*(N)} \quad \text{and} \quad N = N(n) = I\left(\frac{1}{3}n(\log n)^{1+\varepsilon}\right).$$

Let $n_k = 2^k$. Then by the Borel–Cantelli lemma we get (cf. (29.16))

$$M^+(\rho_{n_k}) < I\left(\frac{1}{3}n_k(\log n_k)^{1+\varepsilon}\right) \quad \text{a.s.} \quad (\mathbf{P}_\mathcal{E})$$

for all but finitely many k. If $n_k \leq n < n_{k+1}$ we have

$$M^+(\rho_n) \leq M^+(\rho_{n_{k+1}}) \leq I\left(\frac{1}{3}n_{k+1}(\log n_{k+1})^{1+\varepsilon}\right) \leq I\left(n_k(\log n_k)^{1+\varepsilon}\right)$$
$$\leq I\left(n(\log n)^{1+\varepsilon}\right) \quad \text{a.s.} \quad (\mathbf{P}_\mathcal{E}) \text{ for all but finitely many } n.$$

Hence we have the upper part of (30.9).

Now we turn to the proof of the other inequality of (30.9).

By Lemma 30.4 and (29.26) we have

$$\mathbf{P}_\mathcal{E}\{M^+(\rho_n) < I(n(\log n)^{-1-\varepsilon})\} = \left(1 - \frac{E_0}{D(I(n(\log n)^{-1-\varepsilon}))}\right)^n$$
$$\leq \left(1 - \frac{E_0(\log n)^{1+\varepsilon}}{n}\right)^n$$
$$\leq \exp(-E_0(\log n)^{1+\varepsilon}).$$

Hence we have (30.9) by the Borel–Cantelli lemma.

The proof of (30.10) is identical. (30.11) is a trivial consequence of (30.9) and (30.10).

In order to get some estimates of $M(n)$ (resp. $M^+(n), M^-(n)$) the Lord is interested to estimate ρ_n or equivalently $\xi(0,n)$. To study this problem in a more general form we present a few results describing the behaviour of the local time $\xi(x,n)$.

LEMMA 30.5 *For any integer $k = 1, 2, \ldots$ and any environment $\mathcal{E}$ we have*

$$\mathbf{P}_\mathcal{E}\{\xi(k,\rho_1) = l\} = \begin{cases} 1 - \dfrac{E_0}{D(k)} & \text{if } l = 0, \\ \dfrac{E_0(1-E_k)}{D(k)D^*(k)}\left(1 - \dfrac{1-E_k}{D^*(k)}\right)^{l-1} & \text{if } l = 1, 2, \ldots \end{cases}$$

Consequently

$$\mathbf{P}_\mathcal{E}\{\xi(k,\rho_1) \geq L\} = \frac{E_0}{D(k)}\left(1 - \frac{1-E_k}{D^*(k)}\right)^{L-1} \quad (L = 1, 2, \ldots).$$

Proof. Clearly we have

$$\mathbf{P}_{\mathcal{E}}\{\xi(k,\rho_1)=0\} = 1 - E_0 + E_0 p(0,1,k)$$
$$= 1 - E_0 + E_0\left(1 - \frac{1}{D(k)}\right) = 1 - \frac{E_0}{D(k)}.$$

In case $l = 1, 2, \ldots$ we get

$$\mathbf{P}_{\mathcal{E}}\{\xi(k,\rho_1)=l\} = E_0(1-p(0,1,k))(1-E_k)p(0,k-1,k)$$
$$\times \sum_{j=0}^{l-1} \binom{l-1}{j} E_k^j ((1-E_k)(1-p(0,k-1,k)))^{l-1-j}$$
$$= \frac{E_0(1-E_k)}{D(k)D^*(k)}\left(1 - \frac{1-E_k}{D^*(k)}\right)^{l-1}.$$

A trivial calculation gives

LEMMA 30.6 (Csörgő–Horváth–Révész, 1987). *For any* $k = 1, 2, \ldots$

$$m_k = \mathbf{E}_{\mathcal{E}}\xi(k,\rho_1) = \frac{E_0}{1-E_k}\frac{D^*(k)}{D(k)} = \frac{E_0}{1-E_k}e^{-T_{k-1}}, \tag{30.12}$$

$$\sigma_k^2 = \mathbf{E}_{\mathcal{E}}(\xi(k,\rho_1) - m_k)^2$$
$$= \frac{E_0}{(1-E_k)^2}\frac{(D^*(k))^2}{D(k)}\left(2 - \frac{1-E_k}{D^*(k)} - \frac{E_0}{D(k)}\right), \tag{30.13}$$

$$\varphi(\lambda) = \mathbf{E}_{\mathcal{E}}\exp(\lambda\xi(k,\rho_1))$$
$$= 1 - \frac{E_0}{D(k)} + \frac{E_0(1-E_k)}{D(k)D^*(k)}\frac{e^\lambda}{1 - e^\lambda\left(1 - \frac{1-E_k}{D^*(k)}\right)} \tag{30.14}$$

for any $\lambda < -\log(1 - (1-E_k)/D^*(k))$. *Especially*

$$\varphi\left(\frac{1-E_k}{2D^*(k)}\right) = 1 + \frac{E_0}{D(k)}\left(\frac{2\lambda e^\lambda}{1 - e^\lambda(1-2\lambda)} - 1\right). \tag{30.15}$$

Observe that

$$0 < \lambda = \lambda_k = \frac{1-E_k}{2D^*(k)} < \frac{1}{2}$$

and

$$\sqrt{e} \leq \frac{2\lambda e^\lambda}{1 - e^\lambda(1-2\lambda)} \leq 2 \quad \text{if} \quad 0 < \lambda < 1/2. \tag{30.16}$$

Proof. As an example we prove (30.14). By Lemma 30.5 we have

$$\mathbf{E}_{\mathcal{E}} \exp(\lambda \xi(k, \rho_1)) = 1 - \frac{E_0}{D(k)} + \sum_{l=1}^{\infty} e^{\lambda l} \frac{E_0(1 - E_k)}{D(k)D^*(k)} \left(1 - \frac{1 - E_k}{D^*(k)}\right)^{l-1}$$

$$= 1 - \frac{E_0}{D(k)} + \frac{e^{\lambda} E_0(1 - E_k)}{D(k)D^*(k)} \frac{1}{1 - e^{\lambda}\left(1 - \frac{1 - E_k}{D^*(k)}\right)}.$$

Remark 1. (30.12) implies that: for any $\varepsilon > 0$

$$m_k \geq \frac{\beta}{1 - \beta} \exp((1 - \varepsilon)\sigma(2k \log \log k)^{1/2}) \quad \text{i.o. a.s.} \quad (\mathbf{P}_1)$$

and

$$m_k \leq \frac{1 - \beta}{\beta} \exp(-(1 - \varepsilon)\sigma(2k \log \log k)^{1/2}) \quad \text{i.o. a.s.} \quad (\mathbf{P}_1).$$

Compare these inequalities and (9.6).

LEMMA 30.7 *For any $k = 1, 2, \ldots$ and any environment $\mathcal{E}$ we have*

$$\lim_{n \to \infty} \mathbf{P}_{\mathcal{E}} \left\{ \frac{\xi(k, \rho_n) - nm_k}{\sqrt{n}\sigma_k} < x \right\} = \Phi(x), \tag{30.17}$$

$$\limsup_{n \to \infty} \frac{\xi(k, \rho_n) - nm_k}{\sigma_k \sqrt{2n \log \log n}} = 1 \quad a.s. \quad (\mathbf{P}_{\mathcal{E}}). \tag{30.18}$$

Proof is trivial.

Now we give a somewhat deeper consequence of (3.14).

LEMMA 30.8 (Csörgő–Horváth–Révész, 1987). *For any*

$$-1 < \lambda < \min\left\{1, -\log\left(1 - \frac{1 - E_k}{D^*(k)}\right)\right\}$$

and any $k = 1, 2, \ldots$, we have

$$\mathbf{E}_{\mathcal{E}} \exp(\lambda(\xi(k, \rho_1) - m_k)) = 1 + \frac{\lambda^2}{2}\sigma_k^2 + \lambda^3 \Theta_k$$

where

$$|\Theta_k| \leq A \left(\frac{D^*(k)}{1 - E_k}\right)^5 \quad \text{and } A \text{ is a positive constant.}$$

AFTER THE SIXTH DAY 319

Proof. By Taylor expansion we get

$$\frac{e^\lambda}{1 - e^\lambda\left(1 - \frac{1 - E_k}{D^*(k)}\right)} = \frac{D^*(k)}{1 - E_k} \frac{1}{1 - \frac{D^*(k)}{1 - E_k}(1 - e^{-\lambda})}$$
$$= \mathcal{D}(k)(1 + \mathcal{D}(k)h(\lambda) + (\mathcal{D}(k)h(\lambda))^2 + \eta(\mathcal{D}(k)h(\lambda))^3)$$

with $|\Theta| \leq 1, |\eta| \leq 1$, where

$$\mathcal{D}(k) = \frac{D^*(k)}{1 - E_k} \quad \text{and} \quad h(\lambda) = \lambda - \frac{\lambda^2}{2} + \Theta\frac{\lambda^3}{6}.$$

Consequently

$$\left|\frac{e^\lambda}{1 - e^\lambda(1 - \frac{1 - E_k}{D^*(k)})} - \mathcal{D}(k)\left(1 + \lambda\mathcal{D}(k) + \lambda^2(\mathcal{D}(k))^2 - \frac{1}{2}\mathcal{D}(k)\right)\right|$$

$$\leq A|\lambda|^3(\mathcal{D}(k))^4$$

if A is big enough. Hence by (30.14) we have

$$\left|\mathbf{E}_\mathcal{E} \exp \lambda\xi(k, \rho_1) - \left(1 + \lambda m_k + \lambda^2\left(m_k \frac{D^*(k)}{1 - E_k} - \frac{1}{2}m_k\right)\right)\right|$$

$$\leq A|\lambda|^3 \frac{E_0}{D(k)}\left(\frac{D^*(k)}{1 - E_k}\right)^3.$$

Multiplying the above inequality by

$$\exp(-\lambda m_k) = 1 - \lambda m_k + \frac{\lambda^2}{2}m_k^2 + \eta\frac{\lambda^3}{6}m_k^3$$

one gets the Lemma.

LEMMA 30.9 *Let*

$$0 < x < \sigma_k \left(\min\left\{\sqrt{n}, -\sqrt{n}\log\left(1 - \frac{1 - E_k}{D^*(k)}\right)\right\}\right)^{1/2}.$$

Then for any $k = 1, 2, \ldots$ and $n = 1, 2, \ldots$ we have

$$\mathbf{P}_\mathcal{E}\{|\xi(k, \rho_n) - nm_k| \geq x\sqrt{n}\}$$
$$\leq 2\exp\left(A\left(\frac{D^*(k)}{1 - E_k}\right)^5 x^3\sigma_k^{-6}n^{-1/2}\right)\exp\left(-\frac{x^2}{2\sigma_k^2}\right).$$

Proof. Apply Lemma 30.8 with

$$\lambda = xn^{-1/2}\sigma_k^{-2} \quad \text{and} \quad \lambda = -xn^{-1/2}\sigma_k^{-2}.$$

Then we get

$$\mathbf{P}_{\mathcal{E}}\{|\xi(k,\rho_n) - nm_k| \geq x\sqrt{n}\} \leq \frac{2\mathbf{E}_{\mathcal{E}}\exp(xn^{-1/2}\sigma_k^{-2}(\xi(k,\rho_n) - nm_k))}{\exp(x^2\sigma_k^{-2})}$$

$$\leq 2\left(1 + \frac{x^2}{2n\sigma_k^4}\sigma_k^2 + \Theta_k\frac{x^3}{n^{3/2}\sigma_k^6}\right)^n \exp\left(-\frac{x^2}{\sigma_k^2}\right)$$

and we have the Lemma.

This last inequality gives a very sharp result for $\xi(k,\rho_n)$ when k is not too big. In cases where k can be very big it is worthwhile to give another consequence of Lemma 30.6. In fact we prove

LEMMA 30.10 *For any $K > 0$ there exists a $C = C(K) > 0$ such that*

$$\mathbf{P}_{\mathcal{E}}\{\xi(k,\rho_n) \geq 2nm_k + C\log n D^*(k)\} \leq n^{-K} \tag{30.19}$$

$(k = 1, 2, \ldots; n = 1, 2, \ldots).$

Proof. Let $\lambda = \lambda_k = (1 - E_k)/2D^*(k)$. Then by (30.15) and (30.16) we get

$$\mathbf{P}_{\mathcal{E}}\{\xi(k,\rho_n) \geq 2nm_k + CD^*(k)\log n\}$$
$$= \mathbf{P}_{\mathcal{E}}\{\exp\lambda\xi(k,\rho_n) \geq \exp(2\lambda nm_k + \lambda CD^*(k)\log n)\}$$
$$\leq \mathbf{E}_{\mathcal{E}}(\exp\lambda\xi(k,\rho_1))^n \exp(-2\lambda nm_k - \lambda CD^*(k)\log n)$$
$$\leq \exp\left(n\frac{E_0}{D(k)} - 2n\frac{1-E_k}{2D^*(k)}\frac{E_0}{1-E_k}\frac{D^*(k)}{D(k)} - \frac{1-E_k}{2D^*(k)}CD^*(k)\log n\right)$$
$$= \exp\left(-\frac{1-E_k}{2}C\log n\right)$$

which proves (30.19).

A very similar result is the following:

LEMMA 30.11 *For any $C_1 > 2/\beta$ (cf. (C.1)) we have*

$$\mathbf{P}_{\mathcal{E}}\left\{\xi(k,\rho_n) \geq C_1 n\frac{D^*(k)}{D(k)}\right\} \leq \exp\left(-\frac{n}{D(k)}\right) \tag{30.20}$$

$(k = 1, 2, \ldots; n = 1, 2, \ldots).$

Proof. Let $\lambda = \lambda_k = (1 - E_k)/2D^*(k)$. Then by (30.15) and (30.16) we get

$$\mathbf{P}_{\mathcal{E}}\left\{\xi(k,\rho_n) \geq C_1 n \frac{D^*(k)}{D(k)}\right\}$$

$$= \mathbf{P}_{\mathcal{E}}\left\{\exp(\lambda\xi(k,\rho_n)) \geq \exp\left(\lambda C_1 n \frac{D^*(k)}{D(k)}\right)\right\}$$

$$\leq \exp\left(-\lambda C_1 n \frac{D^*(k)}{D(k)}\right) \mathbf{E}_{\mathcal{E}}(\exp \lambda\xi(k,\rho_n))$$

$$= \left[\exp\left(-\lambda C_1 \frac{D^*(k)}{D(k)}\right) \mathbf{E}_{\mathcal{E}}(\exp \lambda\xi(k,\rho_1))\right]^n$$

$$= \left[\exp\left(-\lambda C_1 \frac{D^*(k)}{D(k)}\right)\left(1 + \frac{E_0}{D(k)}\left(\frac{2\lambda e^\lambda}{1 - e^\lambda(1 - 2\lambda)} - 1\right)\right)\right]^n$$

$$\leq \exp\left(-\frac{n}{D(k)}\right).$$

Hence we have (30.20).

An analogue result describes the behaviour of $\xi(k,\rho_1)$ when k is a big positive number.

LEMMA 30.12 *There exist positive constants C and C_1 such that*

$$\mathbf{P}_{\mathcal{E}}\{\xi(k,\rho_1) \geq C_1 D^*(k) \log k \mid \xi(k,\rho_1) > 0\} \leq C_1 k^{-2} \qquad (30.21)$$

and

$$\mathbf{P}_{\mathcal{E}}\{\xi(k,\rho_1) \leq k^{-2} D^*(k) \mid \xi(k,\rho_1) > 0\} \leq C k^{-2}. \qquad (30.22)$$

Proof. Let μ_k be the number of negative excursions away from k between 0 and ρ_1. Clearly, we have

$$\mathbf{P}_{\mathcal{E}}\{\mu_k = l \mid \xi(k,\rho_1) > 0\} = p(0, k-1, k)(1 - p(0, k-1, k))^{l-1}$$
$$= (D^*(k))^{-1}(1 - (D^*(k))^{-1})^{l-1}.$$

Consequently

$$\mathbf{P}_{\mathcal{E}}\{\mu_k > L \mid \xi(k,\rho_1) > 0\} = (1 - (D^*(k))^{-1})^L. \qquad (30.23)$$

Hence

$$\mathbf{P}_{\mathcal{E}}\{\xi(k,\rho_1) \leq k^{-2} D^*(k) \mid \xi(k,\rho_1) > 0\}$$
$$< \mathbf{P}_{\mathcal{E}}\{\mu_k \leq k^{-2} D^*(k) \mid \xi(k,\rho_1) > 0\} = 1 - (1 - (D^*(k))^{-1})^{k^{-2}D^*(k)}$$
$$\leq C k^{-2}$$

and we have (30.22). In order to prove (30.21) observe that for any $0 < \delta < \varepsilon$ there exists a $C_2 = C_2(\delta) > 0$ such that

$$\mathbf{P}_{\mathcal{E}}\{\mu_k < \hat{E}_k \xi(k, \rho_1) \mid \xi(k, \rho_1)\} \leq \exp(-C_2 \xi(k, \rho_1)),$$

where $\hat{E}_k = 1 - E_k - \delta$. Hence by (30.23) we have

$$\mathbf{P}_{\mathcal{E}}\{\xi(k, \rho_1) \geq C_1 D^*(k) \log k \mid \xi(k, \rho_1) > 0\}$$
$$= \mathbf{P}_{\mathcal{E}}\{\xi(k, \rho_1) \geq C_1 D^*(k) \log k, \mu_k \geq \hat{E}_k \xi(k, \rho_1) \mid \xi(k, \rho_1) > 0\}$$
$$+ \mathbf{P}_{\mathcal{E}}\{\xi(k, \rho_1) \geq C_1 D^*(k) \log k, \mu_k < \hat{E}_k \xi(k, \rho_1) \mid \xi(k, \rho_1) > 0\}$$
$$\leq \mathbf{P}_{\mathcal{E}}\{\mu_k \geq C_1 \hat{E}_k D^*(k) \log k \mid \xi(k, \rho_1) > 0\}$$
$$+ \sum_{l=C_1 D^*(k) \log k}^{\infty} \mathbf{P}_{\mathcal{E}}\{\mu_k < \hat{E}_k l \mid \xi(k, \rho_1) = l\} \mathbf{P}_{\mathcal{E}}\{\xi(k, \rho_1) > 0\}$$
$$\leq (1 - (D^*(k))^{-1})^{C_1 \hat{E}_k D^*(k) \log k} + \exp(-C_2 D^*(k) \log k)$$

which proves (30.21).

In the following lemma we investigate the probability of the event that $\xi(k, \rho_n)$ is very small.

LEMMA 30.13

$$\mathbf{P}_{\mathcal{E}}\left\{\xi(k, \rho_n) \leq \frac{1}{4} n m_k = \frac{n}{4} \frac{E_0}{1 - E_k} \frac{1 - p(0, 1, k)}{p(0, k-1, k)}\right\} \leq C \exp\left(-\frac{n E_0}{12 D(k)}\right)$$

with some constant $C > 0$.

Proof. Let

$$\zeta_i = \begin{cases} 1 & \text{if } \max_{1 \leq j \leq \rho_{i+1} - \rho_i} R_{\rho_i + j} \geq k, \\ 0 & \text{otherwise}, \end{cases}$$
$$S = S_n = \zeta_0 + \zeta_1 + \cdots + \zeta_{n-1}.$$

Then

$$\mathbf{P}_{\mathcal{E}}\{\zeta_i = 1\} = E_0(1 - p(0, 1, k)) = p$$

and by the Bernstein inequality (Theorem 2.3)

$$\mathbf{P}_{\mathcal{E}}\left\{S \leq \frac{1}{2} np\right\} \leq C \exp\left(-\frac{np}{10}\right).$$

Let $1 \leq i_1 < i_2 < \ldots < i_S \leq n$ be the sequence of those i's for which

$$\max_{1 \leq l \leq \rho_{i+1} - \rho_1} R_{\rho_i + l} \geq k$$

AFTER THE SIXTH DAY

and let
$$\nu_j = \xi(k, \rho_{i_j+1}) - \xi(k, \rho_{i_j}) \quad (j = 1, 2, \ldots, S),$$
i.e. ν_j is the number of excursions away from k between ρ_{i_j} and ρ_{i_j+1}. Further, let ν_j^- resp. ν_j^+ be the number of the corresponding negative resp. positive excursions. Then
$$\nu_j = \nu_j^+ + \nu_j^-,$$
$$\mathbf{P}_\varepsilon\{\nu_j^- = m\} = (1-q)^{m-1}q, \quad q = (1 - E_k)p(0, k-1, k),$$
and using again the Bernstein inequality we obtain
$$\mathbf{P}_\varepsilon\left\{\nu_1^- + \nu_2^- + \cdots + \nu_s^- \leq \frac{S}{2q} \mid S\right\} \leq Ce^{-S/6}.$$

Hence
$$\mathbf{P}_\varepsilon\left\{\nu_1^- + \nu_2^- + \cdots + \nu_s^- \leq \frac{1}{4}nm_k\right\}$$
$$= \mathbf{P}_\varepsilon\left\{\nu_1^- + \nu_2^- + \cdots + \nu_s^- \leq \frac{1}{4}nm_k, S \leq \frac{1}{2}np\right\}$$
$$+ \mathbf{P}_\varepsilon\left\{\nu_1^- + \nu_2^- + \cdots + \nu_s^- \leq \frac{1}{4}nm_k, S > \frac{1}{2}np\right\}$$
$$\leq C\exp\left(-\frac{np}{10}\right) + \mathbf{P}_\varepsilon\left\{\nu_1^- + \nu_2^- + \cdots + \nu_s^- \leq \frac{1}{4}nm_k, S = \frac{1}{2}np\right\}$$
$$\leq C\exp\left(-\frac{np}{10}\right) + C\exp\left(-\frac{np}{12}\right).$$

Since $\nu_j \geq \nu_j^-$, we have the Lemma.

Now we give an upper bound for ρ_n.

THEOREM 30.3 *For any $\varepsilon > 0$ and for all but finitely many n we have*

$$\rho_n \leq 2n \sum_{k=-I(-n(\log n)^{1+\varepsilon})}^{I(n(\log n)^{1+\varepsilon})} m_k + C\log n \sum_{k=-I(n(\log n)^{1+\varepsilon})}^{I(n(\log n)^{1+\varepsilon})} D^*(k) \quad a.s. \quad (\mathbf{P}_\varepsilon) \tag{30.24}$$

where C is a big enough positive constant.

Proof. By Theorem 30.2 we have
$$\rho_n = \sum_{k=-I(-n(\log n)^{1+\varepsilon})}^{I(n(\log n)^{1+\varepsilon})} \xi(k, \rho_n).$$

Lemma 30.10 and the Borel–Cantelli lemma imply

$$\sum_{k=0}^{I(n(\log n)^{1+\varepsilon})} \xi(k,\rho_n) \leq 2n \sum_{k=0}^{I(n(\log n)^{1+\varepsilon})} m_k + C\log n \sum_{k=0}^{I(n(\log n)^{1+\varepsilon})} D^*(k).$$

Analogous inequality can be obtained for negative k's. Hence we have (30.24).

A somewhat weaker but simpler upper bound of ρ_n is given in the following:

THEOREM 30.4 *For any $\varepsilon > 0$ and for all but finitely many n we have*

$$\rho_n \leq n(\log n)^{2+\varepsilon} \sum_{k=-I(-n(\log n)^{1+\varepsilon})}^{I(n(\log n)^{1+\varepsilon})} m_k \quad a.s. \quad (\mathbf{P}_\varepsilon).$$

Proof. By (30.12), (29.23) and (C.1) of Chapter 29 for any $0 \leq k \leq I(n(\log n)^{1+\varepsilon})$, we have

$$D^*(k) = \frac{1-E_k}{E_0} D(k) m_k$$

$$\leq \frac{1-E_k}{E_0} D(I(n(\log n)^{1+\varepsilon})) m_k \leq \frac{1-\beta}{\beta} n(\log n)^{1+\varepsilon} m_k.$$

Hence

$$C\log n \sum_{k=0}^{I(n(\log n)^{1+\varepsilon})} D^*(k) \leq n(\log n)^{2+2\varepsilon} \sum_{k=0}^{I(n(\log n)^{1+\varepsilon})} m_k.$$

Since analogous inequality can be obtained for negative k's we have the Theorem.

A lower bound for ρ_n is the following:

THEOREM 30.5

$$\rho_n \geq \frac{n}{4} \max_{k \in A} m_k \quad a.s. \quad (\mathbf{P}_\varepsilon) \tag{30.25}$$

where $A = A_n = \{k : 0 < D(k) < \kappa n/\log n\}$, $\kappa < \beta/12$ and β is defined in (C.1) of Chapter 29.

Proof. Since

$$\rho_n \geq \max_k \xi(k,\rho_n),$$

(30.25) follows by Lemma 30.13.

Remark 2. Remark 1 easily implies that

$$\lim_{n\to\infty} \max_{k\in A} m_k = \infty \text{ a.s. } (\mathbf{P}_1).$$

Hence (30.25) is much stronger than the trivial inequality $\rho_n \geq 2n$.

Clearly having the upper bound (30.11) of $M(\rho_n)$ and the lower bound (30.25) of ρ_n we can obtain an upper bound of $M(n)$. Similarly having the lower bound (30.11) of $M(\rho_n)$ and the upper bound (30.24) of ρ_n a lower bound of $M(n)$ can be obtained. In fact we have

THEOREM 30.6 *Let*

$$f_\varepsilon^+(n) = \max\{I(n(\log n)^{1+\varepsilon}), I(-n(\log n)^{1+\varepsilon})\},$$
$$f_\varepsilon^-(n) = \max\{I(n(\log n)^{-1-\varepsilon}), I(-n(\log n)^{-1-\varepsilon})\},$$

$$g(n) = n(\log n)^{2+\varepsilon} \sum_{k=-I(-n(\log n)^{1+\varepsilon})}^{I(n(\log n)^{1+\varepsilon})} m_k,$$

$$h(n) = \frac{n}{4} \max_{k\in A} m_k.$$

Then for all but finitely many n

$$f_\varepsilon^-(g^{-1}(n)) \leq M(n) \leq f_\varepsilon^+(h^{-1}(n)) \quad \text{a.s.} \quad (\mathbf{P}\varepsilon) \tag{30.26}$$

where $g^{-1}(\cdot)$ *resp.* $h^{-1}(\cdot)$ *are the inverse functions of* $g(\cdot)$ *resp.* $h(\cdot)$.

Proof. Theorems 30.2 and 30.4 combined imply

$$f_\varepsilon^- \leq M(\rho_n) \leq M(g(n)),$$

which, in turn, implies the lower inequality of (30.26). Similarly by Theorems 30.2 and 30.5 we get

$$f_\varepsilon^+(n) \geq M(\rho_n) \geq M(h(n))$$

and we have the upper inequality of (30.26).

Remark 3. Note that knowing the environment $\mathcal{E}$ the lower and upper bounds of (30.26) can be evaluated.

30.4 A prediction of the physicist

Having Theorem 30.2 and Lemma 29.4 the physicist can say

$$\frac{1-\varepsilon}{2\sigma^2}\frac{\log^2 n}{\log_3 n} \leq I(n(\log n)^{-1-\varepsilon}) \leq M^+(\rho_n)$$
$$\leq I(n(\log n)^{1+\varepsilon}) \leq (\log n \log\log n \cdots \log_{p-1} n(\log_p n)^{1+\varepsilon})^2 \quad (30.27)$$

and the analogue inequalities are true for $M^-(\rho_n)$ and $M(\rho_n)$. Theorem 30.2 and Lemma 29.4 also suggest that (30.27) and the corresponding inequalities for $M^-(\rho_n)$ and $M(\rho_n)$ are the best possible ones. It is really so. In fact we have

THEOREM 30.7 (Deheuvels–Révész, 1986). *For any $\varepsilon > 0$ and $p = 1, 2, \ldots$ we have*

$$M^+(\rho_n) \leq (\log n \log\log n \cdots \log_{p-1} n(\log_p n)^{1+\varepsilon})^2$$
$$\text{a.s.}(\mathbf{P}) \text{ if } n \text{ is big enough,} \quad (30.28)$$

$$M^+(\rho_n) \geq (\log n \log\log n \cdots \log_{p-1} n \log_p n)^2 \text{ i.o. a.s. } (\mathbf{P}), \quad (30.29)$$

$$M^+(\rho_n) \leq \frac{1+\varepsilon}{2\sigma^2}\frac{\log^2 n}{\log_3 n} \text{ i.o. a.s. } (\mathbf{P}), \quad (30.30)$$

$$M^+(\rho_n) \geq \frac{1-\varepsilon}{2\sigma^2}\frac{\log^2 n}{\log_3 n} \text{ a.s. } (\mathbf{P}) \text{ if } n \text{ is big enough.} \quad (30.31)$$

The same inequalities hold for $M^-(\rho_n)$ and $M(\rho_n)$.

Proof. (30.27) gives the proofs of (30.28) and (30.31). Since by Theorem 30.2 for all but finitely many n

$$M^+(\rho_n) \geq I(n(\log n)^{-1-\varepsilon}) \quad \text{a.s.} \quad (\mathbf{P}_\varepsilon)$$

and by (29.20)

$$I(n(\log n)^{-1-\varepsilon}) \geq ((\log n - (1+2\varepsilon)\log_2 n)\log_2 n \cdots \log_{p+1} n)^2$$
$$\geq (\log n \log_2 n \cdots \log_p n)^2 \quad \text{i.o. a.s.} \quad (\mathbf{P}_1)$$

we have (30.29). Similarly, by Theorem 30.2

$$M^+(\rho_n) \leq I(n(\log n)^{1+\varepsilon}) \quad \text{a.s.} \quad (\mathbf{P}_\varepsilon),$$

and by (29.21)

$$I(n(\log n)^{1+\varepsilon}) \leq \frac{1+2\varepsilon}{2\sigma^2}\frac{\log^2 n}{\log_3 n} \quad \text{i.o. a.s.} \quad (\mathbf{P}_1),$$

hence we get (30.30). Clearly, the physicist is more interested in the behaviour of $M^+(n), M^-(n), M(n)$ than those of $M^+(\rho_n), M^-(\rho_n), M(\rho_n)$. Since $\rho_n \geq 2n$ by (30.28) and (30.30) we have

THEOREM 30.8 (Deheuvels–Révész, 1986). *For any $\varepsilon > 0$ and $p = 1, 2, \ldots$ we have*

$$M^+(n) \leq (\log n \log \log n \cdots \log_{p-1} n (\log_p n)^{1+\varepsilon})^2$$
$$a.s. \ (\mathbf{P}) \ \text{if } n \ \text{is big enough}, \tag{30.32}$$

and

$$M^+(n) \leq \frac{1+\varepsilon}{2\sigma^2} \frac{\log^2 n}{\log_3 n} \quad i.o. \ a.s. \ (\mathbf{P}). \tag{30.33}$$

The same inequalities hold for $M^-(n)$ and $M(n)$.

To get a lower bound for $M^+(n), M^-(n)$ and $M(n)$ is not so easy. However, as a consequence of Theorem 30.6 we prove

THEOREM 30.9 (Deheuvels–Révész, 1986). *For any $\varepsilon > 0$ we have*

$$M^+(n) \geq \frac{\log^2 n}{(\log \log n)^{2+\varepsilon}} \quad a.s. \ (\mathbf{P}) \tag{30.34}$$

for any $\varepsilon > 0$ and for all but finitely many n. The same inequality holds for $M^-(n)$ and $M(n)$.

Proof. Let

$$g^+(n) = n(\log n)^{2+\varepsilon} \sum_{k=0}^{I(n(\log n)^{1+\varepsilon})} m_k.$$

Then by Condition (C.1), (30.12), (29.19) and (29.6)

$$g^+(n) \leq \frac{1-\beta}{\beta} n(\log n)^{2+\varepsilon} \sum_{k=0}^{I(n(\log n)^{1+\varepsilon})} e^{-T_{k-1}}$$
$$\leq \frac{1-\beta}{\beta} n(\log n)^{2+\varepsilon} I(n(\log n)^{1+\varepsilon}) \max_{0 \leq k \leq I(n(\log n)^{1+\varepsilon})} e^{T_{k-1}}$$
$$\leq \frac{1-\beta}{\beta} n(\log n)^{2+\varepsilon} (\log n)^2 (\log_2 n)^{2+2\varepsilon}$$
$$\times \exp((1+2\varepsilon)\sigma(2(\log n)^2(\log_2 n)^{2+3\varepsilon})^{1/2}$$
$$\leq \exp(\log n (\log \log n)^{1+2\varepsilon}). \tag{30.35}$$

It can be shown similarly that for any $\varepsilon > 0$

$$g(n) \leq \exp(\log n (\log \log n)^{1+\varepsilon}) \quad \text{a.s.} \quad (\mathbf{P}_1) \tag{30.36}$$

for all but finitely many n. Consequently

$$g^{-1}(n) \geq \exp\left(\frac{\log n}{(\log \log n)^{1+\varepsilon}}\right) \quad \text{a.s.} \quad (\mathbf{P}_1) \tag{30.37}$$

if n is big enough. Hence by (29.22)

$$I(g^{-1}(n)(\log g^{-1}(n))^{-1-\varepsilon}) \geq \frac{(\log n)^2}{(\log \log n)^{2+\varepsilon}} \quad \text{a.s.} \quad (\mathbf{P}_1). \tag{30.38}$$

(30.26) and (30.38) combined imply the Theorem.

A much stronger theorem is proved by Hu and Shi (1998/B).

THEOREM 30.10 *Assume conditions* (C.1), (C.2), (C.3) *and let* $\{a_n\}$ *be a sequence of positive nondecreasing numbers. Then we have*

$$\mathbf{P}\{R_n \geq a_n (\log n)^2 \ i.o.\} = 1 \quad (\mathbf{P})$$

if and only if

$$\sum_{n=2}^{\infty} \frac{a_n}{n \log n} \exp\left(-\frac{\pi^2 \sigma^2}{8} a_n\right) = \infty,$$

and

$$\mathbf{P}\left\{M_n \leq \frac{(\log n)^2}{a_n} \ i.o.\right\} = 1 \quad (\mathbf{P})$$

if and only if

$$\sum_{n=2}^{\infty} \frac{a_n^{1/2}}{n \log n} \exp\left(-\frac{a_n}{\sigma^2}\right) = \infty,$$

and

$$\mathbf{P}\left\{M_n^+ \leq \frac{(\log n)^2}{a_n} \ i.o.\right\} = 1 \quad (\mathbf{P})$$

if and only if

$$\sum_{n=2}^{\infty} \frac{1}{n a_n^{1/2} \log n} = \infty.$$

Remark 4. In the above theorem conditions (C.1), (C.2), (C.3) can be replaced by the weaker condition:

$$\mathbf{P}\left\{\sup_{1 \leq |m| \leq n} \left|\sum_{j=1}^{m} \log\left(\frac{1-E_j}{E_j}\right) - \sigma W(m)\right| \geq C_1 \log n\right\} \leq \frac{C_2}{n^{C_3}} \tag{30.39}$$

where $0 < C_i < \infty$ $(i = 1, 2, 3)$.

Chapter 31

What Can a Physicist Say About the Local Time $\xi(0, n)$?

31.1 Two further lemmas on the environment

In this section we study a few further properties of $\mathcal{E}$. These results are simple consequences of the corresponding results of Part I.

LEMMA 31.1 *For any $0 < \varepsilon < 1$ and $0 < \delta < \varepsilon/2$ there exists a random sequence of integers $0 < n_1 = n_1(\omega_1; \varepsilon, \delta) < n_2 = n_2(\omega_1; \varepsilon, \delta) < \ldots$ such that*

$$T_{n_k} \leq -(1-\varepsilon)\sigma b_{n_k}^{-1} \quad \text{and} \quad \max_{0 \leq j \leq n_k} T_j \leq n_k^{1/2}(\log n_k)^{-\delta} \quad (31.1)$$

where $b_n = b(n) = (2n \log \log n)^{-1/2}$. Consequently by (29.5) and (29.18)

$$D(n_k) \leq \exp(n_k^{1/2}(\log n_k)^{-\delta}) \quad \text{and}$$
$$D^*(n_k) \geq \exp((1-\varepsilon)\sigma b_{n_k}^{-1}). \quad (31.2)$$

Proof. (31.1) follows from (5.11) and Invariance Principle 2 of Section 6.4.

LEMMA 31.2 *There exist two constants $C_1 > 0$, $C_2 > 0$ and a random sequence $0 < n_1 = n_1(\omega_1) < n_2 = n_2(\omega_1) < \ldots$ such that*

$$T_{n_k} \geq C_2 b_{n_k}^{-1} \quad \text{and}$$

$$\max_{0 \leq i < j \leq n_k}(T_i - T_j) \leq C_1 \left(\frac{n_k}{\log \log n_k}\right)^{1/2}. \quad (31.3)$$

Consequently by (29.5) and (29.18)

$$D(n_k) \geq \exp\left(C_2 b_{n_k}^{-1}\right) \quad \text{and}$$

$$\max_{0 \leq j \leq n_k} D^*(j) \leq \exp\left(C_1 \left(\frac{n_k}{\log \log n_k}\right)^{1/2}\right). \quad (31.4)$$

Proof. (31.3) is a simple consequence of Theorem 10.5 and Invariance Principle 2.

31.2 On the local time $\xi(0, n)$

Since $\xi(0, \rho_n) = n$ Theorem 30.4 and (30.36) imply

THEOREM 31.1 *For any $\varepsilon > 0$ we have*

$$n = \xi(0, \rho_n) \leq \xi(0, \exp(\log n (\log \log n)^{1+\varepsilon}))$$

i.e.

$$\xi(0, N) \geq \exp\left(\frac{\log N}{(\log \log N)^{1+\varepsilon}}\right) \quad a.s. \quad (\mathbf{P}) \qquad (31.5)$$

for all but finitely many N.

Now we prove that (31.5) is nearly the best possible result. In fact we have

THEOREM 31.2 *For any $\varepsilon > 0$ we have*

$$\xi(0, N) \leq \exp\left(\frac{\log N}{(\log \log N)^{1-\varepsilon}}\right) \quad i.o. \; a.s. \quad (\mathbf{P}). \qquad (31.6)$$

Proof. Define the random sequence $\{N_k\}$ as follows: let N_k be the largest integer for which

$$I(N_k(\log N_k)^{-(1+\varepsilon)}) + 1 \leq n_k$$

where n_k is the random sequence of Lemma 31.1. Then by (30.9)

$$M^+(\rho_{N_k}) \geq I(N_k(\log N_k)^{-(1+\varepsilon/2)}) \geq I(N_k(\log N_k)^{-1-\varepsilon}) + 2 \geq n_k \text{ a.s. } (\mathbf{P})$$

for all but finitely many k, i.e. $\xi(n_k, \rho_{N_k}) > 0$. That is to say, there exists a $0 < j = j(k) < N_k$ such that $\xi(n_k, (\rho_j, \rho_{j+1})) = \xi(n_k, \rho_{j+1}) - \xi(n_k, \rho_j) > 0$. Hence by (30.22)

$$\mathbf{P}_\varepsilon\{\xi(n_k, (\rho_j, \rho_{j+1})) \leq n_k^{-2} D^*(n_k)\} \leq C n_k^{-2},$$

and by (31.2) and the Borel–Cantelli lemma

$$\xi(n_k, (\rho_j, \rho_{j+1})) \geq n_k^{-2} D^*(n_k) \geq \exp((1-\varepsilon)\sigma b_{n_k}^{-1}) \quad a.s. \quad (\mathbf{P})$$

for all but finitely many k (where $j = j(k)$). Consequently

$$\rho_{N_k} \geq \rho_{j+1} - \rho_j \geq \xi(n_k, (\rho_j, \rho_{j+1})) \geq \exp((1-\varepsilon)\sigma b_{n_k}^{-1}) \quad a.s. \quad (\mathbf{P}) \quad (31.7)$$

for all k big enough. By (29.23) and (31.2)

$$N_k(\log N_k)^{-(1+\varepsilon)}$$
$$\leq D(I(N_k(\log N_k)^{-(1+\varepsilon)}) + 1) \leq D(n_k) \leq \exp(n_k^{1/2}(\log n_k)^{-\delta})$$

i.e.
$$n_k \geq \frac{1}{2}(\log N_k)^2 (\log\log N_k)^{2\delta}. \tag{31.8}$$

(31.7) and (31.8) combined imply for any $\delta^* < 1$ and for all but finitely many k

$$\rho_{N_k} \geq \exp\left(\log N_k (\log\log N_k)^{\delta^*}\right) \quad \text{a.s.} \quad (\mathbf{P}). \tag{31.9}$$

(31.9) in turn implies Theorem 31.2.

Theorems 31.1 and 31.2 have shown how small $\xi(0, N)$ can be. Essentially we found that $\xi(0, N)$ can be as small as $N^{1/\log\log N}$. In the next two theorems we investigate the question of how big $\xi(0, N)$ can be. In fact we prove

THEOREM 31.3 *There exists a $C = C(\beta) > 0$ such that*

$$\xi(0, N) \geq \exp\left(\left(1 - \frac{C}{\log_3 N}\right) \log N\right) \quad \text{i.o. a.s.} \quad (\mathbf{P})$$

where β is defined in condition (C.1).

Proof. By (30.12) we have

$$m_j = \frac{E_0}{1 - E_j} \frac{D^*(j)}{D(j)} \leq \frac{D^*(j)}{\beta}. \tag{31.10}$$

Hence by Lemma 30.10 for any $K > 0$ there exists a $C = C(K) > 0$ such that

$$\mathbf{P}_{\mathcal{E}}\{\xi(j, \rho_n) \geq CnD^*(j)\} \leq n^{-K} \quad (j = 1, 2\ldots, \ n = 1, 2, \ldots). \tag{31.11}$$

Define the random sequence $\{N_k\}$ as follows: let N_k be the smallest positive integer for which

$$I(N_k(\log N_k)^{1+\varepsilon}) \geq n_k \tag{31.12}$$

where $\{n_k\}$ is the random sequence of Lemma 31.2. Observe that by (31.4) and (29.23) for all but finitely many k

$$\exp\left(C_2 b_{n_k}^{-1}\right) \leq D(n_k) \leq D(I(N_k(\log N_k)^{1+\varepsilon})) \leq N_k(\log N_k)^{1+\varepsilon} \quad \text{a.s.} \quad (\mathbf{P}_1).$$

Hence

$$n_k \leq \frac{1}{C_2^2} \frac{(\log N_k)^2}{\log_3 N_k} \quad \text{a.s.} \quad (\mathbf{P}_1), \tag{31.13}$$

and by (30.9) for all but finitely many k

$$M^+(\rho_{N_k}) \leq I(N_k(\log N_k)^{1+\varepsilon/2}) \leq n_k \quad \text{a.s.} \quad (\mathbf{P}).$$

Consequently
$$\xi(j, \rho_{N_k}) = 0 \text{ if } j > n_k. \tag{31.14}$$

By (31.10), Lemma 30.10 and (31.13) for any $K > 0$ there exists a $C = C(K) > 0$ such that

$$\mathbf{P}_{\mathcal{E}} \left\{ \sum_{j=1}^{n_k} \xi(j, \rho_{N_k}) \geq CN_k \sum_{j=1}^{n_k} D^*(j) \right\} \leq n_k N_k^{-K} \leq \frac{1}{C_2^2} \frac{(\log N_k)^2}{\log_3 N_k} N_k^{-K}. \tag{31.15}$$

Hence by the Borel–Cantelli lemma, (31.13), (31.14), (31.15) and (31.4) we get

$$\sum_{j=1}^{\infty} \xi(j, \rho_{N_k}) = \sum_{j=1}^{n_k} \xi(j, \rho_{N_k}) \leq CN_k n_k \max_{1 \leq j \leq n_k} D^*(j)$$

$$\leq CN_k n_k \exp\left(C_1 \left(\frac{n_k}{\log \log n_k}\right)^{1/2}\right) \leq \frac{C}{C_2^2} N_k \frac{(\log N_k)^2}{\log_3 N_k}$$

$$\leq \frac{C}{C_2^2} N_k \frac{(\log N_k)^2}{\log_3 N_k} \exp\left(\frac{C_1}{C_2} \frac{\log N_k}{\log_3 N_k}\right)$$

$$\leq \exp\left(\left(1 + \frac{C}{\log_3 N_k}\right) \log N_k\right) \quad \text{a.s.} \quad (\mathbf{P})$$

for all but finitely many k. Since similar inequality can be obtained for the sum $\sum_{j=1}^{\infty} \xi(-j, \rho_{N_k})$ and for $\rho_{N_k} = \sum_{j=-\infty}^{\infty} \xi(j, \rho_{N_k})$ we have

$$\rho_{N_k} \leq \exp\left(\left(1 + \frac{C}{\log_3 N_k}\right) \log N_k\right) \quad \text{a.s.} \quad (\mathbf{P})$$

if k is big enough, which implies Theorem 31.3.

Looking through the above proofs of Theorems 31.2 and 31.3 one can realize that somewhat stronger results were proved than stated. In fact we have proved

THEOREM 31.4 *For almost all environment $\mathcal{E}$ and for all $\varepsilon > 0$ and C big enough there exist two random sequences of positive integers*

$$n_1 = n_1(\mathcal{E}, \varepsilon) < n_2 = n_2(\mathcal{E}, \varepsilon) \ldots \text{ and}$$
$$m_1 = m_1(\mathcal{E}, C) < m_2 = m_2(\mathcal{E}, C) < \ldots$$

such that

$$\xi(0, n_k) \leq \exp\left(\frac{\log n_k}{(\log \log n_k)^{1-\varepsilon}}\right)$$

and
$$\xi(0, m_k) \geq \exp\left(\left(1 - \frac{C}{\log_3 m_k}\right) \log m_k\right).$$

Remark 1. Theorems 31.1 – 31.3 are, as we call them, theorems of the physicist. However, Theorem 31.4 can be considered as a theorem of the Lord. Knowing the environment $\mathcal{E}$ the Lord can find the time-points where $\xi(0, \cdot)$ will be very big or very small while the physicist can only say that there are infinitely many points where $\xi(0, \cdot)$ takes very big resp. very small values but he does not know the location of these points.

In the last theorem of this chapter we prove that $\xi(0, n)$ cannot be very close to n, i.e. Theorem 31.4 is not far from the best possible one.

THEOREM 31.5 *For any $C > 0$ we have*
$$\xi(0, n) \leq \exp((1 - \theta_n) \log n) \quad a.s. \quad (\mathbf{P})$$
for all but finitely many n where
$$\theta_n = (\log n)^{-C \log_4 n (\log_3 n)^{-1/2}}.$$

Proof. Introduce the following notations:
$$N = N(n, \varepsilon) = [(\log n)^2 (\log_2 n)^{-(2+\varepsilon)}],$$
$$M^+(\rho_j, \rho_{j+1}) = \max_{\rho_j \leq k \leq \rho_{j+1}} R_k \quad (j = 1, 2, \ldots),$$
$$\psi^*(N) = \max\{n : 0 \leq n \leq N, T(n) \leq -\sigma b_n^{-1}\},$$
$$\xi^*(x, n) = \#\{j : 1 \leq j \leq \xi(0, n), M^+(\rho_{j-1}, \rho_j) \geq x\}.$$

Note that by Theorem 5.8 (especially Example 3) and by Invariance Principle 2 we obtain
$$\max_{0 \leq k \leq \psi^*(N)} T_k \leq \varepsilon (b(\psi^*(N)))^{-1} \quad a.s. \quad (\mathbf{P}) \quad (31.16)$$

for any $\varepsilon > 0$ and for all but finitely many n. Hence by Lemma 30.1, (29.5) and (31.16)
$$\mathbf{P}_{\mathcal{E}}\{M^+(\rho_1) \geq \psi^*(N)\} = \frac{E_0}{D(\psi^*(N))} \geq n \exp(-\varepsilon (b(\psi^*(N)))^{-1}).$$

Consequently if C is big enough then for any $n = 1, 2, \ldots$ we have
$$\mathbf{P}_{\mathcal{E}}\left\{\xi^*(\psi^*(N), n) \leq \frac{E_0}{2D(\psi^*(N))} \xi(0, n)\right\} \leq C n^{-2}$$

and by the Borel–Cantelli lemma

$$\xi^*(\psi^*(N), N) \geq \frac{E_0}{2D(\psi^*(N))}\xi(0,n) \quad \text{a.s.} \quad (\mathbf{P}) \qquad (31.17)$$

for all but finitely many n. Applying Lemma 30.1 and the definition of $\psi^*(N)$ we obtain

$$\begin{aligned}&D(\psi^*(N))p(0, \psi^*(N) - 1, \psi^*(N)) \\ &= D(\psi^*(N)) - D(\psi^*(N) - 1) = \exp(T_{\psi^*(N)-1}) \\ &\leq \exp\left(-\frac{\sigma}{2}(b(\psi^*(N) - 1))^{-1}\right).\end{aligned} \qquad (31.18)$$

Hence by Lemma 30.1, (31.17) and (31.18) we get

$$\begin{aligned}n \geq \xi(\psi^*(N), n) &\geq \frac{\xi^*(\psi^*(N), n)}{2} \frac{1 - E_{\psi^*(N)}}{p(0, \psi^*(N) - 1, \psi^*(N))} \\ &\geq \frac{E_0(1 - E_{\psi^*(N)})}{4} \frac{\xi(0,n)}{D(\psi^*(N))} \frac{1}{p(0, \psi^*(N) - 1, \psi^*(N))} \\ &\geq \frac{\beta(1-\beta)}{4}\xi(0,n)\exp\left(\frac{\sigma}{2}(b((\psi^*(N) - 1)))^{-1}\right).\end{aligned} \qquad (31.19)$$

Applying Theorem 5.3 we obtain

$$\psi^*(N) \geq N \exp\left(-2^{14}\frac{\log N \log_3 N}{(\log\log N)^{1/2}}\right) \quad \text{a.s.} \quad (\mathbf{P}) \qquad (31.20)$$

for all but finitely many N. (31.19) and (31.20) imply that with some $C > 0$

$$n \geq \frac{\beta(1-\beta)}{4}\xi(0,n)\exp\left(\frac{\sigma}{4}(b(\psi^*(N)))^{-1}\right) \geq \xi(0,n)\exp\left(\frac{\log n}{\Delta_n^C}\right) \quad \text{a.s.} \quad (\mathbf{P})$$

for all but finitely many n where

$$\Delta_n = (\log n)^{\log_4 n (\log_3 n)^{-1/2}}.$$

Hence we have Theorem 31.5.

Remark 2. Since

$$\theta_n \ll C(\log_3 n)^{-1}$$

there is an essential gap between the statements of Theorems 31.3 and 31.5. This is filled in by Hu and Shi (1998/A).

THEOREM 31.6 *Assume condition* (30.39) *and let* $\{a_n\}$ *be a nondecreasing sequence of positive numbers. Then for any* $x \in \mathbb{Z}^1$ *fixed, we have*

$$\mathbf{P}\left\{\frac{\log \xi(x,n)}{\log n} \geq 1 - \frac{1}{a_n} \text{ i.o.}\right\} = 1$$

if and only if

$$\sum_{n=3}^{\infty} \frac{1}{na_n^2 \log n} = \infty,$$

and

$$\mathbf{P}\left\{\frac{\log \xi(x,n)}{\log n} \leq b_n \text{ i.o.}\right\} = 1$$

if and only if

$$\sum_{n=3}^{\infty} \frac{1}{nb_n \log n} = \infty.$$

They also proved

THEOREM 31.7 *Assume condition* (30.39). *Then for any* $x \in \mathbb{Z}^1$ *fixed, we have*

$$\lim_{n \to \infty} \mathbf{P}\left\{\frac{\log \xi(x,n)}{\log n} < y\right\} = \mathbf{P}\{\min(U_1, U_2) < y\}$$

where U_1 *and* U_2 *are independent r.v.'s uniformly distributed in* $(0,1)$.

Chapter 32

On the Favourite Value of the RWIRE

In this chapter we investigate the properties of the sequence $\xi(n) = \max_k \xi(k,n)$
A trivial result can be obtained as a

Consequence of (30.32). For any $\varepsilon > 0$ we have

$$\lim_{n\to\infty} \frac{(\log n)^2 (\log \log n)^{2+\varepsilon}}{n} \xi(n) = \infty \quad \text{a.s.} \quad (\mathbf{P}). \tag{32.1}$$

We also get

Consequence of (30.33).

$$\limsup_{n\to\infty} \frac{(\log n)^2}{(2\sigma^2 \log_3 n)n} \xi(n) \geq 1 \quad \text{a.s.} \quad (\mathbf{P}). \tag{32.2}$$

It looks obvious that much stronger results than those of (32.1) and (32.2) should exist. In fact we prove in the next theorem that (under some extra condition on $\mathcal{E}$)

$$\limsup_{n\to\infty} \frac{\xi(n)}{n} > 0 \quad \text{a.s.} \quad (\mathbf{P}).$$

THEOREM 32.1 (Révész, 1988). *Assume that*

$$\mathbf{P}_1\{E_i = p\} = \mathbf{P}_1\{E_i = 1 - p\} = \frac{1}{2} \quad (0 < p < 1/2). \tag{32.3}$$

Then there exists a constant $g = g(p) > 0$ such that

$$\limsup_{n\to\infty} \frac{\xi(n)}{n} \geq g(p) \quad a.s. \quad (\mathbf{P}).$$

Remark 1. Very likely Theorem 32.1 remains true replacing condition (32.3) by the usual conditions (C.1), (C.2), (C.3). Note that (32.3) implies (C.1), (C.2) and (C.3).

In order to prove this theorem at first we introduce a few notations. Let N be a positive integer and define the random variables on Ω_1:

$$\rho_1^+(N) = \min\{k: \ k > 0, \ T_k = \Delta N\},$$
$$\rho_1^-(N) = \min\{k: \ k > 0, \ T_{-k} = -\Delta N\},$$
$$\mu_N^+ \Delta = -\min\{T_k: \ 0 \le k \le \rho_1^+(N)\},$$
$$\mu_N^- \Delta = \max\{T_{-k}: \ 0 \le k \le \rho_1^-(N)\},$$
$$\mu_N \Delta = \max\{\mu_N^+ \Delta, \mu_N^- \Delta\},$$

where
$$\Delta = \log \frac{1-p}{p}.$$

For the sake of simplicity from now on we assume that $\mu_N^+ \ge \mu_N^-$. Continue the notations as follows:

$$\alpha_N = \max\{k: \ 0 \le k \le \rho_1^+(N), \ T_k = -\mu_N \Delta\},$$
$$\tau_N^- = \begin{cases} \max\{k: \ 0 \le k \le \alpha_N, \ T_k + \mu_N \Delta = \Delta N\} \\ \quad \text{if such a } k \text{ exists,} \\ \max\{k: \ k \le 0, \ -T_k + \mu_N \Delta = \Delta N\} \text{ otherwise,} \end{cases}$$
$$\tau_N^+ = \min\{k: \ \alpha_N \le k \le \rho_1^+(N), \ T_k + \mu_N \Delta = \Delta N\},$$
$$L_N(j) = L(-\Delta(\mu_N - j), (\tau_N^-, \tau_N^+))$$
$$= \#\{k: \ \tau_N^- \le k \le \tau_N^+, \ T_k = -\mu_n \Delta + j\Delta\},$$
$$\Delta \mathcal{F}^+(N) = \max\{T_j - T_i: \ \tau_N^- \le i < j \le \alpha_N\},$$
$$\Delta \mathcal{F}^-(N) = \max\{T_i - T_j: \ \alpha_N \le i < j \le \tau_N^+\};$$

and on Ω:
$$F_N = \min\{k: \ k > 0, \ R_k = \alpha_N\},$$
$$G_N = \min\{k: \ k > 0, \ R_k = \rho_1^-(N)\},$$
$$H_N = \min\{k: \ k > F_N, \ R_k = \tau_N^- \text{ or } \tau_N^+\} - F_N.$$

For the sake of simplicity from now on we assume that $\tau_N^- \le 0$.

The above notations can be seen in Figure, where instead of the process T_k the process
$$\bar{T}_k = \begin{cases} T_k & \text{if } k \ge 0, \\ -T_k & \text{if } k \le 0 \end{cases}$$
is shown.

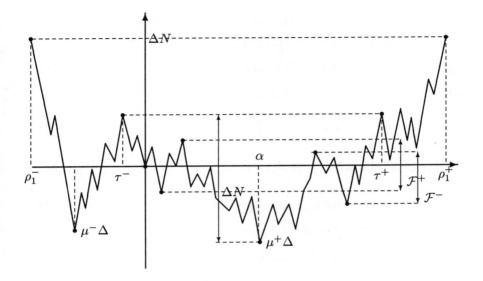

Figure

Now we present a few simple lemmas.

LEMMA 32.1 *There exists an absolute constant θ $(0 < \theta < 1)$ such that*

$$\mathbf{P}_1\{L_N(j) \leq 6j^2 + 4j + 2 \ (j = 0, 1, \ldots, N-1), A\} \geq \theta$$

where

$$A = A_N = \left\{\mathcal{F}^+(N) \leq \frac{N}{2}, \mathcal{F}^-(N) \leq \frac{N}{2}\right\}.$$

Proof. Consequence 1 of Section 12.4 easily implies that

$$\mathbf{P}_1\{L_N(j) \leq 6j^2 + 4j + 2, \ (j = 0, 1, \ldots, N-1)\} \qquad (32.4)$$

is larger than an absolute positive constant independent from N. It is easy to see that

$$\mathbf{P}_1\{A\} = \mathbf{P}_1\left\{\mathcal{F}^+(N) \leq \frac{N}{2}, \mathcal{F}^-(N) \leq \frac{N}{2}\right\} \qquad (32.5)$$

is larger than an absolute positive constant independent from N (cf. Consequence 4 of Section 10.2) and the events involved in (32.4) and (32.5) are asymptotically independent as $N \to \infty$. Hence we have Lemma 32.1.

Let $N = N_k(\mathcal{E})$ be a sequence of positive integers for which
$$L_N(j) \leq 6j^2 + 4j + 2 \quad (j = 0, 1, \ldots, N-1), \quad \text{and} \quad A \quad \text{holds}$$
(by Lemma 32.1 for almost all $\mathcal{E}$ there exists such an infinite sequence).

LEMMA 32.2 *For almost all $\mathcal{E}$ and for any $\varepsilon > 0$ we have*

$$\xi(0, F_N) \leq \exp\left(\frac{(1+\varepsilon)N}{2}\right), \quad G_N \geq \exp((1-\varepsilon)N), \tag{32.6}$$

$$H_N \geq \xi(\alpha_N, H_N + F_N) \geq \exp((1-\varepsilon)N), \quad F_N \leq \exp\left(\frac{(1+\varepsilon)N}{2}\right), \tag{32.7}$$

$$\max_{0 \leq j \leq \alpha_N} \xi(j, F_N) \leq \exp\left(\frac{(1+\varepsilon)N}{2}\right) \tag{32.8}$$

$(N = N_k)$ *a.s.* $(\mathbf{P}_\mathcal{E})$ *for all but finitely many k and*

$$F_N = o(G_N) \quad a.s. \quad (\mathbf{P}_\mathcal{E}). \tag{32.9}$$

Proof. By Lemma 30.1, (29.5) and the definition of $N = N_k$ we have

$$\mathbf{P}_\mathcal{E}\{M^+(\rho_1) \geq \alpha_N\} = E_0(1 - p(0, 1, \alpha_N)) = \frac{E_0}{D(\alpha_N)}$$
$$\geq \frac{E_0}{\alpha_N} \exp(-\max_{0 \leq k \leq \alpha_N - 1} T_k) \geq \frac{E_0}{\alpha_N} \exp(-\mathcal{F}^+(N))$$
$$\geq \frac{E_0}{\alpha_N} \exp\left(-\frac{N}{2}\right)$$

and

$$\mathbf{P}_\mathcal{E}\{M^-(\rho_1) \geq \rho_1^-(N)\} = (1 - E_0)p(\rho_1^-(N), -1, 0)$$
$$\leq (1 - E_0)\exp(-\max_{\rho_1^-(N) \leq k \leq 0} T_k) = (1 - E_0)e^{-N}$$

(cf. the notations in Section 30.2). Hence by the Borel–Cantelli lemma we easily obtain (32.6). The above two inequalities also imply that more excursions are required to arrive at $\rho_1^-(N)$ than at α_N. Hence we get (32.9). The first inequality of (32.7) and (32.8) can be obtained similarly. In order to prove the second inequality of (32.7) observe that by (32.8) and (32.9)

$$F_N = \sum_{j=-\infty}^{\alpha_N - 1} \xi(j, F_N) = \sum_{j=\rho_1^-(N)}^{\alpha_N - 1} \xi(j, F_N)$$
$$\leq (\alpha_N - 1 - \rho_1^-(N))\exp\left((1+\varepsilon)\frac{N}{2}\right) \quad a.s. \quad (\mathbf{P}_\mathcal{E})$$

for any $\varepsilon > 0$ and for all but finitely many k. Hence the lemma is proved.

Introduce the following further notations:

$$\hat{\rho}_1 = \hat{\rho}_1(\alpha_N) = \min\{n : n > 0, R_{F_N+n} = \alpha_N\},$$
$$\hat{\rho}_2 = \hat{\rho}_2(\alpha_N) = \min\{n : n > \hat{\rho}_1, R_{F_N+n} = \alpha_N\},\ldots$$
$$\hat{\xi}(j,\hat{\rho}_n,N) = \hat{\xi}(j,\hat{\rho}_n) = \xi(j,F_N+\hat{\rho}_n) - \xi(j,F_N),$$
$$\hat{D}(j,N) = (p(j,\alpha_N-1,\alpha_N))^{-1} = \frac{D(j,\alpha_N)}{D(j,\alpha_N) - D(j,\alpha_N-1)}$$
$$= 1 + \sum_{k=j}^{\alpha_N-2} \exp(-(T_{\alpha_N-1} - T_k))$$

and

$$\hat{D}^*(j,N) = (1 - p(j,j+1,\alpha_N))^{-1} = D(j,\alpha_N).$$

Observe that

$$\frac{\hat{D}^*(j,N)}{\hat{D}(j,N)} = \frac{p(j,\alpha_N-1,\alpha_N)}{1 - p(j,j+1,\alpha_N)} = D(j,\alpha_N) - D(j,\alpha_N-1)$$
$$= U_{j+1}U_{j+2}\cdots U_{\alpha_N-1} = \exp(T_{\alpha_N-1} - T_j).$$

Clearly Lemmas 30.11 and 30.10 can be reformulated as follows:

LEMMA 32.3 *For any $j < \alpha_n$ we have*

$$\mathbf{P}_{\mathcal{E}}\left\{\hat{\xi}(j,\hat{\rho}_n) \geq C_1 n \frac{\hat{D}^*(j,N)}{\hat{D}(j,N)}\right\} \leq \exp\left(-\frac{n}{\hat{D}(j,n)}\right) \quad (32.10)$$

where $C_1 > 2p^{-1}$. Further, for any $K > 0$ there exists a $C = C(K) > 0$ such that

$$\mathbf{P}_{\mathcal{E}}\left\{\hat{\xi}(j,\hat{\rho}_n) \geq 2n\frac{1 - E_{\alpha_N}}{E_j}\frac{\hat{D}^*(j,N)}{\hat{D}(j,N)} + C\hat{D}^*(j,N)\log n\right\} \leq n^{-K}. \quad (32.11)$$

Proof of Theorem 32.1. In order to simplify the notations from now on we assume that $\tau_N^- > 0$. The case $\tau_N^- \leq 0$ can be treated similarly.

Let $1/2 + \varepsilon < \psi_1 \Delta < \psi_2 < 1 - \varepsilon$ with some $\varepsilon > 0$ and introduce the following notations:

$$n = [\exp(\psi_2 N)] \quad \text{where} \quad N = N_k = N_k(\mathcal{E})$$

and

$$\chi(j) = \min\{k : \tau_N^- < k < \alpha_N, T_k = -\mu_N \Delta + j\Delta\}.$$

Consider any integer $l \in (\chi(\psi_1 N), \alpha_N)$. Then
$$\hat{D}(l,N) \leq \alpha_N \exp(\max_{\chi(\psi_1 N) \leq j \leq \alpha_N}(T_j - T_{\alpha_N - 1})) \leq 2\alpha_N \exp(\psi_1 \Delta N)$$
a.s. ($\mathbf{P}_\mathcal{E}$) for all but finitely many k and by (32.10)

$$\mathbf{P}_\mathcal{E}\left\{\hat{\xi}(l,\hat{\rho}_n) \geq C_1 n \frac{\hat{D}^*(l,N)}{\hat{D}(l,N)}\right\} \leq \exp\left(-\frac{n}{\hat{D}(l,N)}\right)$$
$$\leq \exp\left(-\frac{n}{2\alpha_N \exp(\psi_1 \Delta N)}\right)$$
$$\leq \exp\left(-\exp\left(\frac{\psi_2 - \psi_1 \Delta}{2}N\right)\right).$$

The last inequality of the above inequalities follows from the fact that $\alpha_N < \sqrt{N}$. Consequently by (32.9) and Lemma 32.1

$$\sum_{j=\chi(\psi_1 N)}^{\alpha_N - 1} \hat{\xi}(j, \hat{\rho}_n)$$
$$\leq C_1 n \sum_{j=\chi(\psi_1 N)}^{\alpha_N - 1} \frac{\hat{D}^*(j,N)}{\hat{D}(j,N)} = C_1 n \sum_{j=\chi(\psi_1 N)}^{\alpha_N - 1} \exp(T(\alpha_N - 1) - T(j))$$
$$\leq C_1 n \sum_{l=0}^{\infty}(4l+2)^2 \exp(-l\Delta) = f(\Delta)n \quad \text{a.s.} \quad (\mathbf{P}) \qquad (32.12)$$

for all but finitely many k.

Let $l \in (\tau_N^-, \chi(\psi_1 N))$. Then by (32.11)

$$\mathbf{P}_\mathcal{E}\left\{\hat{\xi}(l,\hat{\rho}_n) \geq 2n\frac{1-E_{\alpha_N}}{E_l}\frac{\hat{D}^*(l,N)}{\hat{D}(l,N)} + C\hat{D}^*(l,N)\log n\right\} \leq n^{-K}.$$

Consequently

$$\sum_{l=\tau_n^-}^{\chi(\psi_1 N)} \hat{\xi}(l,\hat{\rho}_n) \leq 2(1-E_{\alpha_N})n \sum_{l=\tau_N^-}^{\chi(\psi_1 N)} \frac{\hat{D}^*(l,N)}{E_l \hat{D}(l,N)} + C\log n \sum_{l=\tau_N^-}^{\chi(\psi_1 N)} \hat{D}^*(l,N)$$
$$\leq 2\frac{1-p}{p}n\alpha_N \exp(-\psi_1 \Delta N) + C(\log n)\alpha_N \exp\left(\frac{N}{2}\right) = o(n). \qquad (32.13)$$

(32.12) and (32.13) combined imply

$$\sum_{l=\tau_N^-}^{\alpha_N - 1} \hat{\xi}(l,\hat{\rho}_n) \leq 2f(\Delta)n \quad \text{a.s.} \quad (\mathbf{P}) \qquad (32.14)$$

for all but finitely many k. Similarly one can see that

$$\sum_{l=\alpha_N+1}^{\tau_N^+} \hat\xi(l,\hat\rho_n) \leq 2f(\Delta)n.$$

Hence by (32.7)

$$\hat\rho_n = \sum_{l=-\infty}^{+\infty} \hat\xi(l,\hat\rho_n) = \sum_{\tau_N^-}^{\tau_N^+} \hat\xi(l,\hat\rho_n) \leq 4f(\Delta)n.$$

Let $m = 4f(\Delta)n$. Then applying again (32.7) we get

$$\xi((1+\varepsilon)m) \geq \xi(F_N + m) \geq \xi(F_N + \hat\rho_n) \geq \xi(\alpha_N, F_N + \hat\rho_n) = n = \frac{m}{4f(\Delta)}$$

which proves the Theorem.

Note that we have proved a stronger result than Theorem 32.1. In fact we have

THEOREM 32.2 *For almost all environment there exists a sequence of positive integers $n_1 = n_1(\mathcal{E}) < n_2 = n_2(\mathcal{E}) < \ldots$ such that*

$$\xi(n_k) \geq \frac{n_k}{4f(\Delta)}$$

provided that the condition of Theorem 32.1 is fulfilled.

Remark 2. On the connection of Theorems 32.1 and 32.2 the message of Remark 1 of Section 31.2 can be repeated here as well.

Another simple consequence of the proof of Theorem 32.1 is

THEOREM 32.3 *Assume that the condition of Theorem 32.1 is fulfilled. Then there exists an $\varepsilon = \varepsilon(p) > 0$ such that*

$$\liminf_{n\to\infty} \mathbf{P}\left\{\frac{\xi(n)}{n} \geq \varepsilon\right\} > 0.$$

On the liminf behaviour of $\xi(n)$ we present only a

Conjecture 1.

$$\liminf_{n\to\infty} \frac{\xi(n)}{n} \log\log n = 0 \quad \text{a.s.} \quad \mathbf{P}$$

and

$$\liminf_{n\to\infty} \frac{\xi(n)}{n} (\log\log n)^3 = \infty \quad \text{a.s.} \quad \mathbf{P}.$$

Conjecture 2. For any $\varepsilon > 0$ there exists a $K = K(\varepsilon) > 0$ such that

$$\limsup_{n\to\infty} \max_{x\in\mathbb{Z}^1} \frac{1}{n} \sum_{y=x-K}^{x+K} \xi(y,n) \geq 1 - \varepsilon \quad \text{a.s.}$$

Chapter 33

A Few Further Problems

33.1 Two theorems of Golosov

Theorems 30.8 and 30.9 claim that $M(n) \sim (\log n)^2$. As we have already mentioned, this fact was observed first by Sinai (1982). The result of Sinai suggested to Golosov to investigate the limit distributions of the sequences

$$R_n^* = \frac{R_n}{\sigma^2 (\log n)^2}$$

and

$$M_n^* = \frac{M_n}{\sigma^2 (\log n)^2}$$

(for the definition of σ^2, cf. (C.3) of Chapter 29). In order to study the limit distributions of R_n^* and M_n^* he modified a bit the original model. In fact he assumed that $E_0 = 1$, i.e. the random walk is concentrated on the positive half-line. Having this modified model he proved that the limit distributions of the sequences $\{R_n^*\}$ and $\{M_n^*\}$ existed and he evaluated those. In fact we have

THEOREM 33.1 (Golosov, 1983). *For any $u > 0$*

$$\lim_{n \to \infty} \mathbf{P}\{R_n^* < u\} = \frac{1}{2} \int_0^u h_2(v) dv \qquad (33.1)$$

where

$$h_2(v) = \sum_{n=0}^{\infty} \exp\left(-\frac{1}{2} z_n^2 v\right) \quad \text{and} \quad z_n = \frac{(2n+1)\pi}{2},$$

and

$$\lim_{n \to \infty} \mathbf{P}\{M_n^* < u\} = \int_0^u h_1(v) dv \qquad (33.2)$$

where

$$h_1(v) = \sum_{n=0}^{\infty} (-1)^n z_n \exp\left(-\frac{1}{2} z_n^2 v\right).$$

Considering the original model Kesten (1986) proved

THEOREM 33.2

$$\lim_{n\to\infty} \mathbf{P}\{R_n^* < u\} = \frac{2}{\pi} \int_{-\infty}^{u} h_3(v) dv$$

where

$$h_3(v) = \sum_{n=0}^{\infty} \frac{(-1)^n}{2n+1} \exp\left(-\frac{(2n+1)^2 \pi^2}{8}|v|\right).$$

Sinai (1982) also proved that there exists a sequence of random variables $\alpha_1, \alpha_2, \ldots$ defined on Ω_1 such that $R_n - \alpha_n = o((\log n)^2)$ in probability (**P**). This means that knowing the environment $\mathcal{E}$ we can evaluate the sequence $\{\alpha_n\}$ and having the sequence $\{\alpha_n\}$ we can get a much better estimate of the location of the particle R_n than that of Theorem 30.6. Golosov proved a much stronger theorem. His result claims that $R_n - \alpha_n$ has a limit distribution (without any normalising factor), which means that knowing α_n the location of the particle can be predicted with a finite error term with a big probability. The model used by Golosov is a little bit different from the one discussed up to now. He considered a random walk on the right half-line only and he assumed that the particle can stay where it is located. His model can be formulated as follows.

Let $\mathcal{E} = \{p_{-1}(n), p_0(n), p_1(n), \}$ $(n = 0, 1, 2, \ldots)$ be a sequence of independent random three-dimensional vectors whose components are non-negative and $p_{-1}(0) = 0$, $p_{-1}(n) + p_0(n) + p_1(n) = 1$ $(n = 0, 1, 2 \ldots)$. Assume further that

(i) $(p_{-1}(n), p_1(n))$ $(n = 1, 2 \ldots)$ are identically distributed,

(ii) $p_0(n)$ $(n = 0, 1, 2, \ldots)$ are identically distributed,

(iii) the sequences $\{p_0(n), n = 0, 1, 2, \ldots\}$ and $\{p_{-1}(n)/p_1(n), n = 1, 2, \ldots\}$ are independent,

(iv) $\mathbf{E} \log(p_{-1}(n)/p_1(n)) = 0$ and $0 < \mathbf{E}(\log(p_{-1}(n)/p_1(n)))^2 = \sigma^2 < \infty$,

(v) $\mathbf{E}(1 - p_0(n))^{-1} < \infty$ and $\mathbf{P}(p_0(n) > 0) > 0$.

Having the environment $\mathcal{E}$ we define the random walk $\{R_n\}$ by $R_0 = 0$ and

$$\mathbf{P}_{\mathcal{E}}\{R_{n+1} = i + \theta \mid R_n = i, R_{n-1}, \ldots, R_0\} = \begin{cases} p_{-1}(i) & \text{if } \theta = -1, \\ p_0(i) & \text{if } \theta = 0, \\ p_1(i) & \text{if } \theta = 1. \end{cases}$$

Then we have

THEOREM 33.3 (Golosov, 1984). *There exists a random sequence $\{\alpha_n\}$ defined on Ω_1 such that for any $-\infty < y < \infty$*

$$\lim_{n\to\infty} \mathbf{P}\{R_n - \alpha_n \leq y\} = F(y)$$

where the exact form of the distribution function $F(y)$ is unknown.

Remark 1. Clearly Theorems 33.1 and 33.2 can be considered as theorems of the physicist. However, Theorem 33.3 is a theorem of a mixed type. The physicist knows about the existence of α_n but he cannot evaluate it. The Lord can evaluate α_n but He cannot use His further information on $\mathcal{E}$. In fact, He would like to evaluate the distribution $\mathbf{P}_\mathcal{E}\{R_n - \alpha_n \leq y\}$. It is not clear at all whether the $\lim_{n\to\infty} \mathbf{P}_\mathcal{E}\{R_n - \alpha_n \leq y\}$ exists for any given $\mathcal{E}$.

Remark 2. Theorems 32.1 and 32.2 also suggest that R_n should be close to α_n.

33.2 Non-nearest-neighbour random walk

The model studied in the previous chapters of this Part is a nearest-neighbour model, i.e. the particle moves in one step to one of its neighbours. In the last model of Golosov the particle keeps its place or moves to one of its neighbours. In a non-nearest-neighbour model the particle can move farther. Such a model can be formulated as follows.

Let $\mathcal{E} = \{p_{-1}(n), p_1(n), p_2(n)\}$ $(n = 0, \pm 1, \pm 2, \ldots)$ be a sequence of independent, identically distributed three-dimensional random vectors whose components are non-negative and $p_{-1}(n) + p_1(n) + p_2(n) = 1$ $(n = 0, \pm 1, \ldots)$. Then we define a random walk $\{R_n\}$ by $R_0 = 0$ and

$$\mathbf{P}_\mathcal{E}\{R_{n+1} = i + \theta \mid R_n = i, R_{n-1}, \ldots, R_0\} = \begin{cases} p_{-1}(i) & \text{if } \theta = -1, \\ p_1(i) & \text{if } \theta = 1, \\ p_2(i) & \text{if } \theta = 2. \end{cases}$$

Studying the properties of $\{R_n\}$ is much, much harder than in the nearest-neighbour case. Even the question of the recurrence is very hard. In fact, the question is to find the necessary and sufficient condition for the distribution function

$$\mathbf{P}_1\{p_{-1}(i) < u_{-1}, p_1(i) < u_1, p_2(i) < u_2\} = F(u_{-1}, u_1, u_2)$$

which guarantees that

$$\mathbf{P}\{R_n = 0 \text{ i.o.}\} = 1. \tag{33.3}$$

This question was studied in a more general form by Key (1984), who in the above formulated case obtained the required condition.

THEOREM 33.4 (Key, 1984). *Let*

$$\alpha = p_1(0) + p_2(0) + ((p_1(0) + p_2(0))^2 + 4p_{-1}(0)p_2(0))^{1/2}(2p_{-1}(0))^{-1}$$

and

$$m = \mathbf{E}\left(\log\left(-\alpha - 1 + \frac{1}{p_{-1}(0)}\right)\right).$$

Then

$$\begin{aligned}
\mathbf{P}\{R_n = 0 \text{ i.o.}\} = 1 & \quad \text{if} \quad m = 0, \\
\mathbf{P}\{\lim_{n\to\infty} R_n = \infty\} = 1 & \quad \text{if} \quad m > 0, \\
\mathbf{P}\{\lim_{n\to\infty} R_n = -\infty\} = 1 & \quad \text{if} \quad m < 0.
\end{aligned}$$

Remark 1. Clearly this Theorem gives the necessary and sufficient condition of (33.3) if the expectation m exists. If m does not exist then the necessary and sufficient condition is unknown, just as in the nearest-neighbour case.

Remark 2. The general non-nearest-neighbour case (i.e. when the environment $\mathcal{E}$ is defined by an i.i.d. sequence

$$\{p_{-L}(n), p_{-L+1}(n), \ldots, p_R(n)\}$$

where L and R are positive integers) was also investigated by Key. However, he cannot obtain an explicit condition for (33.3), but he proves a general zero-one law which implies that

$$\mathbf{P}\{R_n = 0 \text{ i.o.}\} = 0 \quad \text{or} \quad 1.$$

His zero-one law was generalized by Andjel (1988).

33.3 RWIRE in $\mathbb{Z}^d$

The model of the RWIRE can be trivially extended to the multivariate case. For the sake of simplicity here, we formulate the model in the case $d = 2$. Let $U_{ij} = (U_{ij}^{(1)}, U_{ij}^{(2)}, U_{ij}^{(3)}, U_{ij}^{(4)})$ $(i, j = 0, \pm 1, \pm 2, \ldots)$ be an array of i.i.d.r.v.'s with $U_{ij}^{(k)} \geq 0$, $U_{ij}^{(1)} + U_{ij}^{(2)} + U_{ij}^{(3)} + U_{ij}^{(4)} = 1$. The array $\mathcal{E} = \{U_{ij}, i, j = 0, \pm 1, \pm 2, \ldots\}$ is called a two-dimensional random environment. Having an environment $\mathcal{E}$ a random walk $\{R_n, n = 0, 1, 2, \ldots\}$ can be defined by $R_0 = 0$ and

$$\mathbf{P}\{R_{n+1} = (i+1, j) \mid R_n = (i, j), R_{n-1}, \ldots, R_0\} = U_{ij}^{(1)},$$
$$\mathbf{P}\{R_{n+1} = (i, j+1) \mid R_n = (i, j), R_{n-1}, \ldots, R_0\} = U_{ij}^{(2)},$$

$$\mathbf{P}\{R_{n+1} = (i-1, j) \mid R_n = (i,j), R_{n-1}, \ldots, R_0\} = U_{ij}^{(3)},$$
$$\mathbf{P}\{R_{n+1} = (i, j-1) \mid R_n = (i,j), R_{n-1}, \ldots, R_0\} = U_{ij}^{(4)}.$$

No non-trivial, sufficient condition is known for the recurrence

$$\mathbf{P}\{R_n = 0 \text{ i.o.}\} = 1$$

in the case $d \geq 2$. Kalikow (1981) gave necessary conditions. In fact, he gave a class of environments where $\mathbf{P}\{R_n = 0 \text{ i.o.}\} = 0$. As a consequence of his result he proves

THEOREM 33.5 *Define the environment $\mathcal{E}$ by*

$$\mathbf{P}_1\left\{(U_{ij}^{(1)}, U_{ij}^{(2)}, U_{ij}^{(3)}, U_{ij}^{(4)}) = (a_1, b_1, c_1, d_1)\right\} = p$$

and

$$\mathbf{P}_1\left\{(U_{ij}^{(1)}, U_{ij}^{(2)}, U_{ij}^{(3)}, U_{ij}^{(4)}) = (a_2, b_2, c_2, d_2)\right\} = 1 - p.$$

Assume that

$$\frac{p(a_1 - c_1)}{(1-p)(c_2 - a_2)} > \max\left(\frac{a_1}{a_2}, \frac{b_1}{b_2}, \frac{c_1}{c_2}, \frac{d_1}{d_2}\right).$$

Then

$$\mathbf{P}\{R_n = 0 \text{ i.o. }\} = 0,$$

moreover

$$\lim_{n \to \infty} R_n^{(1)} = \infty \quad \text{a.s.} \quad (\mathbf{P})$$

where $R_n^{(1)}$ is the first coordinate of R_n.

Kalikow also proves a zero-one law, i.e. he can prove under some regularity conditions that $\mathbf{P}\{R_n = 0 \text{ i.o.}\} = 0$ or 1. This zero-one law was extended by Andjel (1988).

Kalikow also formulated some unsolved problems. Here we quote two of them.

Problem 1. Is every three-dimensional RWIRE transient?

Problem 2. Let $0 < p < 1/2$ and define the random environment $\mathcal{E}$ by

$$\mathbf{P}_1\left\{(U_{ij}^{(1)}, U_{ij}^{(2)}, U_{ij}^{(3)}, U_{ij}^{(4)}) = \left(p, \frac{1}{4}, \frac{1}{2} - p, \frac{1}{4}\right)\right\}$$
$$= \mathbf{P}\left\{(U_{ij}^{(1)}, U_{ij}^{(2)}, U_{ij}^{(3)} U_{ij}^{(4)}) = \left(\frac{1}{2} - p, \frac{1}{4}, p, \frac{1}{4}\right)\right\} = \frac{1}{2}.$$

Is this RWIRE recurrent?

33.4 Non-independent environments

In Chapter 28 we mentioned the magnetic fields as possible applications of the RWIRE. However, up to now it was assumed that the environment $\mathcal{E}$ consists of i.i.d.r.v.'s. Clearly the condition of independence does not meet with the properties of the magnetic fields and most of the possible physical applications. In most cases it can be assumed that the environment is a stationary field. A lot of papers are devoted to studying the properties of the RWIRE in case of a stationary environment $\mathcal{E}$.

In the multivariate case it turns out that having some natural conditions on the stationary environment $\mathcal{E}$ (which exclude the case of independent environments) one can prove the recurrence and a central limit theorem with a normalizing factor $(\log n)^2$.

33.5 Random walk in random scenery

Let $\zeta = \{\zeta_i = \zeta(i),\ i = 0, \pm 1, \pm 2, \ldots\}$ be a sequence of i.i.d.r.v.'s with

$$\mathbf{E}\zeta_i = 0, \quad \mathbf{E}\zeta_i^2 = 1, \quad \mathbf{E}(\exp t\zeta_i) < \infty$$

for some $|t| < t_0$ ($t_0 > 0$). ζ is called random scenery. Further, let $\{S_k\}$ (independent from $\{\zeta_k\}$) be a simple symmetric random walk. Kesten and Spitzer (1979) were interested in the sum

$$K_n = n^{-3/4} \sum_{k=0}^{n} \zeta(S_k).$$

If the particle has to pay ζ_i guilders whenever it visits i, then the amount paid by the particle during the first n steps of the random walk is $n^{3/4} K_n$. Clearly

$$K_n = n^{-3/4} \sum_{k=-\infty}^{+\infty} \zeta_k \xi(k, n)$$

where $\xi(\cdot, \cdot)$ is the local time of $\{S_k\}$.

Studying the sequence $\{K_n\}$ Kesten and Spitzer are arguing heuristically as follows: let $\sum_{k=0}^{n} \zeta_k = L_n$ ($n = 0, \pm 1, \pm 2, \ldots$) then one can define independent Wiener processes $\{W_1(n)\ (n = 0, 1, \ldots)\}$ and $\{W_2(n)\ (n = 0, \pm 1, \ldots)\}$ such that $W_2(n)$ should be near enough to L_n and simultaneously $\xi(k, n)$ should be near to the local time $\eta_1(k, n)$ of the Wiener process $W_1(\cdot)$. Hence

$$K_n \sim n^{-3/4} \sum_{k=-\infty}^{\infty} (W_2(k+1) - W_2(k)) \eta_1(k, n)$$

$$\sim n^{-3/4} \int_{-\infty}^{+\infty} \eta_1(x,n) dW_2(x). \tag{33.4}$$

Since it is not very hard to prove that $n^{-3/4}\int_{-\infty}^{+\infty}\eta_1(x,n)dW_2(x)$ has a limit distribution, the above heuristic approach suggests that K_n has a limit distribution.

Applying Invariance Principle 2. (Section 6.2) and Theorem 10.1 it is not hard to get a precise form of (33.4).

We note that Kesten and Spitzer investigated a much more general situation than the above one and they initiated an extended research of random sceneries. They conjectured that in the case when $\{S_k\}$ is a simple random walk on the plane then $n^{3/4}K_n$ can be approximated by a Wiener process. This conjecture was proved by Bolthausen (1989) (see also Borodin, 1980). The strongest form of this statement is

THEOREM 33.6 (Csáki–Révész–Shi, 2001) *Let $d = 2$ and assume that $\mathbf{E}|\zeta_i|^q < \infty$ for some $q > 2$. Possibly in an enlarged probability space, there exists a version of K_n and a standard one-dimensional Wiener process $\{W(t), t \geq 0\}$ such that for any $\varepsilon > 0$ as $n \to \infty$*

$$\left|\sum_{k=0}^{n} \zeta(S_k) - (2/\pi)^{1/2} W(n \log n)\right| = o(n^{1/2}(\log n)^{3/8}) \quad a.s.$$

A similar result can be obtained in case $d \geq 3$.

THEOREM 33.7 (Révész–Shi, 2000) *Let $d \geq 3$ and assume that $\mathbf{E}|\zeta_i|^q < \infty$ for some $q > 2$. Possibly in an enlarged probability space there exists a Wiener process $\{W(t), t \geq 0\}$ such that for any $\varepsilon > 0$*

$$\left|\sum_{k=0}^{n} \zeta(S_k) - \left(\frac{2-\gamma}{\gamma}\right)^{1/2} W(n)\right| = o(n^{\vartheta+\varepsilon}) \quad a.s.$$

where

$$\vartheta = \max\left(\frac{1}{q}, \frac{5}{12}\right).$$

Consequence.

$$\lim_{n \to \infty} \mathbf{P}\left\{\left(\frac{\gamma}{2-\gamma}\right)^{1/2} n^{-1/2} \sum_{k=1}^{n} \zeta(S_k) < x\right\} = \Phi(x). \tag{33.5}$$

Here we present only the

Proof of (33.5). Let
$$A(i,n) = \{x \in \mathbb{Z}^d : \xi(x,n) = i\} = \{x_1(i,n), x_2(i,n), \ldots, x_{Q(i,n)}(i,n)\}$$
where $Q(i,n)$ is the number of sites visited exactly i times up to n. Then
$$\sum_{k=0}^{n} \zeta(S_k) = \sum_{x \in \mathbb{Z}^d} \zeta(x)\xi(x,n) = \sum_{i=1}^{\infty} i \sum_{x \in A(i,n)} \zeta(x)$$
$$= \sum_{i=1}^{t_n} i \sum_{x \in A(i,n)} \zeta(x) + \sum_{i=t_n+1}^{\infty} i \sum_{x \in A(i,n)} \zeta(x).$$

It is easy to see (cf. Theorem 22.1) that
$$\sum_{i=t_n+1}^{\infty} i \sum_{x \in A(i,n)} \zeta(x) = o(n^{1/2-\varepsilon}) \quad \text{a.s.}$$

Assume that $i \leq t_n$ and $Q(i,n) \geq n\gamma^2(1-\gamma)^{i-1}$. Then
$$\sum_{x \in A(i,n)} \zeta(x) = \sum_{l=1}^{Q(i,n)} \zeta(x_l(i,n))$$
$$= \sum_{l=1}^{n\gamma^2(1-\gamma)^{i-1}} \zeta(x_l(i,n)) + \sum_{l=n\gamma^2(1-\gamma)^{i-1}+1}^{Q(i,n)} \zeta(x_l(i,n)).$$

It is easy to see that
$$\sum_{l=n\gamma^2(1-\gamma)^{i-1}+1}^{Q(i,n)} \zeta(x_l(i,n)) = o(n^{1/2-\varepsilon}) \quad \text{a.s.}$$

and
$$\lim_{n\to\infty} \mathbf{P}\left\{\frac{\sum_{l=1}^{n\gamma^2(1-\gamma)^{i-1}} \zeta(x_l(i,n))}{(n\gamma^2(1-\gamma)^{i-1})^{1/2}} < x\right\} = \Phi(x).$$

The case $Q(i,n) < n\gamma^2(1-\gamma)^{i-1}$ can be treated similarly. Since the r.v.'s
$$\sum_{l=1}^{n\gamma^2(1-\gamma)^{i-1}} \zeta(x_l(i,n)) \quad (i = 1, 2, \ldots)$$
are independent we have (33.5).

A FEW FURTHER PROBLEMS

33.6 Random environment and random scenery

It is a natural idea to introduce a common generalization of the random environment and of the random scenery.

Let R_n be a RWIRE (Section 33.3) and let $\zeta(x)$ ($x \in \mathbb{Z}^d$) be an array of i.i.d.r.v.'s. Then we are interested in the properties of the sequence $\sum_{k=0}^{n} \zeta(R_k)$. Here we present only a

Conjecture. Assume that there exists an $a > 0$ such that $\mathbf{P}\{\zeta(x) \geq a\} > 0$. Then

$$\limsup_{n \to \infty} \frac{\sum_{k=0}^{n} \zeta(R_k)}{an} = 1 \quad \text{a.s.}$$

33.7 Reinforced random walk

Construct a random environment on $\mathbb{Z}^1$ by the following procedure. Let $R_0 = 0$, $\mathbf{P}\{R_1 = 1\} = \mathbf{P}\{R_1 = -1\} = 1/2$ and let the weight of each interval $(i, i+1)$ ($i = 0, \pm 1, \pm 2, \ldots$) be initially 1 and increased by 1 at each time when the process jumps across it, so that its weight at time n is one plus the number of indices $k \leq n$ such that (R_k, R_{k+1}) is either $(i, i+1)$ or $(i+1, i)$. Given $\{R_0 = 0, R_1 = i_1, \ldots, R_n = i_n\}$ R_{n+1} is either $i_n + 1$ or $i_n - 1$ with probabilities proportional to the weights at time n of $(i_n, i_n + 1)$ and $(i_n - 1, i_n)$ where $i_1, i_2, \ldots, i_n$ is a sequence of integers with $|i_{j+1} - i_j| = 1$ ($j = 1, 2, \ldots, n$). Hence if $R_1 = 1$, the weight of $[0, 1]$ at time $n = 1$ is 2. Consequently

$$\mathbf{P}\{R_2 = 2 \mid R_1 = 1\} = \frac{1}{3} \quad \text{and} \quad \mathbf{P}\{R_2 = 0 \mid R_1 = 1\} = \frac{2}{3}.$$

Similarly

$$\mathbf{P}\{R_2 = -2 \mid R_1 = -1\} = \frac{1}{3} \quad \text{and} \quad \mathbf{P}\{R_2 = 0 \mid R_1 = -1\} = \frac{2}{3}.$$

Further, in the case $R_1 = 1$, $R_2 = 2$ the weights of $[0, 1]$ and $[1, 2]$ at time $n = 2$ are equal to 2. Hence

$$\mathbf{P}\{R_3 = 3 \mid R_1 = 1, R_2 = 2\} = 1 - \mathbf{P}\{R_3 = 1 \mid R_1 = 1, R_2 = 2\} = \frac{1}{3}.$$

In the case $R_1 = 1$, $R_2 = 0$ the weight of $[0, 1]$ at time $n = 2$ is equal to 3. Hence

$$\mathbf{P}\{R_3 = 1 \mid R_1 = 1, R_2 = 0\} = 1 - \mathbf{P}\{R_3 = -1 \mid R_1 = 1, R_2 = 0\} = \frac{3}{4}.$$

Similarly

$$\mathbf{P}\{R_3 = -3 \mid R_1 = -1, \ R_2 = -2\}$$
$$= 1 - \mathbf{P}\{R_3 = -1 \mid R_1 = -1, \ R_2 = -2\} = \frac{1}{3}$$

and

$$\mathbf{P}\{R_3 = -1 \mid R_1 = -1, \ R_2 = 0\} = 1 - \mathbf{P}\{R_3 = 1 \mid R_1 = -1, \ R_2 = 0\} = \frac{3}{4}.$$

This model was introduced by Coppersmith and Diaconis (cf. Davis (1989)) and studied by Davis (1989, 1990).

Intuitively it is clear that the random walk generated by this model is "more recurrent" than the simple symmetric random walk. However, to prove that it is recurrent is not easy at all. This was done by Davis (1989, 1990), who studied the recurrence in more general models as well.

Note that in this model the random environment is changing in time and depends on the random walk itself. Situations where the random environment is changing in time look very natural in different practical models.

A more general model is the following: the random walk $\{R_i\}$ starts from the origin of $\mathbb{Z}^1$ and at time $i+1$ it jumps to one of the two neighbouring sites of R_i, so that the probability of jumping along a bond of the lattice is proportional to

$$\omega(\text{number of previous jumps along that bond})$$

where ω is a weight function. More formally, for a nearest neighbour walk $x^{(i)} = (x_0, x_1, \ldots, x_i)$ let

$$r(x^{(i)}) = \#\{j: \ 0 \le j \le i, \ (x_j, x_{j+1}) = (x_i, x_i + 1) \text{ or } (x_i + 1, x_i)\},$$
$$l(x^{(i)}) = \#\{j: \ 0 \le j \le i, \ (x_j, x_{j+1}) = (x_i, x_i - 1) \text{ or } (x_i - 1, x_i)\}$$

that is r (resp. l) shows how many times had the walk visited the edge adjacent from the right (resp. left) to the terminal site x_i. Then the random walk $\{R_i\}$ is governed by the law

$$\mathbf{P}\{R_{i+1} = R_i + 1 \mid R_j = x_j, \ j = 0, 1, 2, \ldots, i\}$$
$$= \frac{\omega(r(x^{(i)}))}{\omega(r(x^{(i)})) + \omega(l(x^{(i)}))}$$
$$= 1 - \mathbf{P}\{R_{i+1} = R_i - 1 \mid R_j = x_j, \ j = 0, 1, 2, \ldots, i\}.$$

Clearly the properties of the random walk $\{R_i\}$ depend strongly on the choice of the weight function $\omega(\cdot)$. The most complete description of this

A FEW FURTHER PROBLEMS

type of walks is due to Tóth (1995, 1996, 1997, 1999). He studied for example the cases when

$$(\omega(n))^{-1} = 2^{-\alpha}(\alpha+1)n^{\alpha} + 2^{1-\alpha}Bn^{\alpha-1} + O(n^{\alpha-2}) \quad (\alpha \in \mathbb{R},\ B \in \mathbb{R})$$

and
$$\omega(n) = e^{-gn} \quad (g > 0).$$

For example in the first case with $\alpha \geq 0$ he proved that

$$\lim_{A \to \infty} A^{1/2} \mathbf{P}\{R_{[As]} = [A^{1/2}x]\}$$

exists and its value is evaluated.

References

ADELMAN, O. – BURDZY, K. – PEMANTLE, R.
(1998) Sets avoided by Brownian motion. *The Annals of Probability* **26**, 429–464.

ANDJEL, E. D.
(1988) A zero or one law for one-dimensional random walks in random environments. *The Annals of Probability* **16**, 722–729.

AUER, P.
(1989) Some hitting probabilities of random walks on $\mathbb{Z}^2$. *Coll. Math. Soc. J. Bolyai: Limit Thorems in Probability and Statistics* (ed. I. Berkes, E. Csáki, P. Révész) North-Holland, **57**, 9–25.

(1990) The circle homogeneously covered by random walk on $\mathbb{Z}^2$. *Statistics & Probability Letters* **9**, 403–407.

AUER, P. – RÉVÉSZ, P.
(1990) On the relative frequency of points visited by random walk on $\mathbb{Z}^2$. *Coll. Math. Soc. J. Bolyai: Limit Theorems in Probability and Statistics* (ed. I. Berkes, E. Csáki, P. Révész) North-Holland, **57**, 27–33.

BÁRTFAI, P.
(1966) Die Bestimmung der zu einem wiederkehrenden Prozess gehörenden Verteilungsfunktion aus den mit Fehlern behafteten Daten einer Einzigen Realisation. *Studia Sci. Math. Hung.* **1**, 161–168.

BASS, R. F. – GRIFFIN, P. S.
(1985) The most visited site of Brownian motion and simple random walk. *Z. Wahrscheinlichkeitstheorie verw. Gebiete* **70**, 417–436.

BASS, R. F. – KUMAGAI, T.
(2002) Laws of the iterated logarithm for the range of randow walks in two and three dimensions. *The Annals of Probability* **30**, 1369–1396.

BENJAMINI, I. – HÄGGSTRÖM, O. – PERES, Y. – STEIF, J. E.
(2003) Which properties of a random sequence are dinamically sensitive. *The Annals of Probability* **31**, 1–134.

BERKES, I.
(1972) A remark to the law of the iterated logarithm. *Studia Sci. Math. Hung.* **7**, 189–197.

BICKEL, P. J. – ROSENBLATT, M.
(1973) On some global measures of the deviations of density function estimates. *The Annals of Statistics* **1**, 1071–1095.

BILLINGSLEY, P.
(1968) *Convergence of Probability Measures.* J. Wiley, New York.

BINGHAM, N. H.
(1989) The work of A. N. Kolmogorov on strong limit theorems. *Theory of Probability and its Applications* **34**, 152–164.

BOLTHAUSEN, E.
(1978) On the speed of convergence in Strassen's law of the iterated logarithm. *The Annals of Probability* **6**, 668–672.

(1989) A central limit theorem for two-dimensional random walk in random sceneries. *The Annals of Probability* **17**, 108–115.

BOOK, S. A. – SHORE, T. R.
(1978) On large intervals in the Csörgő – Révész Theorem on Increments of a Wiener Process. *Z. Wahrscheinlichkeitstheorie verw. Gebiete* **46**, 1–11.

BOREL, E.
(1909) Sur lés probabilités dénombrables et leurs applications arithmétiques. *Rendiconti del Circolo Mat. di Palermo* **26**, 247–271.

BORODIN, A. N.
(1982) Distribution of integral functionals of Brownian motion. *Zap. Nauchn. Semin. Leningrad Otd. Mat. Inst. Steklova* **119**, 13–38.

(1986/A,B) On the character of convergence to Brownian local time I, II. *Probab. Th. Rel. Fields* **72**, 231–250, 251–277.

BROSAMLER, G. A.
(1988) An almost everywhere central limit theorem. *Math. Proc. Camb. Phil. Soc.* **104**, 561–574.

CHEN, X.
(2004) Exponential asymptotics and law of the iterated logarithm for intersection local times of random walks. *The Annals of Probability* **32**, 3248–3300.

CHEN, X. – LI, W.
(2004) Large and moderate deviations for intersection local times. *Probab. Th. Rel. Fields* **128**, 213–254.

CHUNG, K. L.
(1948) On the maximum partial sums of sequences of independent random variables. *Trans. Am. Math. Soc.* **64**, 205–233.

CHUNG, K. L. - ERDŐS, P.
  (1952) On the application of the Borel – Cantelli lemma. *Trans. Am. Math. Soc.* **72**, 179–186.

CHUNG, K. L. – HUNT, G. A.
  (1949) On the zeros of $\sum_1^n \pm 1$. *Annals of Math.* **50**, 385–400.

COPPERSMITH, D. – DIACONIS, P.
  (1987) Random walks with reinforcement. *Stanford Univ.* Preprint.

CSÁKI, E.
  (1978) On the lower limits of maxima and minima of Wiener process and partial sums. *Z. Wahrscheinlichkeitstheorie verw. Gebiete* **43**, 205–221.

  (1980) A relation between Chung's and Strassen's law of the iterated logarithm. *Z. Wahrscheinlichkeitstheorie verw. Gebiete* **54**, 287–301.

  (1989) An integral test for the supremum of Wiener local time. *Probab. Th. Rel. Fields* **83**, 207–217.

  (1990) A liminf result in Strassen's law of the iterated logarithm. *Coll. Math. Soc. J. Bolyai: Limit Theorems in Probability and Statistics* (ed. I. Berkes, E. Csáki, P. Révész) North-Holland, **57**, 83–93.

CSÁKI, E. – CSÖRGŐ, M. – FÖLDES, A. – RÉVÉSZ, P.
  (1983) How big are the increments of the local time of a Wiener process? *The Annals of Probability* **11**, 593–608.

  (1989) Brownian local time approximated by a Wiener-sheet. *The Annals of Probability* **17**, 516–537.

CSÁKI, E. – ERDŐS, P. – RÉVÉSZ, P.
  (1985) On the length of the longest excursion. *Z. Wahrscheinlichkeitstheorie verw. Gebiete* **68**, 365–382.

CSÁKI, E. – FÖLDES, A.
  (1984/A) On the narrowest tube of a Wiener process. *Coll. Math. Soc. J. Bolyai: Limit Theorems in Probability and Statistics* (ed. P. Révész) North-Holland, **36**, 173–197.

  (1984/B) The narrowest tube of a recurrent random walk. *Z. Wahrscheinlichkeitstheorie verw. Gebiete* **66**, 387–405.

  (1984/C) How big are the increments of the local time of a simple symmetric random walk? *Coll. Math. Soc. J. Bolyai: Limit Theorems in Probability and Statistics* (ed. P. Révész) North-Holland, **36**, 199–221.

  (1986) How small are the increments of the local time of a Wiener process? *The Annals of Probability* **14**, 533–546.

(1987) A note on the stability of the local time of a Wiener process. *Stochastic Processes and their Applications* **25**, 203–213.

(1988/A) On the length of the longest flat interval. *Proc. of the 5th Pannonian Symp. on Math. Stat.* (ed. Grossmann, W. – Mogyoródi, J. – Vincze, I. – Wertz, W.). 23–33.

(1988/B) On the local time process standardized by the local time at zero. *Acta Mathematica Hungarica* **52**, 175–186.

CSÁKI, E. – FÖLDES, A. – KOMLÓS, J.
(1987) Limit theorems for Erdős – Rényi type problems. *Studia Sci. Math. Hung.* **22**, 321–332.

CSÁKI, E. – FÖLDES, A. – RÉVÉSZ, P.
(1987) On the maximum of a Wiener process and its location. *Probab. Th. Rel. Fields* **76**, 477–497.

(2005) Heavy points of a d-dimensional simple random walk. To appear.

CSÁKI, E. – FÖLDES, A. – RÉVÉSZ, P. – ROSEN, J. – SHI, Z.
(1998) A strong invariance principle for the local time difference of a symple symmetric planar random walk. *Studia Sci. Math. Hung.* **34**, 25–39.

(2005) Frequently visited sets for random walks. To appear.

CSÁKI, E. – FÖLDES, A. – RÉVÉSZ, P. – SHI, Z.
(1999) On the excursions of two-dimensional random walk and Wiener process. *Bolyai Soc. Math. Studies 9.* (ed. P. Révész, B. Tóth) 43–58.

CSÁKI, E. – GRILL, K.
(1988) On the large values of the Wiener process. *Stochastic Processes and their Applications* **27**, 43–56.

CSÁKI, E. – RÉVÉSZ, P.
(1979) How big must be the increments of a Wiener process? *Acta Math. Acad. Sci. Hung.* **33**, 37–49.

(1983) A combinatorial proof of P. Lévy on the local time. *Acta Sci. Math. Szeged* **45**, 119–129.

CSÁKI, E. – RÉVÉSZ, P. – ROSEN, J.
(1998) Functional law of iterated logarithm for local times of recurrent random walks on $\mathbb{Z}^2$. *Ann. Inst. Henri Poincaré, Probabilités et statistiques* **34**, 545–463.

REFERENCES

CSÁKI, E. – RÉVÉSZ, P. – SHI, Z.
    (2000) Favourite sites, favourite values and jump sizes for random walk and Brownian motion. *Bernoulli* **6**, 951–975.

    (2001/A) A strong invariance principle for two-dimensional random walk in random scenery. *Stochastic Processes and their Applications* **92**, 181–200.

    (2001/B) Long excursions of a random walk. *J. of Theoretical Probability* **14**, 821–844.

CSÁKI, E. – VINCZE, I.
    (1961) On some problems connected with the Galton test. *Publ. Math. Inst. Hung. Acad. Sci.* **6**, 97–109.

CSÖRGŐ, M. – HORVÁTH, L.
    (1989) On best possible approximations of local time. *Statistics & Probability Letters* **8**, 301–306.

CSÖRGŐ, M. – HORVÁTH, L. – RÉVÉSZ, P.
    (1987) Stability and instability of local time of random walk in random environment. *Stochastic Processes and their Applications* **25**, 185–202.

CSÖRGŐ, M. – RÉVÉSZ, P.
    (1979/A) How big are the increments of a Wiener process? *The Annals of Probability* **7**, 731–737.

    (1979/B) How small are the increments of a Wiener process? *Stochastic Processes and their Applications.* **8**, 119–129.

    (1981) *Strong Approximations in Probability and Statistics.* Akadémiai Kiadó, Budapest and Academic Press, New York.

    (1985/A) On the stability of the local time of a symmetric random walk. *Acta Sci. Math.* **48**, 85–96.

    (1985/B) On strong invariance for local time of partial sums. *Stochastic Processes and their Applications* **20**, 59–84.

    (1986) Mesure du voisinage and occupation density. *Probab. Th. Rel. Fields* **73**, 211–226.

    (1992) Long random walk excursions and local time. *Stochastic Processes and their Applications* **41**, 181–190.

DARLING, D. A. – ERDŐS, P.
    (1956) A limit theorem for the maximum of normalized sums of independent random variables. *Duke Math. J.* **23**, 143–155.

DAVIS, B.
    (1989) Loss of recurrence in reinforced random walk. Technical Report, Purdue University.

(1990) Reinforced random walk. *Probab. Th. Rel. Fields.* **84**, 203–229.

DE ACOSTA, A.
(1983) Small deviations in the functional central limit theorem with applications to functional laws of the iterated logarithm. *The Annals of Probability* **11**, 78–101.

DEHEUVELS, P.
(1985) On the Erdős – Rényi theorem for random fields and sequences and its relationship with the theory of runs and spacings. *Z. Wahrscheinlichkeitstheorie verw. Gebiete* **70**, 91–115.

DEHEUVELS, P – DEVROYE, L. – LYNCH, I.
(1986) Exact convergence rates in the limit theorem of Erdős – Rényi and Shepp. *The Annals of Probability* **14**, 209–223.

DEHEUVELS, P. – ERDŐS, P. – GRILL, K. – RÉVÉSZ, P.
(1987) Many heads in a short block. *Mathematical Statistics and Probability Theory, Vol. A., Proc. of the 6th Pannonian Symp.* (ed. Puri, M. L. – Révész, P. – Wertz, W.) 53–67.

DEHEUVELS, P. – RÉVÉSZ, P.
(1986) Simple random walk on the line in random environment. *Probab. Th. Rel. Fields* **72**, 215–230.

(1987) Weak laws for the increments of Wiener processes, Brownian bridges, empirical processes and partial sums of i.i.d.r.v.'s. *Mathematical Statistics and Probability Theory, Vol. A., Proc. of the 6th Pannonian Symp.* (ed. Puri, M. L. – Révész, P. – Wertz, W.) 69–88.

DEHEUVELS, P. – STEINEBACH, J.
(1987) Exact convergence rates in strong approximation laws for large increments of partial sums. *Probab. Th. Rel. Fields* **76**, 369–393.

DEMBO, A. – PERES, Y. – ROSEN, J. – ZEITOUNI, O.
(2000) Thick points for spatial Brownian motion: multifractal analysis of occupation measure. *The Annals of Probability* **28**, 1–35.

(2001) Thick points for planar Brownian motion and the Erdős – Taylor conjecture on random walk. *Acta Mathematica* **186**, 239–270.

(2005/A) The largest disc covered by a planar random walk. To appear.

(2005/B) Cover time for Brownian motion and random walks in two dimensions. To appear.

DOBRUSHIN, R. L.
(1955) Two limit theorems for the simplest random walk on a line. *Uspehi Math. Nauk* (N. N) **10**, 139–146. In Russian.

DONSKER, M. D. - VARADHAN, S. R. S.
   (1977) On laws of the iterated logarithm for local times. *Comm. Pure Appl. Math.* **30**, 707–753.

   (1979) On the number of distinct sites visited by a random walk. *Comm. Pure Appl. Math.* **27**, 721–747.

DURRETT, R.
   (1991) *Probability: Theory and Examples.* Wadsworth Brooks/Cole, Pacific Grove, California.

DVORETZKY, A. - ERDŐS, P.
   (1951) Some problems on random walk in space. *Proc. Second Berkeley Symposium* 353–367.

DVORETZKY, A. - ERDŐS, P. - KAKUTANI, S.
   (1950) Double points of Brownian paths in n-space. *Acta Sci. Math. Szeged* **12**, 75–81.

ERDŐS, P.
   (1942) On the law of the iterated logarithm. *Annals of Math.* **43**, 419–436.

ERDŐS, P. - CHEN, R. W.
   (1988) Random walks on Z_2^n. *J. Multivariate Analysis* **25**, 111–118.

ERDŐS, P. - RÉNYI, A.
   (1970) On a new law of large numbers. *J. Analyse Math.* **23**, 103–111.

ERDŐS, P. - RÉVÉSZ, P.
   (1976) On the length of the longest head-run. *Coll. Math. Soc. J. Bolyai: Topics in Information Theory* (ed. Csiszár, I. - Elias, P.) **16**, 219–228

   (1984) On the favourite points of a random walk. *Mathematical Structures - Computational Mathematics - Mathematical Modelling* 2. Sofia, 152–157.

   (1987) Problems and results on random walks. *Math. Statistics and Probability Theory, Vol. B., Proc. 6th Pannonian Symp.* (ed. Bauer, P. - Konecny, F. - Wertz, W.) D. Reidel, Dordrecht. 59–65.

   (1988) On the area of the circles covered by a random walk. *Journal of Multivariate Analysis* **27**, 169–180.

   (1990) A new law of the iterated logarithm. *Acta Math. Hung.* **55**, 125–131.

(1991) Three problems on the random walk in $\mathbb{Z}^d$. *Studia Sci. Math. Hung.* **26**, 309–320.

(1997) On the radius of the largest ball left empty by a Wiener process. *Studia Sci. Math. Hung.* **33**, 117–125.

ERDŐS, P. – TAYLOR, S. J.
(1960/A) Some problems concerning the structure of random walk paths. *Acta Math. Acad. Sci. Hung.* **11**, 137–162.

(1960/B) Some intersection properties of random walk paths. *Acta Math. Acad. Sci. Hung.* **11**, 231–248.

FELLER, W.
(1943) The general form of the so-called law of the iterated logarithm. *Trans. Am. Math. Soc.* **54**, 373–402.

(1966) *An Introduction to Probability Theory and Its Applications,.* Vol. II. J. Wiley, New York.

FISHER, A.
(1987) Convex – invariant means and a pathwise central limit theorem. *Advances in Mathematics* **63**, 213–246.

FLATTO, L.
(1976) The multiple range of two-dimensional recurrent walk. *The Annals of Probability* **4**, 229–248.

FÖLDES, A.
(1975) On the limit distribution of the longest head-run. *Matematikai Lapok* **26**, 105–116. In Hungarian.

(1989) On the infimum of the local time of a Wiener process. *Probab. Th. Rel. Fields* **82**, 545–563.

FÖLDES, A. – PURI, M. L.
(1993) The time spent by the Wiener process in a narrow tube before leaving a wide tube. *Proc. Amer. Math. Soc.* **117**, 529–537.

FÖDES, A. – RÉVÉSZ, P.
(1992) On hardly visited points of the Brownian motion. *Probability Theory and Related Fields* **91**, 71–80.

(1993) Quadratic variation of the local time of a random walk. *Statistics & Probability Letters* **17**, 1–12.

GNEDENKO, B. V. – KOLMOGOROV, A. N.
(1954) *Limit Distributions for Sums of Independent Random Variables.* Addison – Wesley, Reading, Massachusetts.

GOLOSOV, A. O.
   (1983) Limit distributions for random walks in random environments. *Soviet Math. Dokl.* **28**, 18–22.

   (1984) Localization of random walks in one-dimensional random environments. *Commun. Math. Phys.* **92**, 491–506.

GONCHAROV, V. L.
   (1944) From the domain of Combinatorics. *Izv. Akad. Nauk SSSR Ser. Math.* **8**(1), 3–48.

GOODMAN, V. – KUELBS, J.
   (1988) Rates of convergence for increments of Brownian motion. *J. Theor. Probab.* **1**, 27–63.

GORN, N. L. – LIFSCHITZ, M. A.
   (1998) Chung's law and Csáki function. *J. Theor. Probab.* **12**, 399 – 420.

GRIFFIN, P.
   (1990) Accelerating beyond the third dimension: Returning to the origin in simple random walk. *The Mathematical Scientist.* **15**, 24–35.

GRILL, K.
   (1987/A) On the rate of convergence in Strassen's law of the iterated logarithm. *Probab. Th. Rel. Fields* **74**, 583–589.

   (1987/B) On the last zero of a Wiener process. *Mathematical Statistics and Probability Theory, Vol. A., Proc. 6th Pannonian Sump.* (ed. Puri, M. L. – Révész, P. – Wertz, W.) D. Reidel, Dordrecht. 99–104.

   (1991) On the increments of the Wiener processes. *Studia Sci. Math. Hung.* **26**, 329–354.

GUIBAS, L. J. – ODLYZKO, A. M.
   (1980) Long repetitive patterns in random sequences. *Z. Wahrscheinlichkeitstheorie verw. Gebiete* **53**, 241–262.

HAMANA, Y.
   (1995) On the central limit theorem for multiple point range of random walk. *J. Fac. Sci. Univ. Tokyo Sect. I A Math.* **39**, 339–363.

   (1997) The fluctuation result for the multiple point range of two-dimensional recurrent random walks. *The Annals of Probability* **25**, 598–639.

   (1998) An almost sure invariance principle for the range of random walks. *Stochastic Process. Appl.* **78**, 131–143.

HANSON, D. L. - RUSSO, R. P.
(1983/A) Some results on increments of the Wiener process with applications to lag sums of I.I.D. random variables. *The Annals of Probability* **11**, 609–623.

(1983/B) Some more results on increments of the Wiener process. *The Annals of Probability* **11**, 1009–1015.

HARTMAN, P. - WINTNER, A.
(1941) On the law of iterated logarithm. *Amer. J. Math.* **63**, 169–176.

HAUSDORFF, F.
(1913) *Grundzüge der Mengenlehre*. Leipzig.

HIRSCH, W. M.
(1965) A strong law for the maximum cumulative sum of independent random variables. *Comm. Pure Appl. Math.* **18**, 109–217.

HOUGH, J. B. - PERES, Y.
(2005) An LIL for cover times of discs by planar random walk and Wiener sousage. To appear.

HU, Y. - SHI, Z.
(1998/A) The local time of simple random walk in random environment. *J. Theoretical Probability* **11**, 765–793.

(1998/B) The limits of Sinai's simple random walk in random environment. *The Annals of Probability* **26**, 1477–1521.

IMHOF, I. P.
(1984) Density factorizations for Brownian motion meander and the three-dimensional Bessel process. *J. Appl. Probab.* **21**, 500–510.

ITÔ, K.
(1942) Differential equations determining a Markoff process. *Kiyosi Itô Selected Papers*. Springer-Verlag, New York (1986), 42–75.

ITÔ, K. - MCKEAN Jr., H. P.
(1965) Diffusion processes and their sample paths. *Die Grundlagen der Mathematischen Wissenschaften Band 125*. Springer-Verlag, Berlin.

JAIN, N. C. - PRUITT, W. E.
(1971) The range of transient random walk. *J. Analyse Math.* **24**, 369–373.

(1972/A) The law of iterated logarithm for the range of random walk. *Ann. Math. Statist.* **43**, 1692–1697.

(1972/B) The range of random walk. *Proc. Sixth Berkeley Symp. Math. Statist. Probab.* **3**, Univ. California Press, Berkeley, 31–50.

(1974) Further limit theorems for the range of random walk. *J. Analyse Math.* **27**, 94–117.

KALIKOW, S. A.
    (1981) Generalized random walk in a random environment. *The Annals of Probability* **9**, 753–768.

KARLIN, S. – OST, F.
    (1988) Maximal length of common words among random letter sequences. *The Annals of Probability* **16**, 535–563.

KESTEN, H.
    (1965) An iterated logarithm law for the local time. *Duke Math. J.* **32**, 447–456.

    (1980) The critical probability of band percolation on $\mathbb{Z}^2$ equals $1/2$. *Comm. Math. Phys.* **74**, 41–59.

    (1986) The limit distribution of Sinai's random walk in random environment. *Comm. Math. Phys.* **138**, 299–309.

    (1987) Hitting probabilities of random walks on $\mathbb{Z}^d$. *Stochastic Processes and their Applications* **25**, 165–184.

    (1988) Recent progress in rigorous percolation theory. *Astérisque* **157–158**, 217–231.

KESTEN, H. – SPITZER, F.
    (1979) A limit theorem related to a new class of self similar processes. *Z. Wahrscheinlichkeitstheorie verw. Gebiete* **50**, 5–25.

KEY, E. S.
    (1984) Recurrence and transience criteria for random walk in a random environment. *The Annals of Probability* **12**, 529–560.

KHINCHINE, A.
    (1923) Über dyadische Brüche. *Math. Zeitschrift* **18**, 109–116.

KHOSHNEVISAN, D.
    (1994) Exact rates of convergence to Brownian local time. *The Annals of Probability* **22**, 1295–1330.

    (2002) *Multiparameter Processes.* Springer-Verlag, New York, Berlin, Heidelberg.

KNIGHT, F. B.
    (1981) *Essentials of Brownian Motion and Diffusion* Am. Math. Soc., Providence, R.I.

(1986) On the duration of the longest excursion. *Seminar on Stochastic Processes, 1985.* Birkhäuser, Boston. 117–147.

KOLMOGOROV, A. N.
(1933) *Grundbegriffe der Wahrscheinlichkeitsrechnung.* Springer, Berlin.

KOMLÓS, J. – MAJOR, P. – TUSNÁDY, G.
(1975) An approximation of partial sums of independent R.V.'s and the sample DF. I. *Z. Wahrscheinlichkeitstheorie verw. Gebiete* **32**, 111–131.

(1976) An approximation of partial sums of independent R.V.'s and the sample DF. II. *Z. Wahrscheinlichkeitstheorie verw. Gebiete* **34**, 33–58.

LACEY, M. T. – PHILIPP, W.
(1990) A note on the almost sure central limit theorem. *Statistics & Probability Letters* **9**, 201–205.

LAMPERTI, J.
(1977) *Stochastic Processes. A Survey of the Mathematical Theory.* Springer – Verlag, New York.

LAWLER, G. F.
(1980) A self-avoiding random walk. *Duke Mathematical Journal* **47**, 655–692.

(1991) *Intersections of Random Walks.* Birkhäuser, Boston.

(1993) On the covering time of a disc by simple random walk in two-dimensionals. *Seminar on Stochastic Processes 1992.* Birkhäuser, Boston. **33**, 189–208.

LE GALL, J.-F.
(1986) Propriétés d' intersections des marches aléatoires 1. Convergence rers le temps local d' intersection. *Comm. Math. Phys.* **104**, 471–507.

(1988) Fluctuation results for the Wiener sausage. *The Annals of Probability* **16**, 991–1018.

LE GALL, J.-F. – ROSEN, J.
(1991) The range of stable random walks. *The Annals of Probability* **19**, 650–705.

LÉVY, P.
(1948) *Processu Stochastique et Mouvement Brownien.* Gauthier – Villars, Paris.

MAJOR, P.
(1988) On the set visited once by a random walk. *Probab. Th. Rel. Fields* **77**, 117–128.

MARCUS, M. B. – ROSEN, J.
   (1997) Laws of iterated logarithm for intersections of random walks on $\mathbb{Z}^4$. *Ann. Inst. H. Poincaré Probab. Statist.* **33**, 37–63.

MCKEAN Jr, H. P.
   (1969) *Stochastic Integrals*. Academic Press, New York.

MOGUL'SKII, A. A.
   (1979) On the law of the iterated logarithm in Chung's form for functional spaces. *Th. of Probability and its Applications* **24**, 405–412.

MÓRI, T.
   (1989) More on the waiting time till each of some given patterns occurs as a run. Preprint.

MUELLER, C.
   (1983) Strassen's law for local time. *Z. Wahrscheinlichkeitstheorie verw. Gebiete* **63**, 29–41.

NEMETZ, T – KUSOLITSCH, N.
   (1982) On the longest run of coincidences. *Z. Wahrscheinlichkeitstheorie verw. Gebiete* **61**, 59–73.

NEWMAN, D.
   (1984) In a random walk the number of "unique experiences" is two on the average. *SIAM Review* **26**, 573–574.

OREY, S. – PRUITT, W. E.
   (1973) Sample functions of the N-parameter Wiener process. *The Annals of Probability* **1**, 138–163.

ORTEGA, I. – WSCHEBOR, M.
   (1984) On the increments of the Wiener process. *Z. Wahrscheinlichkeitstheorie verw. Gebiete* **65**, 329–339.

PERKINS, E.
   (1981/A) A global instrinsic characterization of Brownian local time. *The Annals of Probability* **9**, 800–817.

   (1981/B) On the iterated logarithm law for local time. *Proc. Amer. Math. Soc.* **81**, 470–472.

PERKINS, E. – TAYLOR, S. J.
   (1987) Uniform measure results for the image of subsets under Brownian motion. *Probab. Th. Rel. Fields* **76**, 257–289.

PETROV, V. V.
(1965) On the probabilities of large deviations for sums of independent random variables. *Th. of Probability and its Applications* **10**, 287–298.

PETROWSKY, I. G.
(1935) Zur ersten Randwertaufgabe der Warmleitungsgleichung. *Comp. Math.* **B. 1**, 383–419.

PITT, J. H.
(1974) Multiple points of transient random walk. *Proc. Amer. Math. Soc.* **43**, 195–199.

PÓLYA, G.
(1921) Über eine Aufgabe der Wahrscheinlichkeitsrechnung betreffend die Irrfahrt im Strassennetz. *Math. Ann.* **84**, 149–160.

QUALLS, G. – WATANABE, H.
(1972) Asymptotic properties of Gaussian processes. *Annals Math. Statistics* **43**, 580–596.

RÉNYI, A.
(1970/A) *Foundations of Probability*. Holden-Day, San Francisco.

(1970/B) *Probability Theory*. Akadémiai Kiadó, Budapest and North-Holland, Amsterdam.

RÉVÉSZ, P.
(1978) Strong theorems on coin tossing. *Proc. Int. Cong. of Mathematicians*, Helsinki.

(1979) A generalization of Strassen's functional law of iterated logarithm. *Z. Wahrscheinlichkeitstheorie verw. Gebiete* **50**, 257-264.

(1981) Local time and invariance. *Lecture Notes in Math.: Analytical Methods in Probab. Th.* **861**, 128–145.

(1982) On the increments of Wiener and related process. *The Annals of Probability* **10**, 613-622.

(1988) In random environment the local time can be very big. Société Mathématique de France, *Astérisque* **157-158**, 321-339.

(1989) Simple symmetric random walk in $\mathbb{Z}^d$. *Almost Everywhere Convergence. Proceedings of the Int. Conf. on Almost Everywhere Convergence* (ed. G. A. Edgar, L. Sucheston) Academic Press, Boston. 369–392.

(1990/A) Estimates of the largest disc covered by a random walk. *The Annals of Probability* **18**, 1784–1789.

(1990/B) On the volume of spheres covered by a random walk. *A tribute to Paul Erdős* (ed. A. Baker, B. Bollobás, A. Hajnal) Cambridge Univ. Press. 341–347.

(1991) Waiting for the coverage of the Strassen's set. *Studia. Sci. Math. Hung.* **26**, 379–391.

(1992) Black holes on the plane drawn by a Wiener process. *Probability Theory and Related Fields* **93**, 21–37.

(1993/A) Clusters of a random walk on the plane. *The Annals of Probability* **21**, 318–328.

(1993/B) Covering problems. *Theory of Probability and its Applications* **38**, 367–379.

(1993/C) A homogenity property of the $\mathbb{Z}^2$ random walk. *Acta Sci. Math. (Szeged)* **57**, 477–484.

(1996) Balls left empty by a critical branching Wiener process. *J. of Applied Math. and Stochastic Analysis* **9**, 531–549.

(2000) On the inverse local time process of a plane random walk. *Periodica Math. Hung.* **41**, 227–236.

(2004) The maximum of the local time of a transient random walk. *Studia Sci. Math. Hung.* **41**, 379–390.

RÉVÉSZ, P. – SHI, Z.
(2000) Strong approximation of spatial random walk in random scenery. *Stochastic Processes and their Applications* **88**, 329–345.

RIESZ, F. – SZ. NAGY, B.
(1953) *Functional Analysis.* Frederick Ungar, New York.

ROSEN, J.
(1997) Laws of the iterated logarithm for triple intersections of three-dimensional random walks. *Electron. J. Probab.* **2**, 1–32.

SAMAROVA, S. S.
(1981) On the length of the longest head-run for a Markov chain with two states. *Th. of Probability and its Applications* **26**, 489–509.

SCHATTE, P.
(1988) On strong versions of the central limit theorem. *Math. Nachr.* **137**, 249–256.

SHAO, Q. M.
(1995) On a conjecture of Révész. *Proc. Am. Math. Soc.* **123**, 575–582.

SHI, Z. – TÓTH, B.
(2000) Favourite sites of simple random walks. *Periodica Math. Hung.* **41**, 237–249.

SIMONS, G.
> (1983) A discrete analogue and elementary derivation of "Lévy's equivalence" for Brownian motion. *Statistics & Probability Letters* **1**, 203–206.

SINAĬ, JA. G.
> (1982) Limit behaviour of one-dimensional random walks in random environment. *Th. of Probability and its Applications* **27**, 247–258.

SKOROHOD, A. V.
> (1961) *Studies in the Theory of Random Processes.* Addison – Wesley, Reading, Mass.

SOLOMON, F.
> (1975) Random walks in random environment. *The Annals of Probability* **3**, 1–31.

SPITZER, F.
> (1958) Some theorems concerning 2-dimensional Brownian motion. *Transactions of the Am. Math. Soc.* **87**, 187–197.

> (1964) *Principles of Random Walk.* Van Nostrand, Princeton, N.J.

STRASSEN, V.
> (1964) An invariance principle for the law of iterated logarithm. *Z. Wahrscheinlichkeitstheorie verw. Gebiete* **3**, 211–226.

> (1966) A converse to the law of the iterated logarithm. *Z. Wahrscheinlichkeitstheorie verw. Gebiete* **4**, 265–268.

SZABADOS, T.
> (1989) A discrete Itô formula. *Coll. Math. Soc. J. Bolyai: Limit Theorems in Probability and Statistics* (ed. I. Berkes, E. Csáki, P. Révész) North-Holland 491–502.

SZÉKELY, G. – TUSNÁDY, G.
> (1979) Generalized Fibonacci numbers, and the number of "pure heads". *Matematikai Lapok* **27**, 147–151. In Hungarian.

TÓTH, B.
> (1985) A lower bound for the critical probability of the square lattice site percolation. *Z. Wahrscheinlichkeitstheorie verw. Gebiete* **69**, 19–22.

> (1995) The "true" self-avoiding walk with bond repulsion on $\mathbb{Z}$: limit theorems. *The Annals of Probability* **23**, 1523–1556.

> (1996/A) Multiple covering of the range of a random walk on $\mathbb{Z}$ (On a question of P. Erdős and P. Révész). *Studia Sci. Math. Hung.* **31**, 355–359.

(1996/B) Generalized Ray – Knight theory and limit theorems for self-interacting random walks on $\mathbb{Z}^1$. *The Annals of Probability* **24**, 1324–1367.

(1997) Limit theorems for weakly reinforced random walks on $\mathbb{Z}$. *Studia Sci. Math. Hung.* **33**, 321–337.

(1999) Self-interacting random motions – A survey. *Bolyai Soc. Math. Studies 9.* (ed: P. Révész, B. Tóth) 349–384.

(2001) No more than three favorite sites for simple random walk. *The Annals of Probability* **29**, 484–503.

TÓTH, B. – WERNER, W.
(1997) Tied favourite edges for simple random walk. *Combinatorics, Probability and Computing* **6**, 359–396.

TROTTER, H. F.
(1958) A property of Brownian motion paths. *Illinois J. of Math.* **2**, 425–433.

WEIGL, A.
(1989) *Zwei Sätze über die Belegungszeit beim Random Walk.* Diplomarbeit, TU Wien.

WICHURA, M.
(1977) Unpublished manuscript.

ZIMMERMANN, G.
(1972) Some sample function properties of the two-parameter Gaussian process. *Ann. Math. Statistics* **43**, 1235–1246.

Author Index

Adelman, O. 279
Andjel, E. D. 348, 349
Auer, P. 245, 251, 264, 266, 297

Bártfai, P. 53, 77
Bass, R. F. 157, 222, 225
Benjamini, I. 60
Berkes, I. 32
Bickel, P. J. 180
Billingsley, P. 16
Bingham, N. H. 38
Bolthausen, E. 92, 351
Book, S. A. 71
Borel, E. 28, 29
Borodin, A. N. 107, 110, 351
Brosamler, G. A. 140, 141
Burdzy, K. 279

Chen, R. W. 62
Chen, X. 243, 244
Chung, K. L. 72, 112, 118, 121, 135, 136
Coppersmith, N. 354
Csáki, E. 18, 41, 45, 71, 72, 74, 81, 82, 88, 93, 94, 107, 112–115, 118–123, 125, 127, 137–139, 152, 159, 166, 167, 169, 172–176, 179, 227, 236, 238, 281, 283, 285, 351
Csörgő, M. 66, 73, 110, 119, 120, 126, 141, 163, 164, 166, 167, 185, 317, 318

Darling, D. A. 180
Davis, B. 354
De Acosta, A. 93
Deheuvels, P. 18, 64, 75, 79, 80, 326, 327
Dembo, A. 219, 240, 246, 263, 277
Devroye, L. 80
Diaconis, P. 354
Dobrushin, R. L. 129, 130

Donsker, M. D. 126, 224
Durett, R. 14
Dvoretzky, A. 209, 211, 221, 242

Erdős, P. 17, 18, 34, 39, 53, 59, 62, 65, 67, 77, 79, 121, 135–139, 157, 158, 160, 180, 194, 209, 211, 213, 215, 219, 220, 221, 241, 242, 243, 245, 272, 278

Feller, W. 19, 20, 34
Fisher, A. 140
Flatto, L. 220
Földes, A. 18, 21, 74, 81, 82, 107, 118–123, 127, 134, 139, 155, 166, 167, 169, 172–175, 227, 236, 238

Gnedenko, B. V. 19
Golosov, A. O. 345, 346, 347
Goncharov, V. L. 21
Goodman, V. 94
Gorn, N. L. 94
Griffin, P. 157, 199, 233
Grill, K. 18, 45, 68, 70, 71, 79, 92, 176, 178, 179
Guibas, L. J. 59, 62

Häggström, O. 60
Hamana, Y. 220, 225, 235
Hanson, D. L. 72, 181
Hartman, P. 53
Horváth, L. 110, 317, 318
Hough, J. B. 245
Hu, Y. 328, 334
Hunt, G. A. 112, 118

Imhof, I. P. 175
Itô, K. 141, 183

Jain, N. C. 222, 225

Kakutani, S. 242

AUTHOR INDEX

Kalikow, S. A. 349
Karlin, S. 66
Kesten, H. 112, 118, 296–298, 346, 350, 351
Key, E. S. 347, 348
Khinchine, A. 30
Khoshnevisan, D. 107, 108, 146, 164
Knight, F. B. 52, 111, 139, 204, 205
Kolmogorov, A. N. 19, 25, 34, 38
Komlós, J. 18, 53, 54
Kuelbs, J. 94
Kumagai, T. 222, 225
Kusolitsch, N. 66

Lacey, M. T. 141
Lamperti, J. 293
Lawler, G. F. 195, 242, 244, 246
Le Gall, J. F. 225, 226
Lévy, P. 33, 104, 111, 140, 141
Li, W. 243
Lifschitz, M. A. 94
Lynch, I. 80

Major, P. 53, 54, 161
Marcus, M. B. 244
McKean Jr, H. P. 141, 185
Mogul'skii, A. A. 116
Móri, T. 62, 63
Mueller, C. 95, 126

Nemetz, T. 66
Newman, D. 161

Odlyzko, A. M. 59, 62
Orey, S. 208
Ortega, I. 68, 72
Ost, F. 66

Pemantle, R. 279
Peres, Y. 60, 219, 240, 245, 246, 263
Perkins, E. 119, 141, 240
Petrov, V. V. 66
Petrowsky, I. G. 34
Philipp, W. 141
Pitt, J. H. 220
Pólya, G. 23, 193

Pruitt, W. E. 208, 222, 225
Puri, M. L. 155

Qualls, G. 180, 182

Rényi, A. 13, 14, 19, 20, 28, 53, 67, 77, 101
Révész, P. 17, 18, 39, 59, 63, 65, 66, 71–73, 75, 79, 88, 91, 109, 112–115, 119, 120, 125, 126, 134, 137, 138, 141, 157–160, 163, 164, 166, 167, 172–175, 185, 227, 235, 236, 238, 245, 246, 256, 263, 264, 266, 269, 272, 278, 281, 283, 285, 317, 318, 326, 327, 337, 351
Riesz, F. 85
Rosen, J. 219, 236, 238, 240, 244, 246, 263, 281, 283
Rosenblatt, M. 180
Russo, R. P. 72, 181

Samarova, S. S. 59
Schatte, P. 140
Shao, Q. M. 182
Shi, Z. 158, 159, 236, 238, 263, 283, 285, 328, 334, 351
Shore, T. R. 71
Simons, G. 113, 115
Sinaĭ, JA, G. 314, 345, 346
Skorohod, A. V. 53
Solomon, F. 311
Spitzer, F. 27, 205, 206, 248, 350, 351
Steinebach, J. 80
Steif, J. E. 60
Strassen, V. 32, 88, 89
Szabados, T. 183
Székely, G. 18
Sz.–Nagy, B. 85

Taylor, S. J. 139, 194, 209, 211, 213, 215, 219, 220, 240, 241, 242, 243
Tóth, B. 158, 159, 162, 297, 355
Trotter, H. F. 105
Tusnády, G. 18, 53, 54

Varadhan, S. R. 126, 224

Vincze, I. 113

Watanabe, H. 180, 183
Weigl, A. 141
Werner, W. 159
Wichura, M. 95, 126

Wintner, A. 53
Wschebor, M. 68, 72

Zeitouni, O. 219, 240, 246, 263
Zimmermann, G. 163

Subject Index

Arcsine law 104
Asymptotically deterministic sequence 34

Bernstein inequality 13
Borel – Cantelli lemma 27
Brownian motion 9

Central limit theorem 19
Chebyshev inequality 28

Dirichlet problem 293
DLA model 296

EFKP LIL 34

Gap method 29

Invariance principle 52, 109, 203
Itô formula 183
Itô integral 183

Large deviation theorem 14, 19
Lévy classes 33
LIL of Hartman – Wintner 32
LIL of Khinchine 31
Logarithmic density 140
Long head-runs 57

Markov inequality 28
Method of high moments 29

Normal numbers 29

Ornstein – Uhlenbeck process 179

Percolation 297

Quasi asymptotically deterministic sequence 34

Rademacher functions 10, 11

Random walk in random environment
   definition 303, 348
   local time 313, 330, 335, 337
   maximum 313, 325, 327, 345
   recurrence 311
Random walk in random scenery 350
Random walk in $\mathbb{Z}^1$
   definition 9
   excursion 146, 147
   favourite sites 157
   first recurrence 97
   increments 57, 77
   increments of the local time 123
   law of the iterated logarithm 31
   law of the large numbers 28
   local time 98, 117, 146,
   location of the last zero 102, 136, 175
   location of the maximum 102, 104, 136, 171
   longest run 21, 57
   longest zero-free interval 135
   maximum 14, 20, 31, 35, 41
   maximum of the absolute value 20, 31, 35, 41, 171
   mesure du voisinage 141
   number of crossings 100, 113
   range 44
   rarely visited points 157, 161
   recurrence 23
   Strassen type theorems 83, 90, 124
Random walk in $\mathbb{Z}^d$
   almost covered discs 264
   completely covered balls 272
   completely covered discs 245, 263
   definition 192
   excursions 281, 284
   first recurrence 211
   heavy points 227
   heavy balls 236
   law of the iterated logarithm 207

local time 211, 218
maximum 206, 209
range 221
rate of escape 209
recurrence 193
self-crossing 241
speed of escape 288
Strassen type theorems 204
Reflection principle 15
Reinforced random walk 353

Skorohod embedding scheme 53
Stirling formula 19

Tanaka formula 185
Theorem of Borel 29
Theorem of Chung 39
Theorem of Donsker and Varadhan 125
Theorem of Hausdorff 30
Theorem of Hirsch 39

Wichura's theorem 95
Wiener process in $\mathbb{R}^1$
   definition 51

excursion 139, 141
increments 66
increments of the local time 119
local time 104, 109,
location of the last zero 179
location of the maximum 175
longest zero-free interval 121
maximum 53
maximum of the absolute value 53
mesure du voisinage 141
occupation time 155
Strassen type theorems 84, 90, 92, 124
Wiener process in $\mathbb{R}^d$
   definition 203
   law of the iterated logarithm 207
   maximum 206
   rate of escape 209
   self-crossing 241
   Strassen type theorems 204
Wiener sausage 226
Wiener sheet 163

Zero-one law 25